Proceedings of the Ninth Asia-Pacific International Conference on

Gravitation and Astrophysics

Proceedings of the Ninth Asia-Pacific International Conference on

Gravitation and Astrophysics

Wuhan, China, 29 June – 2 July 2009

Editors

Jun Luo

Huazhong University of Science and Technology, China

Ze-Bing Zhou

Huazhong University of Science and Technology, China

Hsien-Chi Yeh

Huazhong University of Science and Technology, China

Jong-Ping Hsu

University of Massachusetts Dartmouth, USA

World Scientific

NEW JERSEY • LONDON • SINGAPORE • BEIJING • SHANGHAI • HONG KONG • TAIPEI • CHENNAI

Published by

World Scientific Publishing Co. Pte. Ltd.
5 Toh Tuck Link, Singapore 596224
USA office: 27 Warren Street, Suite 401-402, Hackensack, NJ 07601
UK office: 57 Shelton Street, Covent Garden, London WC2H 9HE

British Library Cataloguing-in-Publication Data
A catalogue record for this book is available from the British Library.

Back cover: Yellow Crane Tower is a famous and historic tower, and the subject of many famous poems. CUI Hao, a poet in The Tang Dynasty, wrote the most famous poem about it: "Long ago a man rode off a yellow crane, all that remains here is Yellow Crane Tower. Once the yellow crane left it never returned, for one thousand years the clouds wandered without cares. . . "

Front image: "The torsion pendulum for measuring the Newtonian gravitational constant G. It was built at the Key Laboratory for Measurements of Fundamental Physical Quantities at the Huazhong University of Science and Technology (HUST), Ministry of Education, China. The most accurate value obtained so far at HUST is $G = (6.67349 \pm 0.00018) \times 10^{-11} m^3 kg^{-1} s^{-2}$."

GRAVITATION AND ASTROPHYSICS
Proceedings of the 9th Asia-Pacific International Conference

ISBN-13 978-981-4307-66-6
ISBN-10 981-4307-66-1

Printed in Singapore by B & Jo Enterprise Pte Ltd

To

ZHANG Heng (78-139, Han dynasty)

and

SHEN Kua (1031-1095, Song dynasty)

Inventors of Astronomical Instruments

张衡

ZHANG Heng

沈括

SHEN Kua

9th Asia-Pacific International Conference on Gravitation and Astrophysics
第九届亚太引力与天体物理国际会议
HUST. Wuhan, P. R. China 28 June - 2 July, 2009

Preface

The year of 2009 was the International Year of Astronomy, commemorating the 400th anniversary of Galileo's use of a telescope to study the sky. This historical event paved the way for the following development of astrophysics, gravitation and cosmology. As one of the celebration activities in this meaningful year, the Ninth Asia-Pacific International Conference on Gravitation and Astrophysics (ICGA9) was held on June 28th - July 2nd at the Huazhong University of Science and Technology (HUST), Wuhan, China.

The previous ICGAs had been held in Seoul, Korea (1993), Hsinchu, Taiwan (1995), Tokyo, Japan (1997), Beijing, China (1999), Moscow, Russia (2001), Seoul, Korea (2003), Jhongli, Taiwan (2005), and Nara, Japan (2007). At the ICGA9, a growing number of participants representing the researchers and scientists of various countries met together to discuss the current status of Gravitation and Astrophysics and shared their research experiences in this cordial and professional meeting.

The ancient Chinese astronomers developed many precision instruments to observe and record astronomical events in very early periods of Chinese history. The recordings turned out to contain important discoveries of supernovae and novae. According to Prof. T. D. Lee (Nobel laureate), the ancient Chinese jades Pi, Zong and Xuan Ji were artistic expressions of astronomical instruments for the determination of the Fixed Point in the sky (circa 2700, B.C.). Ancient astronomers SHI Shen and GAN De (circa 400, B.C.) started to build up the armillary sphere (浑仪), which was accomplished by the famous astronomer ZHANG Heng (78-139 A.D.) with horizon and meridian rings. Later it was further improved by GUO Shoujing in 1276, using a novel design called abridged armilla (简仪). We hope that scientists in China can continue this glorious tradition of developing excellent precision measurements and instruments, especially for research in gravitation and astrophysics.

Prof. LUO Jun organized the ICGA9 as an open conference for astronomers, physicists and graduate students, based on the foundations of previous ICGA conferences. It aimed to serve as a common platform around the Asia-Pacific region for exchange and communication among all workers in this research field. At ICGA9, there were more than 40 reports presented, and part of these are now collected and published in this special issue. We expect that these papers reflect the current status and active development in Gravitation and Astrophysics around the Asia-Pacific region.

According to the response and feedback received from participants, the ICGA9 was a wonderful and successful meeting. We would like to take this opportunity to express our gratitude to the Asia Pacific Center for Theoretical Physics (APCTP), the National Natural Science Foundation of China (NSFC), the Ministry of Education of P. R. China, and the University of Massachusetts Dartmouth Foundation for their financial support of ICGA9. We also thank all colleagues and graduate students of HUST participating in this conference for their help and effort to make the ICGA9 successful.

The ICGA series of conferences meets a significant need and plays an increasing role as a center of exchange for all scientists and researchers in this field in the Asia-Pacific region. We believe that it will continue growing and gradually become the information hub among researchers involved in Gravitation and Astrophysics all over the world. We wish that the present volume will serve as the first step toward our hope and expectation.

Editors

The 9th ICGA

Contents

Gravitational Experiments

THE NEWTONIAN GRAVITATIONAL CONSTANT: THE HISTORY OF THE DETERMINATION AND THE ENVIRONMENTAL NOISE PROBLEM FOR THE EXPERIMENTAL MEASUREMENT

VADIM MILYUKOV*

Sternberg Astronomical Institute of Moscow State University
Moscow, Universitetsky pr., 13, Russia
**E-mail: milyukov@sai.msu.ru*
www.sai.msu.ru

Due to the weakness of gravity, the accuracy of the Newtonian gravitational constant G is essentially below the accuracy of other fundamental constants. The current value of G, recommended by CODATA in 2006, based on all results available at the end of 2006, is $G = (6.67428 \pm 0.00067) \times 10^{-11}\ m^3 kg^{-1} s^{-2}$ with a relative error of 100 ppm. The modern history of G determination is considered. New experiments at a level of accuracy of 10 to 30 ppm are now in progress in some world gravitational laboratories. One of the problems of improving the accuracy of G is a precision measurement of the period of eigen oscillations of the torsion balance. The torsion balance responses to tiny environmental noises, which accordingly result in changes of the torsional oscillation period. The seismic environmental noise in the underground laboratory condition has been studied on the base of data of the Baksan Laser Interferometer. The dependence of the torsion period value and the error of its estimation on the level of the acting seismic noise has been simulated and studied. To measure the Newtonian gravitational constant at the accuracy level of 10 ppm, the environmental seismic noise must be below 10^{-2} mGal.

Keywords: Newtonian gravitational constant; torsion balance; seismic noise.

1. Introduction

The Newtonian gravitational constant G together with Planck's constant $\hbar$ and the speed of light c are the fundamental constants of nature which represent the fundamental limits: c is the maximal speed, $\hbar$ is the minimal angular momentum and G is the gravitational radius of unit mass (the maximal radius of the sphere for relativistic gravitational collapse).

If absolute values of fundamental constants such as c and $\hbar$ are known with high accuracy, a situation with the gravitational constant G absolutely by others. Due to the weakness and nonshieldability of gravitational interaction accuracy of experimental determination of G is essential below accuracy of other fundamental constants.

Measurement of a gravitational constant is connected with absolute measurements of three physical values - time, mass and length, and consequently it is necessary to produce the absolute measurements on the high technology level in order to have a reliable estimation of G.

The first device for measurement of a mutual gravitational attraction of small laboratory bodies - the horizontal torsion balance - has been made at the end of XVIII century by Henry Cavendish, outstanding English scientist. Hundred years later after Newton's discovering the law of gravitation, the Cavendish experiment (Fig. 1), done in 1797–98, was the first experiment to measure the force of gravity between masses in the laboratory, and the first to give accurate values for the gravitational constant, $G = (6.67 \pm 0.07) \times 10^{-11}\ m^3kg^{-1}s^{-2}$, with a relative uncertainty of 10^4 ppm (a relative uncertainty is expressed in units of *part per million*, i.e. $10^4 \times 10^{-6}$). Cavendish also has determined the mass and the mean density of the Earth. A significance of the Cavendish's experiment was not only limited by the determination of the G value. The main thing is he has proved the validity of the universal law of gravitation for small laboratory bodies.

The modern experimental installations for measurement of gravitational constant are complicated devises, performed on the high technology level, but the main part of them is also the horizontal torsion balance. After 2000 several new results on the measurement of G with a relative error, less than 50 ppm have been published. However these results are also not covered each other within confidential intervals. In 2006 the Committee on Data for Science and Technology (CODATA) recommended for the Newtonian gravitational constant a value $G = (6.67428 \pm 0.00067) \times 10^{-11}\ m^3kg^{-1}s^{-2}$, with an uncertainty of 100 ppm. This new value of G is based on the data accessible at the end of 2006.

Since the first laboratory measurement of Cavendish over 200 years ago, the reduction in uncertainty in G has been only two order of magnitude. Progress in the measurement of G occurs slowly enough: the error value decreases approximately 10 times per century, and the knowledge of the absolute value of G is still rather poor. New experiments on measurement of the Newtonian gravitational constant at the accuracy level of 10–30 ppm are actual and desirable.

Fig. 1. The artist's conception of Cavendish conducting his experiment. He performed the experiment inside a closed shed and observed the result from outside through a telescope. The opening in the wall was added by the artist to show the apparatus.

2. The modern history of G determination

The modern history of the G determination covers 30–35 years and has started from the three experiments, performed in 70-th of last century. There were the experiment of *Observatoire de Recherches de la Meteorologie Nationale* (France), reported in 1972,[1] the experiment of *Sternberg Astronomical Institute of Moscow University*, reported in 1979,[2] and the experiment of *National Bureau of Standards* (USA), reported in 1982.[3] The system of values of fundamental constants CODATA 1986 has contained the G value with relative accuracy 128 ppm, which was based mainly on the value, obtained by Luther and Towler,[3] but with its uncertainty doubled, what reflects the fact, that, historically, measurements of G have been difficult to carry out and the result of Luther and Towler was possibly not final.[4]

Within 90th years of the last century large enough numbers of laboratory experiments on the measurement of the Newtonian Gravitational constant were done with relative accuracy about of 100 ppm and less.[5–7,9,10] The part of these of results is summarized in Table 1. Nevertheless, the discrepancies between the values of a gravitational constant obtained in these experiments remained enough big. In particular, the value $G = 6.7146$ obtained in *Physikalish Technische Bundesanstalt* (Germany),[5] was more than on 40 standard deviations (i.e. more than 5000 ppm) above the G value recommended CODATA in 1986.[4] As a result of such scattering of G values, CODATA should increase significantly an uncertainty and recommended in 1998 value $G = (6.673 \pm 0.010) \times 10^{-11}\ m^3 kg^{-1} s^{-2}$,[10] with a relative error of 1500 ppm. I.e. "uncertainty of knowledge" of G has increased almost in 10 times!

Table 1. The best world experiments on the measurement of G and CODATA values.

Authors, year of publication	G, $\times 10^{-11}$ ($m^3 kg^{-1} s^{-2}$)	STD, $\times 10^{-11}$ ($m^3 kg^{-1} s^{-2}$)	ppm
Facy, Ponticis, 1972[1]	6.6714	0.0006	90
Sagitov, Milyukov, et al., 1979[2]	6.6745	0.0008	120
Luther, Towler, 1982[3]	6.6726	0.0005	75
CODATA, 1986[4]	**6.67259**	**0.00085**	**128**
Michaelis, et al., 1995[5]	6.7154	0.0006	90
Karagioz, Izmailov, 1996[6]	6.6729	0.0005	75
Bagley, Luther, 1997[7]	6.674	0.0007	105
CODATA, 1998[8]	**6.673**	**0.01**	**1500**
Jun Luo, et al., 1999[9]	6.6699	0.0007	105
Fitzgerald, Armstrong, 1999[10]	6.6742	0.0007	105
Gundlach, Merkowich, 2000[11]	6.674215	0.000092	14
Quinn, Speake et all., 2001[12]	6.67559	0.00027	41
Schlamminger et all., 2002[13]	6.67407	0.00022	33
CODATA, 2002[14]	**6.6742**	**0.001**	**150**
Armstrong, Fitzgerald, 2003[15]	6.67387	0.00027	41
Schlamminger et all., 2006[16]	6.67425	0.0001	16
CODATA, 2006	**6.67428**	**0.00067**	**100**
Jun Luo, et al., 2009[17]	6.67349	0.00018	26

During following years (2000–2002) three new results, with relative errors less than 50 ppm, have been published. These are the experiment of *University of Washington* (USA), 2000, with a relative error of 14 ppm,[12] the experiment of *University of Birmingham* (Great Britain), 2001, with a relative error of 41 ppm,[12] and the experiment of University of Zurich (Switzerland), 2002, with a relative error of 33 ppm.[13] Although the situation with G has improved considerably since 1998 adjustment, these new results are not in complete agreement, as can be seen from the table 1 and Fig. 2. These new G values are not crossed inside of confidential intervals. Based on weighted means of results after 1998, all of which round to $G = 6.6742 \times 10^{-11} m^3 kg^{-1} s^{-2}$, as well as their uncertainties, the relatively poor agreement of the data, and the historic and apparently continuing difficulty of assigning an uncertainty to a measured value of G that adequately reflects its true reliability, CODATA has taken as the 2002 recommended value $G = (6.6742 \pm 0.0010) \times 10^{-11} m^3 kg^{-1} s^{-2}$ with relative error of 150 ppm.[14]

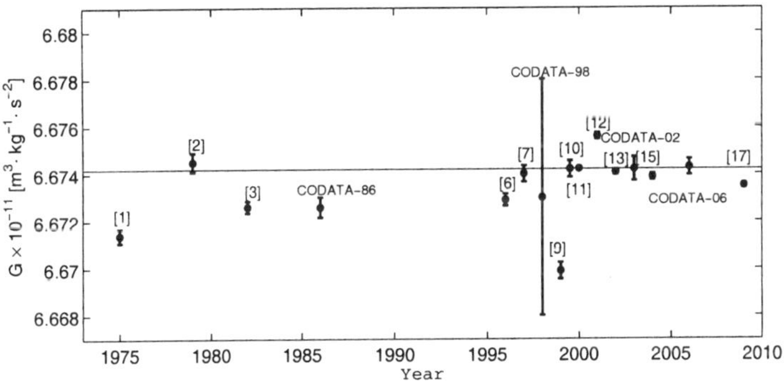

Fig. 2. The results of the best world experiments on measurement of G and CODATA values. The dash line is the CODATA-2006 value.

In 2003 and 2006 the new results on the measurement of G were reported by Armstrong and Fizgerald from *the Measurement Standards Laboratory* (New Zealand) with uncertainty of 40 ppm[15] and by Shlamminger, et al. from *University of Zurich* (Switzerland) with uncertainty of 16 ppm.[16] The uncertainty of the New Zealand experiment is about 2.5 times smaller than the uncertainty of the previous result of these authors,[10] but its standard deviation interval does not intersect formally with CODATA 2002 one. Alternatively, the uncertainty of the Switzerland result coincides practically with CODATA 2002 value. The current value of the Newtonian gravitational constant, recommended by CODATA in 2006, based on all results, available at the end of 2006, and is equal $G = (6.67428 \pm 0.00067) \times 10^{-11} m^3 kg^{-1} s^{-2}$

with relative error of 100 ppm (http://www.codata.org).

Finally the new result of determination of G in *Huazhong University of Science and Technology* (China) was published in 2009.[17] The G value, obtained in this experiment is $G = (6.67349 \pm 0.00018) \times 10^{-11} m^3 kg^{-1} s^{-2}$ with a relative error of 26 ppm. The essential feature of this experiment is consideration of the anelasticity effect of the torsion fiber (so called Kuroda effect) which introduces the systematic error in the G value, proportional to Q factor of the torsion balance. One of the sources of the systematic errors which can explain so big scattering the G values obtained in different experiments was expressed by K. Kuroda.[18] The time-of-swing method of G determination assumes, that the elastic torsion constant K_e remains the same at each orientation of source masses, but this assumption has been called into question. Kuroda has shown that the damping of torsion balance is caused by losses in the suspension fiber (anelasticity of the fiber) and for a Heyl-type experiment the measured value of G will be biased upward by a factor of $(1+1/\pi Q)$, were Q is the quality factor of torsion oscillations. The resultant correction of G value due to the anelastisity in HUST experiment is -211.80 ppm, which was measured experimentally.[17]

Table 1 summarizes the various measurements of G that were considered, and Fig. 2 compares them graphically. The 1986, 1998, 2002 and 2006 CODATA recommended values are also included. Finally we would like to emphasize the following fact. The SAI MSU (*Sternberg Astronomical Institute of Moscow University*) value, $G = (6.6745 \pm 0.0008) \times 10^{-11} m^3 kg^{-1} s^{-2}$, was published in 1979.[2] CODATA-1986 has recommended value of $G = (6.67259 \pm 0.00085) \times 10^{-11} m^3 kg^{-1} s^{-2}$, which is not coincide with SAI MSU value. And 27 years after this experiment, in 2006, CODATA recommended value of G, which is practically coincide with this "old" value. From the position of modern experience, the SAI MSU value, due to the carefulness of preparation and carrying out of experiment, despite of imperfection of measuring technics of that time, has been apparently least subject to systematic errors. Thus, the 27-years-old value of G is actual and significant now.

3. Torsion balance is the principal tool of the experiment

The torsion balance, or torsion pendulum, is the traditional tool for performance of high precision gravitational experiments. The torsion balance is horizontal bar with the test masses on the ends suspended from the center by a thin elastic fiber. The system of equations, describing the motion of torsion balance, is five nonlinear equations, containing also fluctuating terms. Two pairs of the equations describe swing oscillations in two orthogonal vertical planes, and the last equation describes torsion oscillations. Analytic solution of this system of equations was done in some works.[19,20] Then the system of equations has been solved by numerical method and simulation of the motion of the torsion balance has been done.[21]

We have studied the character of excitation of the swing oscillations of the torsion balance. The numerical experiment has shown, that oscillations on swing degrees of freedom are exited by the random noise of a seismic origin and occur with the amplitude, varying in time (Fig. 3). The spectral analysis of swing oscillations has shown, that the oscillations on the each swing degree of freedom are beating of all quasi-harmonic swing modes with random amplitude changing in time. The swing frequencies are defined by geometrical parameters of the torsion balance. The random character of the swing oscillations is defined by the seismic noise affected on the suspension point. It was shown also, that even with damping, due to the action of seismic noise, the swing oscillations are steady process.

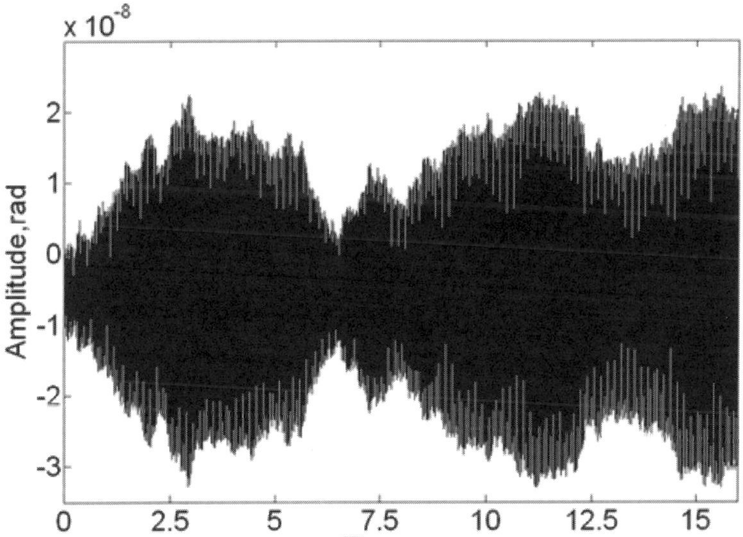

Fig. 3. Swing oscillations on vertical plane, which excited by seismic noise with amplitude about 0.1 mGal.

The torsion balance is the complicated system with many degrees of freedom, and due to the nonlinear couplings between them, new oscillations on so-called coupled modes are appeared. This leads to perturbations of the basic torsional mode and as a consequence, to increase an uncertainty of the determination of its period. The frequencies of coupled modes are simply linear combinations of swing-mode frequencies. If the coupled-mode frequencies are close to the frequency of the torsional oscillations, the latter can be disturbed. Therefore the problem of the high precision measurement of the torsional-mode frequency has to be solved on the condition of maximal suppression of coupled modes.[20] The traditional method to remove the effects of low-frequency torsional mode couplings is employing a magnetic damper in the torsion system to overcome seismic noise and consequently to suppress the intensity of swing modes. This method is called the amplitude suppression and is used in the experimental setup, mentioned above.[17]

There are a number of problems which have to be in attention in order to reach the desirable accuracy of measurement of G. One of them is stability of the torsion oscillations. Stability of torsion oscillations also depends on various external disturbances, among them, on seismic noise. The environmental seismic noise excites the swing oscillations of the torsion balance and due to the nonlinear coupling between different degrees of freedom this leads to the disturbances of the torsion oscillations. To estimate experimentally the level of the seismic noise in the typical underground laboratory condition we have studied the seismic noise on the base of data of the Baksan Laser Interferometer.

4. Influence of environmental noise on the torsional oscillations

A torsion balance being a high sensitive devise is influenced by the environmental noise of different origin, even if it placed in quiet underground condition. We have studied the seismic environmental noise in the underground laboratory condition on the base of data of *the Baksan Laser Interferometer.* The interferometer is mounted at a level of 650 m and at a depth of 400 m along the tunnel of *the Baksan Neutrino Observatory* (the Northern Caucasus, Russia), driven into the massif of Mount Andyrchy to a depth of 4200 m.

The scheme of the interferometer is the two passes ($N = 2$) Michelson-type interferometer with unequal arms. The measurable armlength is 75 m. The long-term monitoring strains of the Earth is provided in several frequency channel: practically from 0 Hz up to 1.62 kHz (acoustic channel). The resolution of the interferometer is 2.3×10^{-13}. Technical details of the interferometer is described in the work.[22]

The spectral amplitude of strain, based on the data of all deformations channels, has a flicker noise character and can be approximated by function $3\times10^{-12} f^{-1}\text{Hz}^{-1/2}$ over the entire frequency range. There are two strongly pronounced maxima corresponding to diurnal and semidiurnal tidal waves. Experimental results based on deformation spectra have been recalculated in the amplitudes of horizontal accelerations in the frequency band of $10^{-6} \div 10^{1}$ Hz (Fig. 4). Spectral amplitudes of horizontal accelerations less depend on frequency. Noise amplitudes vary in a band of 0.001–0.01 mGal. The range of characteristic values of torsion oscillations frequency (10^{-3} Hz) corresponds to the minimum values of noise while the range of characteristic values of swing oscillations (0.5 Hz) corresponds to its maximum values.

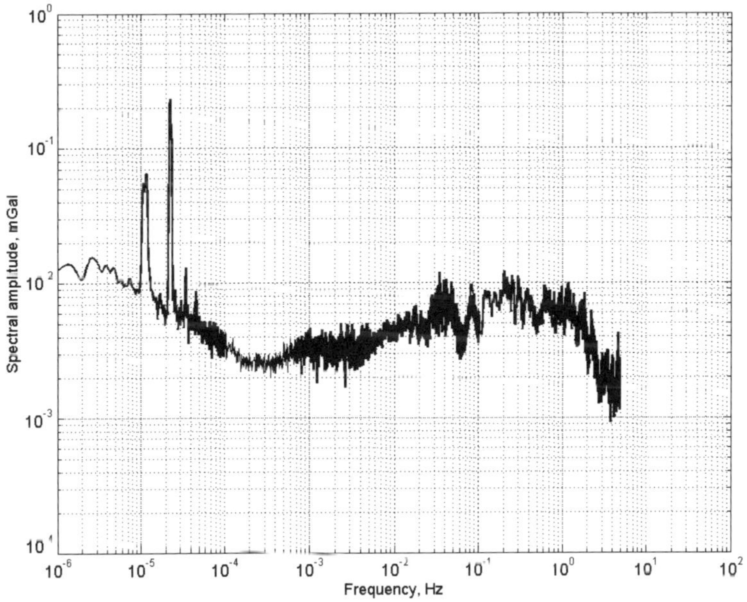

Fig. 4. Spectral amplitude of horizontal accelerations, recalculated from experimental strain spectrum.

Since the environmental noise is stochastic, variations of the period of torsional oscillations are also stochastic. Finally we have studied dependence of the period of torsion oscillations and the accuracy of its estimations from the environmental noise. For estimation of the value of the period of torsion oscillations we use the phase method which was adopted and developed by our colleagues from HUST.[23] Let a time series consist of m torsion oscillations. Following this method, estimation of the experimental frequency ω_{ex} on the base of j oscillations can be done from the formula

$$\omega_{ex} = \omega_0 - \frac{K\omega_0}{2\pi} \,, \tag{1}$$

were ω_0 is initial frequency value, which can be obtained by usual FFT analysis; K is the dependency ratio of the phase angle at the j-th period from period number j.

If the dependence of the phase on the period number has linear character ($K = const$), the frequency ω_{ex} can be estimated using all data (m oscillations) simultaneously. If the torsion oscillations are noticeably disturbed by the environmental noise, the phase behavior can be nonlinear ($K \neq const$). In this case the time series is divided into parts with $K \approx const$, and ω_{ex} is estimated for each such "linear-slope" part. Then the value of ω_{ex} is calculated as a mean value of all "linear-slope" parts.

Numerical simulation of the motion of a torsion balance subject to seismic noise and estimation of the frequency of torsional oscillations have been done for the torsion balance used in the HUST experiment.[17] The time series of data have been obtained as a solution to the set of equations describing the motion of a torsion balance with five degrees of freedom.[24] The length of the time series was 118 h (it means 800 oscillations for the typical period of torsional oscillations of 532 s). The period of torsion oscillations was estimated by the phase method. The level of seismic noise acting on the torsion balance (in terms of acceleration) varied from 10^{-3} mGal to 10^{-1} mGal. The results of a numerical simulation show that, for the seismic noise of 10^{-3} mGal, the dependence of the phase on the number of period has a linear character and the torsion period can be estimated using all data simultaneously. For the seismic noise of 10^{-2} mGal, the phase behavior becomes nonlinear, and the period can be estimated by dividing the data into "linear-slope" parts. Finally, for the seismic noise of 10^{-1} mGal, the phase dependence is a complicated process having nonlinear parts and steps. In this case, the period estimation accuracy is essentially lower than it is required in the experiment (100 ppm against 10–30 ppm). The results of the analysis are summarized in Table 2. Fig. 4 shows estimates of the torsion period obtained for the length of "linear-slope" parts equal to 20 periods.

Table 2. Dependence of the period of torsional oscillations on the environmental seismic noise level.

Seismic noise (mGal)	Period of torsion oscillation (s) (mean for 800 periods)	Period of torsion oscillation (s) (mean for $20 \div 40$ periods)
0.001	532.80882 ± 0.00016 (0.3 *ppm*)	532.80788 ± 0.00036 (0.6 *ppm*)
0.005	532.80851 ± 0.00030 (0.6 *ppm*)	532.80463 ± 0.00068 (1 *ppm*)
0.010		532.82580 ± 0.0010 (2 *ppm*)
0.050		532.6541 ± 0.004 (8 *ppm*)
0.100		532.390 ± 0.062 (100 *ppm*)

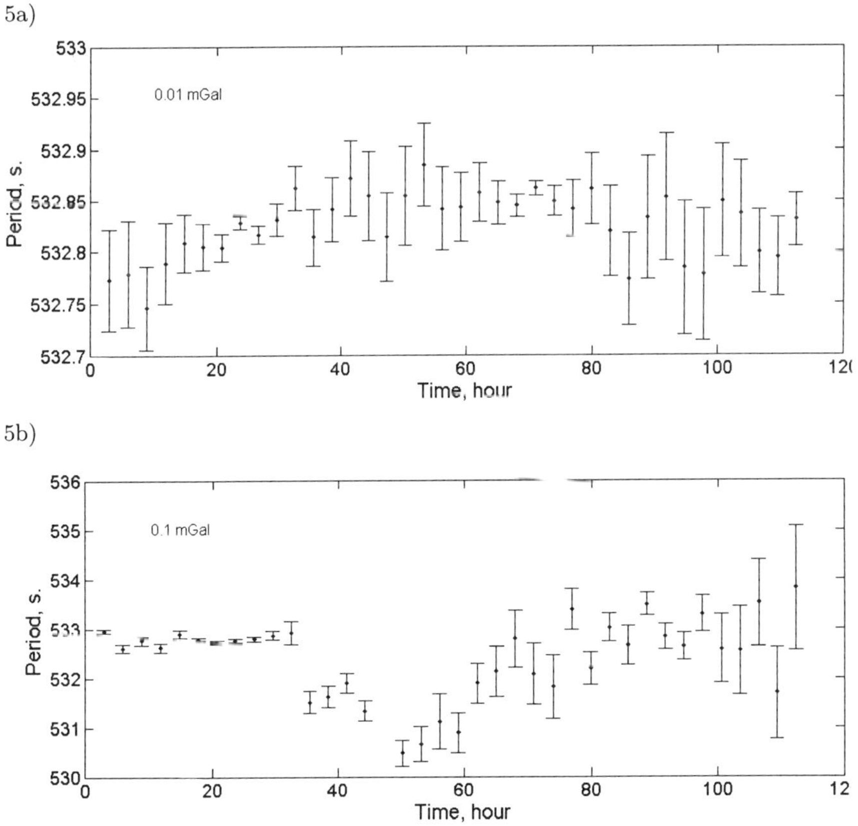

Fig. 5. Estimation of the period of torsional oscillations for different seismic noise levels. Each values was calculated for the "linear-slope" part equal to 20 oscillations. The vertical bars are standard deviations.

5. Conclusion

One of the most important problems which restricts the accuracy of measurement of the Newtonian gravitational constant is the measurement of the period of eigen oscillations of the torsion balance. Due to a very high sensitivity, the torsion balance responses to tiny external noises, which accordingly results in changes of the torsional oscillations period.

The accuracy of period estimations depends on the level of seismic noise which affects the torsion balance. Estimation of the period on the accuracy level of a few ppm is provided by the level of environmental seismic noise with amplitudes smaller than 10^{-2} mGal. The value of the torsional period depends on the seismic noise level. Changing the seismic noise level by one order leads to a change in the torsion period value of a few tens of ppm (table 2). To provide the absence of systematic errors in the period values caused by the seismic noise, the noise conditions should not change during the experiment.

Acknowledgments

This work is supported by *the Russian Foundation for Basic Research*, grant No.08-02-92217.

References

1. C. Pontikis, *C. R. Acad. Sci.* Ser. B **274**, 437 (1972).
2. M. U. Sagitov, V. K. Milyukov, et al, *Dokl. Akad. Nauk SSSR.* **245** (3), 567 (1979). [*Sov. Phys. Dokl.* **245** (1-6), 20 (1981)].
3. G.G. Luther and W.R. Towler, *Phys. Rev. Lett.* **48**, 121 (1982).
4. G. Gilles, *Metrologia.* **24** (Supplement), P. 56. (1987).
5. W. Michaelis, et al, *Metrologia.* **32**, 267 (1995/96).
6. O. V. Karagioz and V. P. Izmailov, *Izmer. Tekh.* **39** (10), 3 (1996). [*Meas. Tech.* **39** (10), 979 (1996)].
7. C. H. Bagley and G. G. Luther, *Phys. Rev. Lett.* **78** (16), 3047 (1997).
8. P. Mohr, *Advances in Quantum Chemistry.* **30**, 77 (1998).
9. J. Luo, et al, *Phys. Rev.* D **59**, 042001 (1999).
10. M. P. Fitzgerald and T. R. Armstrong, *Meas. Sci. Technol.* **10** (6), 439 (1999).
11. J.H. Gundlach and S. M. Merkowich, *Phys. Rev. Lett.* **85**, 2869 (2000).
12. T. J., Quinn C. C. Speake, Richmann S. J. et al, *Phys. Rev. Lett.* **87**, 111101 (2001).
13. St. Schlamminger, E. Holzschuh, Kundig W, *Phys. Rev. Lett.* **89**, 161102 (2002).
14. P.J. Mohr and B.N. Taylor, *Rev. Mod. Phys.* **77** (1), 1 (2005).
15. T.R. Armstrong and M.P. Fitzgerald, *Phys. Rev. Lett.* **91** (20), 201101 (2003).
16. St. Schlamminger, et al., *Phys. Rev.* D **74**, 082001, 25 (2006).
17. J. Luo, et al., *Phys. Rev. Lett.* **102**, 240801 (2009).
18. K. Kuroda, *Phys. Rev. Lett.* **75** (15), 2796 (1995).
19. V. K. Milyukov, *The theory of motion of the torsion balance in inhomogeneous gravitational field under the action of random noise*, in *Problems of gravitation and elementary particles*, Vol. 16 (Moscow, Energoizdat, 1981), pp. 128–136.
20. X.-D. Fan et al., *Physics Letters.* A **372**, 547 (2008).
21. V.K. Milyukov, J. Luo, T. Chen, A.P. Mironov, *Gravitation and Cosmology.* **14**, 368 (2008).
22. V.K. Milyukov, et al., *Instruments and Experimental Techniques.* **48** (6), 780 (2005).
23. J. Luo, D-H. Wang, Q. Liu, Ch-G. Shao, *Chin. Phys. Lett.* **22** (9), 2169 (2005).
24. V.K. Milyukov, T. Chen, and A.P. Mironov, *Gravitation and Cosmology.* **15** (1), 65 (2009).

A NEW DETERMINATION OF G WITH TIME-OF-SWING METHOD

SHAN-QING YANG, QING LI, LIANG-CHENG TU*, CHENG-GANG SHAO, LIN-XIA LIU, QI LIU, QING-LAN WANG and JUN LUO

Department of Physics, Huazhong University of Science and Technology, Wuhan 430074, People's Republic of China

** E-mail: tlc@mail.hust.edu.cn*

A new determination of Newtonian gravitational constant G is presented by using a torsion pendulum with time-of-swing method. Features of the design include: (1) two SS316 stainless steel spheres with more homogeneous density were used as the source masses; (2) a rectangular glass block was used as the main body of the pendulum; (3) both the pendulum and source masses were set in a vacuum chamber to reduce the error of measuring the relative positions conveniently; (4) exchanging of the configurations between "near" and "far" positions was remotely operated by using a stepper motor; (5) the anelastic effect of the torsion fiber was first measured directly by using two disk pendulums with the help of a high-Q factor quartz fiber. We repeated the G measurements twice independently, and two G values have only a difference of 9 ppm. The combined final value of G was found to be $(6.67349 \pm 0.00018) \times 10^{-11}$ $m^3kg^{-1}s^{-2}$ with a relative uncertainty of 26.33 ppm.

Keywords: G; gravitational constant; torsion pendulum; time-of-swing method, precision measurement.

1. Introduction

The Newtonian gravitational constant G holds an important place in physics. Though there have been about 300 measurements of G since the first laboratory measurement by Cavendish[1] over 200 years ago, its measurement precision is the worst among all the fundamental physical constants.[2] In 2006, the Committee on Data for Science and Technology (CODATA), based on the weighted mean of eight values obtained in the past few years, recommended an updated G value of $G_{2006} = 6.67428(67) \times 10^{-11}$ $m^3kg^{-1}s^{-2}$ with a relative standard uncertainty $u_r =$ 100 ppm (part per million).[3] This value exceeds the 2002 recommended value, $G_{2002} = 6.6742(10) \times 10^{-11}$ $m^3kg^{-1}s^{-2}$, by the fractional amount 12 ppm,[4] and the uncertainty of the 2006 value is two-thirds that of the 2002 value. Although the situation of G measurements has been improved considerably since 1998, the values in CODATA 2006 are still in poor agreement. Even for the four most precise values of G with their assigning uncertainties within 50 ppm,[5–8] they are only consistent with each other in the range of about 200 ppm. It seems clear that improving the accuracy of the G measurement, further investigating and depressing more possibly systematic errors is needed. Here we report our new determination of G by means of the time-of-swing method, some of which has already appeared in a concise Letter form,[9] and this paper presents a main discussion on measurements of some systematic errors, such as the density inhomogeneity of the pendulum and the sources masses, the distribution of the coating layer on the pendulum and the fiber's anelasticity, and so on.

The time-of-swing method, developed by Heyl,[10] has been used to measure G commonly.[11–17] In this method, a torsion pendulum is suspended by a very thin fiber, and two source masses are placed aside (as shown in Fig. 1). In their absence, the pendulum's free oscillation frequency squared ω_0^2 is related to the pendulum's moment of inertia I and the fiber's torsion constant K by $\omega_0^2 = K/I$. Due to the gravitational interaction between the pendulum and the source masses, the torsion frequency will be perturbed and it is changed as

$$\omega^2 = \frac{K + GC_g}{I}, \tag{1}$$

where C_g is the gravitational coupling coefficient determined by the mass distributions of the pendulum and source masses. Therefore, the frequency squared of the pendulum in the presence of the source masses at the "near"

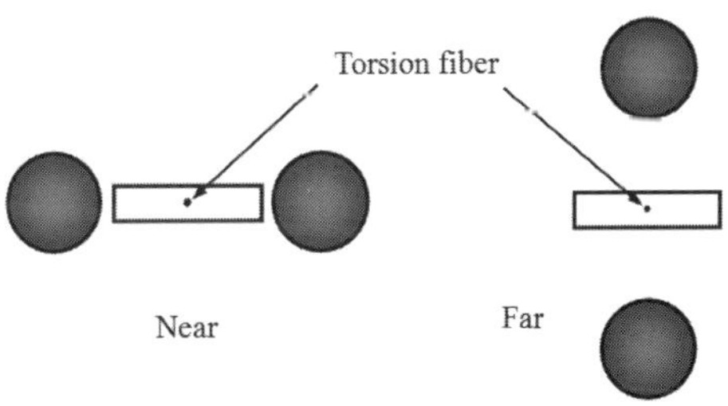

Fig. 1. Top view of the two configurations of the pendulum and the source masses in the time-of-swing method. At "near" configuration, the equilibrium position of the pendulum is in line with the source masses, and at "far" position, it is perpendicular.

or "far" positions (denoted with the subscripts "n" and "f", respectively) is $\omega_{n,f}^2 = (K_{n,f} + GC_{gn,gf})/I$. Then the G can be determined as

$$G = \frac{I(\omega_n^2 - \omega_f^2) - (K_n - K_f)}{C_{gn} - C_{gf}} = \frac{I\Delta(\omega^2)}{\Delta C_g}[1 - \frac{\Delta K}{I\Delta(\omega^2)}], \quad (2)$$

where $\Delta K = (K_n - K_f)$ represents the possible change of the spring constant of the fiber due to the fiber's anelasticity as proposed by Kuroda.[18]

The time-of-swing method has a nice merit that it converts the weak gravitational interaction into a change of frequency which could be measured with a higher precision, and it requires no calibration of the fiber's torsion constant. However, for a non-ideal torsion fiber, the major shortcoming is to need our knowledge of the fiber's properties with sufficient accuracy, such as its response to the ambient temperature, the pendulum's oscillation amplitude and frequency, time, and so on. Another difficulty, just like most of G measurements with other methods, is determining with sufficient accuracy the dimensions and the density distribution of both the torsion pendulum and the source masses. The main progresses of our G measurement with time-of-swing method will be introduced as follows.

2. The experimental design

2.1. *General description*

A schematic drawing of the apparatus is shown in Fig. 2. The pendulum was a gold-coated rectangular quartz block with a length of 91.46546(13) mm, a width of 12.01471(5) mm, a height of 26.21618(7) mm and a mass of 63.38388(21) g, respectively. The pendulum was suspended by a 25-μm-diameter, 890-mm-long tungsten fiber (W99.4/Th 0.6) linked with a cylindrical aluminum clamp, which was glued centrically on the pendulum with the departures of 16(4) μm in X, Y direction. The upper end of the fiber was attached to a magnetic damper to suppress the simple pendulum motions.[19,20] The magnetic damper consisting of a copper disk and an aluminum shaft suspended between two annular permanent magnets was hung from a 50-μm-diameter, 70-mm-long pre-hanger tungsten fiber, which was finally linked to a feedthrough fastened on the top of the vacuum chamber. The attitude angel of the pendulum after suspended by the fiber was $\theta_x = 4.06(5)$ mrad and $\theta_y = 1.91(2)$ mrad (as shown in Fig.2.), measured by an autocollimator when the pendulum was twisted by 180° around the fiber.

The source masses are two SS316 stainless steel spheres with the vacuum masses of 778.1794(9) g and 778.1763(9) g and diameters of 57.15123(24)

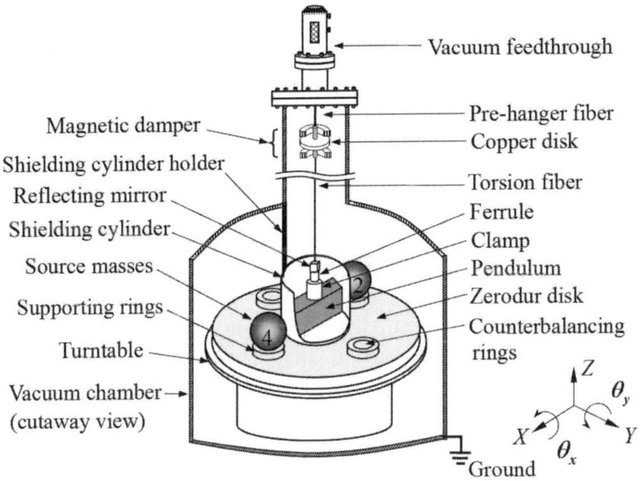

Fig. 2. (Color online) The schematic diagram of the two-stage pendulum system and the source masses with the time-of-swing method in our G measurement. The directions in laboratory frame is also shown in this figure

mm and 57.15074(21) mm, respectively. Four small identical Zerodur rings, two of them used to support the source masses and the others acted as the gravitational counterbalances, are symmetrically adhered on a Zerodur disk with a diameter of 240 mm and a thickness of 20 mm, which is fixed on a turntable driven by a Huber-410 stepper motor with a minimal increment of 0.00005 °/step. Because of the extremely low thermal expansion coefficient for Zerodur of $(0 \pm 1) \times 10^{-7}/\,^{\circ}\mathrm{C}$, the variations of the separation between the mass centers of the source masses caused by the temperature fluctuation is negligible. The rings and the disk are covered by a thin aluminum film to keep the source masses being well electrical-grounded. A thinner hollow gold-coated aluminum cylinder is insertod between the pendulum and the source masses (as shown in Fig. 2), which is treated as the electrostatic shielding and evidently improved the stability of the pendulum's period. The distance between the geometric center of the source mass was measured[21,22] to be 157.16154(37) mm.

The vacuum chamber was maintained at a pressure of $\sim 10^{-5}$ Pa by an ion pump. The apparatus is located in a mu-metal shielded room of $5 \times 3 \times 3.5\ \mathrm{m}^3$ in dimension, which provided the effective magnetic shielding. The experiment site was located in our cave laboratory.

The oscillation of the pendulum was monitored by an optical lever, and the output signals were sampled by a personal computer at a rate of 2 Hz with a frequency accuracy and stability of $\pm 5\times10^{-9}$ Hz and $\leq 3\times10^{-10}$/day, respectively. Six temperature sensors (four inside the chamber), two tiltmeters, and one barometer were also recorded synchronically.

2.2. *Density inhomogeneity of the pendulum and the sources masses*

2.2.1. *The pendulum*

The pendulum is made of fused silica glass, and its impurity content is less than 1 ppm. The density inhomogeneity of the pendulum was investigated by an optical interference method as described in Ref. 23. As we know, the local variations of the refractive index of the glass has a linear relationship with the density fluctuation.[24] Therefore, measuring the changes of the refractive index over the glass block could yield the information on the local density inhomogeneities of the pendulum. The fused silica glass pendulum was placed in one of the optical arms in the Michelson-type interferometer and moved along one direction driven by a one-dimensional translation stage.

If the pendulum's density changed, the refractive index of the quartz glass, in turn, the interference fringe would shift correspondingly. The observed relative variations of the densities over a volume with size of $5 \times 5 \times 5$ mm^3 was $\Delta\rho/\rho = (0.99 \pm 0.92) \times 10^{-5}$. Under an assumption that glass pendulum is subjected to the same density distribution, the density inhomogeneity of the pendulum would introduce u_r <0.21 ppm to G value by performing integral over the whole pendulum body.[23]

2.2.2. *The sources masses*

The density inhomogeneity of the source masses was investigated by different approaches. First, several slices cut from another SS316 stainless steel sphere with the same quality as the source masses were scanned by a scanning electron microscopy method (SEM). The statistical results from 210 samples showed that the relative variation of the density was better than 5.9×10^{-4} over a volume of $0.027 \times 0.034 \times 0.005$ mm^3 and it introduced $u_r \leq 0.034$ ppm to G value.[25] In addition, we exchanged the positions and also changed the orientations of the two source masses and repeated the G measurement. The result showed that the second G value was 9.0 ppm smaller than the first one, so a half of which was chosen as the error due to the density inhomogeneity of the source masses.

2.3. *Coating layer on the pendulum*

To minimize the electrostatic potential variations in the surface electrical potential of the torsion pendulum, its surface was covered by coating two thin metal layers using the magnetron sputtering. During the coating, the metal target was fixed and the sample was rotated at a rate of $\sim$ 20 circles per minute. The sub-layer used a pure copper with the purity of 99.999% to acquire a better adhesive property, and the surface layer was coated with pure gold (purity of 99.999%) to act as a conductive materials. The mass of the coating layers determined from the mass difference before and after coating was 49.863(26) mg, and the mixture density of the coating layer was estimated to be 15.83(1.00) g/cm^3. An identical pendulum coated with the same process was used to investigate the inhomogeneity of the coating layer.[26] The pendulum was cut into 19 small blocks along the length direction, a typical photo of which is shown in Fig. 3.

The coating area of each small block was measured by the 3-dimensional video instrument, and the average thickness of the coating layer was determined by comparing the mass changes of the blocks at different position on the pendulum before and after coating. The coating and measurements were repeated 3 times for each pairs of surface. During each coating, one of the end surfaces was coated, the coating thickness of which was chosen as the reference for comparisons. The statistic results revealed that the thickness of coating layer was almost uniform for each surface, but varied for different surfaces. The measured average ratio of thickness on the lateral (large) surface to that of the end one was 0.824(41), and the bottom surface to that of the end one was 0.782(23). Due to the aluminium clamp adhered

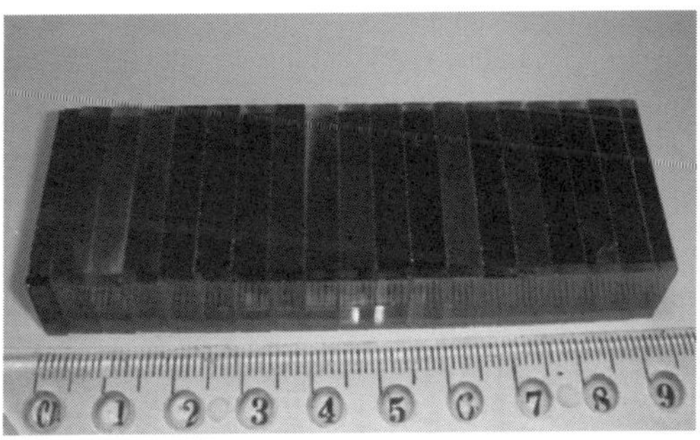

Fig. 3. (Color online) Photo of 19 small blocks after coated by a thin pure copper, which were cut from the glass pendulum with the same dimensions as that used in the G measurement. They are numbered as 1 to 19 from left to right.

in the center of the top surface, the thickness ratio was taken on a symmetrical exponential distribution as $h(x) = 0.524 - 0.299e^{|-x/14.549\text{mm}|}$ with the correlation coefficient of 0.981, where x was the distance from the center to the end. The result showed that the coating layers contributed a downward correction of 24.28(4.33) ppm to our final G value. If the thickness distribution is uniform, it's correction would be only 2.7 ppm.

2.4. *Direct measurement of the anelasticity of the torsion fiber*

According to Eq.(2), the systematical upward bias induced by the anelasticity of the fiber can be expressed in terms of G as

$$\frac{\Delta G}{G} = \frac{1}{I}\frac{K(\omega_n) - K(\omega_f)}{\omega_n^2 - \omega_f^2} = \frac{\Delta K}{I\Delta(\omega_{nf}^2)}. \tag{3}$$

With the typical tungsten fiber's quality factor Q being about 1700 for our G measurement, the anelasticity of the fiber should bring an upward bias about 200 ppm as predicted,[17,18] which was 4 times larger than our goal and it should be determined accurately.

To measure the fiber's anelasticity directly, two additional disk pendulums with different moments of inertia (I_1 and I_2) were used in another experiment.[27] The eigen oscillation periods of the two pendulums, suspended by the tungsten fiber respectively, were about 581.2 s and 488.2 s, covered the range of the periods in our G measurement.[9] By exchanging the two disk pendulums, the upward bias to G value due to the anelasticity could be further expressed as

$$\frac{\Delta G}{G} = \frac{I_1}{I_G[(\omega_2/\omega_1)^2 - 1]}\left[\left(\frac{I_2}{I_1}\right)\left(\frac{\omega_2}{\omega_1}\right)^2 - 1\right], \tag{4}$$

where ω_1 and ω_2 are the eigen frequencies when the two pendulums were suspended by the tungsten fiber, respectively. Obviously, to measure G with an expected accuracy, the parameters ω_1, ω_2, I_1 and I_2 of two pendulums should be measured with the same relative accuracy correspondingly. However, it is more difficult to determine the moments of inertia than the oscillation frequencies of the pendulums. In order to solve the problem, a high-Q quartz fiber ($Q \sim 3.36 \times 10^5$), whose anelastic effect can be negligible, to determine the ratio I_2/I_1 (but not the I_1 or I_2 itself) of the moments of inertia by measuring the the free oscillation periods of two disk pendulums, which were about 29.6 s and 35.3 s, respectively.

Twelve sets of experimental data with two pendulums suspended by the quartz fiber (four sets) and the tungsten fiber (eight sets) alternately were taken. The ratio of two moments of inertia of two pendulums was obtained as $I_2/I_1 = 0.7056832(8)$, and the ratio of squared frequencies for the two disk pendulums suspended by the tungsten fiber is yielded as $(\omega_2/\omega_1)^2 = 1.4171723(92)$. The correction to G due to the anelasticity of the tungsten fiber was

$$\frac{\Delta G}{G} = (211.80 \pm 18.69) \text{ ppm}, \tag{5}$$

with one of the moment of inertia being $I_1 = 5.3318(8) \times 10^{-5}$ kgm^2 obtained by measuring the geometry dimensions and the mass of one pendulum. The anelasticity effect of the fiber was the largest uncertainty attributed to our G measurement.

2.5. $\Delta C_g/I$ *and* $\Delta(\omega^2)$

The moment of inertia of the pendulum was calculated to be $I = 4.505679(35) \times 10^{-5}$ kgm^2, and the ratio $\Delta C_g/I = 25202.85(28)$ kgm^{-3} could be calculated according to the mass distributions of the pendulum system and the sources masses. In order to obtain the value of $\Delta(\omega^2)$, ten sets of experimental data with the source masses in near and far positions were taken alternately, and the duration was about 3 days for each position. The oscillation periods of the pendulum was extracted from the angle-time data by a correlative method,[28,29] and the result is yielded as $\Delta(\omega^2) = 1.682245(31) \times 10^{-6}$ s^{-2}, which has been discussed in detail in Ref. 9.

3. Experimental results

By directly averaging the two values of G yielded from the the two individual experiments, the combined final value of G was obtained to be

$$G = (6.67349 \pm 0.00018) \times 10^{-11} \text{ m}^3\text{kg}^{-1}\text{s}^{-2}, \tag{6}$$

with a relative uncertainty of 26.33 ppm. The 1σ error budget are listed in Table. 1.

The comparison with other results of u_r <50 ppm[5–8] and the CODATA 2002 and 2006 recommended values[3,4] are listed in Fig. 4.

4. Conclusions

By using a flat plane pendulum together with a pair of spheres with the time-of-swing method, the G value is determined in this work with an uncertainty of u_r <26.33 ppm, which is consistent with our previous G value in one σ uncertainty. The largest error source in this experiment is due to the fiber's anelasticity. For our future improvements, we intend to use a higher Q quartz fiber to measure the G value directly, and further improve the stability of the pendulum's periods to suppress statistical uncertainty of $\Delta(\omega^2)$.

Table 1. The 1σ error budget in G measurement. (unit: ppm).

Error source	Correction	$\delta G/G$	
$\Delta C_g/I$		11.85	
Fiber:		18.76	
nonlinearity			<0.70
thermoelasticity	-39.83(8.37)		1.52(0.82)
anelasticity	-211.80		18.69
aging			<0.01
Gravitational nonlinearity	7.73(4.79)	0.30(0.20)	
Magnetic damper	17.54	0.31	
Magnetic field		0.40	
Electrostatic field		0.10	
Statistical $\Delta\omega^2$		18.43(23.31)	
Total		**<28.86(<32.17)**	
Combined		**<26.33**	

Note: The values in the round brackets were yielded from the second experiment, others without the round bracket represent the same values in two experiments.

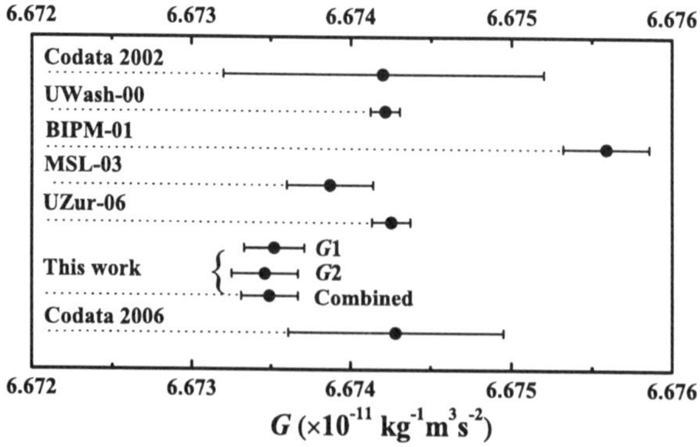

Fig. 4. The comparison of our result with others of uncertainties within 50 ppm[5–8] as well as the CODATA 2002 and 2006 recommended values.[3,4]

Acknowledgements

We are very appreciative of R. D. Newman of UCI and V. Milyukov of MSU for their valuable discussions and suggestions. We are also thank National Institute of Metrology of Peoples Republic of China which provided many certification reports for our experimental equipments and instruments. This research is supported in part by the National Basic Research Program of China (Grant No. 2010CB832801), the National High-Tech Research and Development Plan of China (Grant No 2008AA12A215), and the National Natural Science Foundation of China (Grant No. 10805021).

References

1. H. Cavendish, *Philos. Trans. R. Soc.* (London), **88**, 469 (1798).
2. G. T. Gillies, *Rep. Prog. Phys.* **60**, 151 (1997).
3. P. J. Mohr, B. N. Taylor and D. B. Newell, *Rev. Mod. Phys.* **80**, 633 (2008).
4. P J. Mohr and B. N. Taylor, *Rev. Mod. Phys.* **77**, 1 (2005).
5. J. H. Gundlach and S. M. Merkowitz, *Phys. Rev. Lett.* **85**, 2869 (2000).
6. T. J. Quinn *et al.*, *Phys. Rev. Lett.* **87**, 111101 (2001).
7. T. R. Armstrong and M. P. Fitzgerald, *Phys. Rev. Lett.* **91**, 201101 (2003).
8. S. Schlamminger *et al.*, *Phys. Rev. D.* **74**, 082001 (2006).
9. J. Luo *et al.*, *Phys. Rev. Lett.* **102**, 240801 (2009).
10. P. R. Heyl, *J. Res, Natl. Bur. Stand.* **5**, 1243 (1930).
11. P. R. Heyl and P. Chrzanowski, *J. Res, Natl. Bur. Stand.* **29**, 1 (1942).
12. M. U. Sagitov *et al.*, *Dokl. Akad. Nauk SSSR* **245**, 567(1979) [*Sov. Phys. Dokl.* **245**, 20 (1981)].
13. O. V. Karagioz and V. P. Izmailov, *Izmer. Tekh.* **39**, 3 (1996). (*Meas. Tech.* **39**, 979(1996)).
14. G. G. Luther and W. R. Towler, *Phys. Rev. Lett.* **48**, 121 (1982).
15. C. H. Bagley and G. G. Luther, *Phys. Rev. Lett.* **78**, 3047 (1997).
16. J. Luo *et al.*, *Phys. Rev. D.* **59**, 042001 (1998).
17. R. D. Newman and M. K. Bantel, *Meas. Sci. Technol.* **10**, 445 (1999).
18. K. Kuroda, *Phys. Rev. Lett.* **75**, 2796 (1995).
19. Y. Tu *et al.*, *Phys. Lett. A.* **331**, 354 (2004).
20. X. D. Fan *et al.*, *Phys. Lett. A.* **372**, 547 (2008).
21. J. Luo *et al.*, *Chin. Phys. Lett.* **18**, 1012 (2001).
22. D. C. Chen *et al.*, *Chin. Phys. Lett.* **21**, 33 (2004).
23. L. X. Liu *et al.*, *Chin. Phys. Lett.* **25**, 4203 (2008).
24. B. Marler, *Phys. Chem. Minerals* **16**, 286 (1988).
25. L. X. Liu *et al.*, *Chin. Phys. Lett.* **26**, 010403 (2009).
26. L. X. Liu *et al.*, *Chin. Phys. Lett.* **26**, 090402 (2009).
27. S. Q. Yang *et al.*, *Phys. Rev. D.* **80**, 122005 (2009).
28. Y. L. Tian, Y. Tu and C. G. Shao, *Rev. Sci. Instrum.* **75**, 1971 (2004).
29. F. Chen *et al.*, *Meas. Sci. Technol.* **14**, 619 (2003).

CRYOGENIC TEST OF THE GRAVITATIONAL INVERSE-SQUARE LAW

HO JUNG PAIK*, KRISHNA Y. VENKATESWARA and M. VOL MOODY

Department of Physics, University of Maryland,
College Park, MD 20742, U.S.A
**E-mail: hpaik@umd.edu*

VIOLETA PRIETO

U.S. Army Research Laboratory,2800 Powder Mill Road,
Adelphi, MD 20783, U.S.A
E-mail: violeta.prieto@us.army.mil

In string theories, the law of gravity changes as the distance r is reduced to below the radius of compactification, R. Gravity-only large extra dimensions with $R \sim 100$ μm could solve the gauge hierarchy problem in grand unification. On the other hand, the observed accelerating expansion of the universe is consistent with a vacuum-energy density 10^{120} times smaller than the value predicted from the quantum corrections to it. This enormous discrepancy could be eliminated if gravitons were 'fat' with the size of the order of 50 μm. Both of these possibilities would lead to a violation of the $1/r^2$ law below 100 μm.

We are conducting a sub-millimeter test of the $1/r^2$ law at 4.2 K. To minimize Newtonian errors, the experiment employs a near-null source, a circular disk of large diameter-to-thickness ratio. Two test masses, also disk-shaped, are positioned on the two sides of the source mass at a nominal distance of 260 μm. The test masses are connected to a SQUID to form a superconducting differential accelerometer. As the source is driven sinusoidally, a violation signal should appear at the second harmonic due to symmetry. The detector is rigidly mounted on the housing and the source mass is suspended separately and driven from outside the housing. The source is driven at its resonance frequency. In this paper, we discuss the design, principle, and progress of this experiment.

Keywords: Test of gravity, Cryogenic, Superconducting accelerometer; ICGA 09.

1. Objective

In string theories, the N extra dimensions are compactified, leaving the observed $3 + 1$ space-time dimensions. For $n = 2$ 'large' extra dimensions, the law of gravity changes from $1/r^2$ to $1/r^4$, as r is reduced to below the radius of compactification, R. Different theories that include extra dimensions have been proposed as a solution to the hierarchy problem (the large discrepancy between the electro-weak scale and the gravitational scale). Some theories involve large extra dimensions ($R > 1$ μm), and thus are potentially observable in laboratory-scale experiments.[1–3] This would imply a violation of the inverse-square law at $r \leq R$. A discovery of such a deviation would be ground breaking.

The 'cosmological constant problem' provides further motivation. Present observations favor an accelerating expansion of the universe. This implies a tiny but non-vanishing cosmological constant, which corresponds to a vacuum-energy density 10^{120} times smaller than the value predicted from quantum corrections. This enormous discrepancy could be eliminated if gravity were cut off at ~ 100 μm[4,5] ; hence, the cosmological constant problem strongly motivates sub-millimeter range tests of gravity.

This experiment aims to test the inverse-square law of gravitation to better than 1 part in 100 at a 100 μm range. Figure 1 shows the present limits on the coupling (α) and range (λ) of deviations in a generalized Yukawa potential of the form[6–9] :

$$V(r) = -G\frac{m_1 m_2}{r}[1 + \alpha e^{-r/\lambda}]. \tag{1}$$

The predictions by higher dimensional string theory and the axion theory are also shown, as well as the expected sensitivity of the University of Maryland experiment. The expected resolution of this experiment at 20 μm $< \lambda < 200$ μm represents an improvement by more than an order of magnitude beyond the present limits.[7]

2. Principle and Design

To maximize the masses that can be brought to 100 μm from each other, disk geometry is used for both the source and test masses. An infinite plane slab is a Newtonian null source. We approximate this by using a circular disk of large diameter-to-thickness ratio. Figure 2 shows the configuration of the source and test masses with associated coils. Two disk-shaped test masses are suspended on two sides of the source and are coupled magnetically to form a differential accelerometer. As the source mass is driven at resonance at frequency f along the symmetry axis, the first-order Newtonian fields arising from the finite diameter of the source mass are canceled upon differential measurement, leaving only a second-order error at $2f$. By symmetry, the Yukawa signal also appears at $2f$. As an improvement

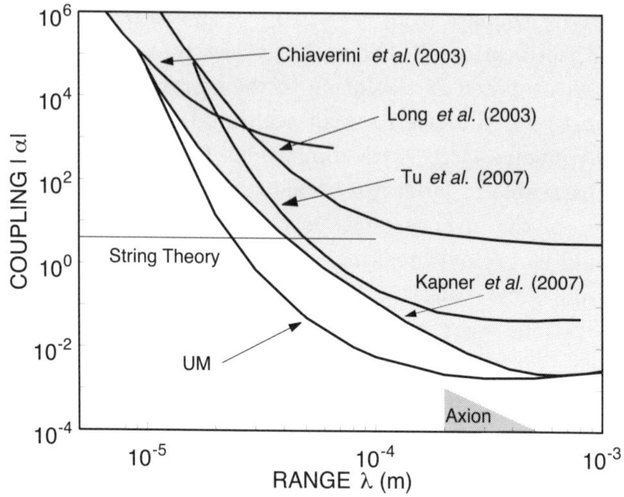

Fig. 1. The expected sensitivity of the University of Maryland experiment (95% confidence) versus the existing limits.

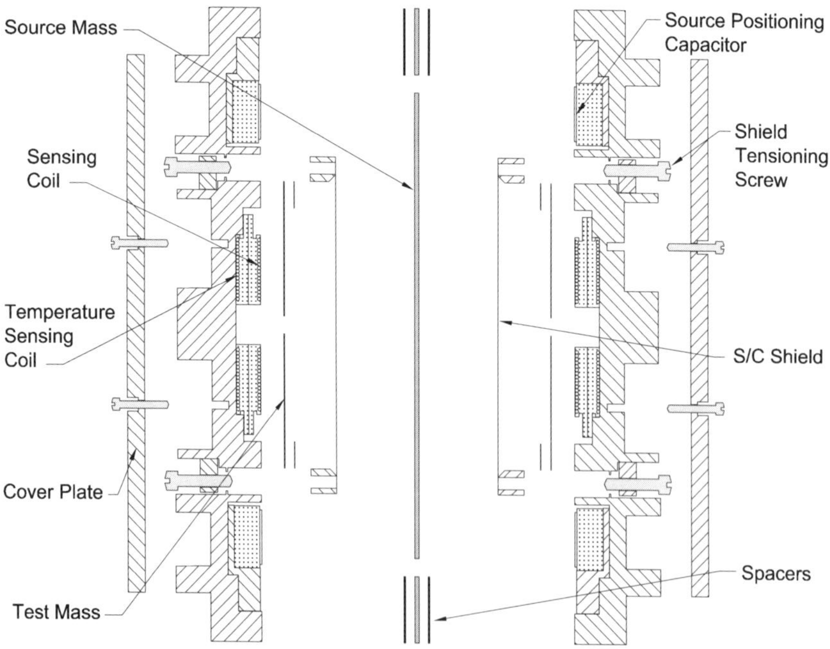

Fig. 2. Expanded cross section of the experiment.

over the previous design,[10] the source mass and the detector are suspended separately from the top of the cryostat, thus minimizing their interactions.

Figure 3 shows a schematic of the test cryostat. To eliminate differential contraction and provide good electromagnetic shielding, the entire housing is fabricated from niobium (Nb). The Nb source mass is suspended by a pair of phosphor bronze wires and driven magnetically with the help of two permanent magnets attached to the suspension wires, and a coil at the top of the cryostat. The test masses are also made out of Nb and suspended by cantilever springs. A 25-μm thick Nb foil provides electrostatic and magnetic shielding between the source and each test mass. The detector

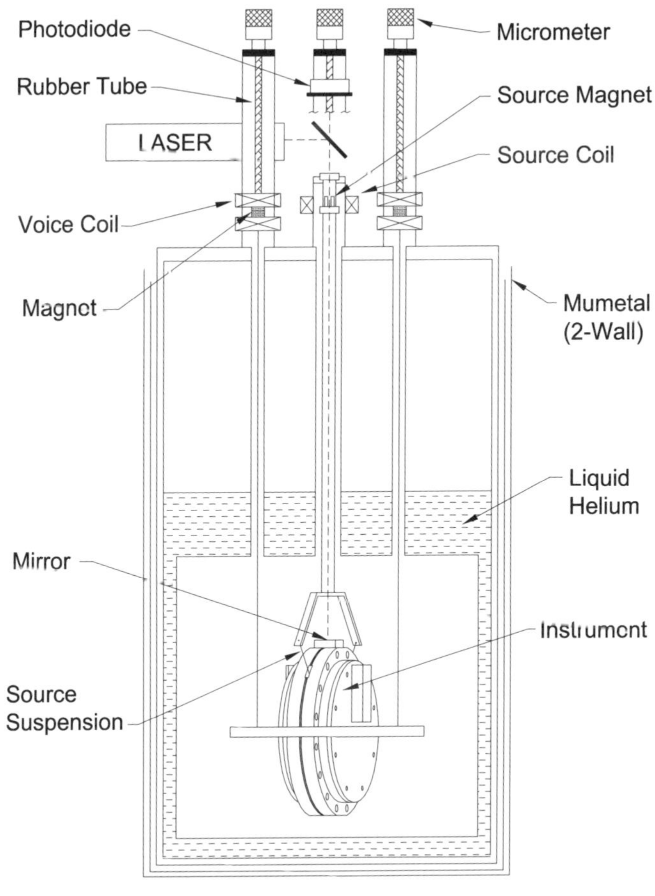

Fig. 3. Schematic of the test cryostat.

is suspended from the top of the cryostat via three rubber tubes. Voice-coil transducers, incorporated into each vertical leg of the suspension, are used to shake the instrument for balance and calibration. By varying the magnitude and phase of the current through the coils, vertical acceleration or tilt in any direction is applied to the instrument. Additional alignment coils are placed around the detector, which allows us to translate it along the experiment axis or rotate it about the vertical. This allows control over the source location with respect to the detector. The tilt of the detector is sensed with a two-dimensional optical lever consisting of a laser, a beam splitter, an x-y photodiode, and a planar mirror mounted at the top of the instrument.

2.1. *Source and test masses*

The source is a disk 2.90-mm thick and 165 mm in diameter, with a mass of 530 g. The source mass was polished to a mirror finish by double disk grinding and then coated with gold (Au) using physical vapor deposition. The test masses are matched Nb disks 250-μm thick by 70 mm in diameter. Their dynamic mass is $m = 8.7$ g. The mechanical resonance frequency of the test mass is about 11 Hz. The equilibrium spacing between the source and each test mass is 260 μm. They are shielded from each other by means of a 25-μm thick Nb shield, which is also coated with gold. The design allows a source amplitude of 130 μm. Figure 4 shows the expected differential acceleration signals as a function of the source position. It compares the signal from the Yukawa force with $\alpha = 10^{-2}$ and $\lambda = 100$ μm, the Yukawa force with $\alpha = 1$ and $\lambda = 35$ μm, and ten times the Newtonian error arising from the finite source mass diameter.

2.2. *Superconducting circuits*

Figure 5(a) is the differential-mode (DM) sensing circuit. A persistent current is stored through the loop comprising the parallel combination of L_{D1} and L_{D2} and the transformer primary, where L_{D1} and L_{D2} are pancake coil inductors facing the two test masses. Another current is stored in the loop comprised of the series combination of L_{D1} and L_{D2}, which in turn tunes the ratio I_{D2}/I_{D1} without modifying the value $I_{D2} + I_{D1}$. For redundancy, the CM circuit, 5(b), is designed to be physically identical but with a different current configuration. Figure 5(c) shows one of the alignment circuits used to center and align the detector with respect to the source mass.

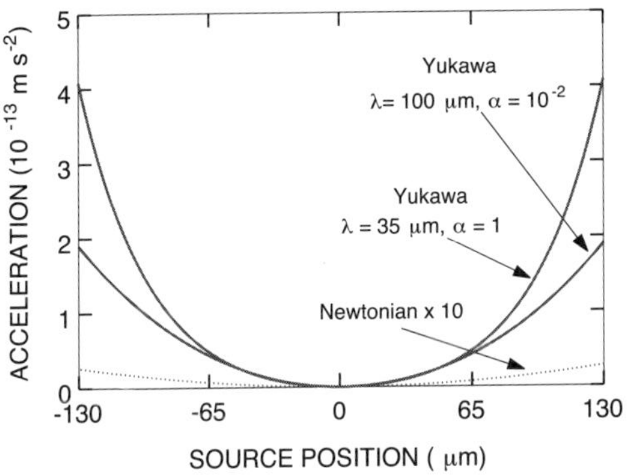

Fig. 4. Newtonian and Yukawa signals versus source position.

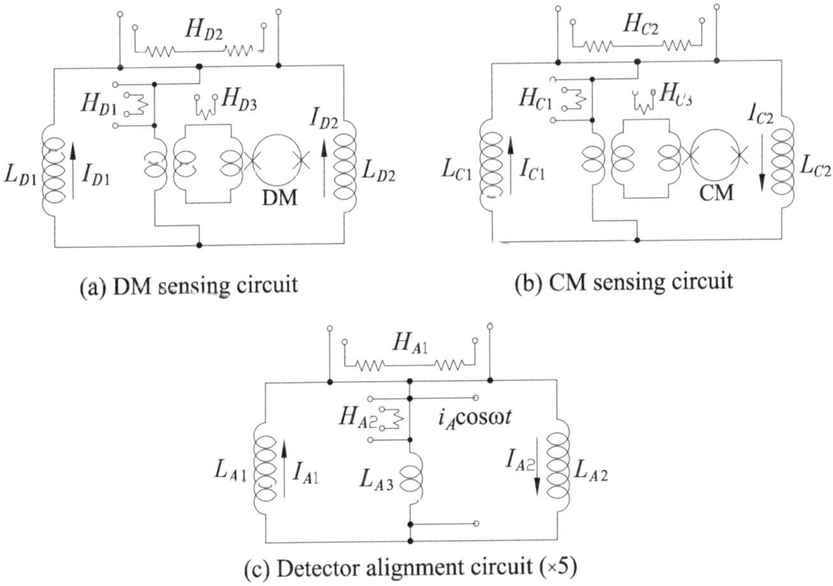

Fig. 5. Superconducting circuits for the detector.

2.3. *Capacitor plates*

The source position and amplitude are sensed by four capacitor plates located on each side of the source. The capacitor plates were cut out of 125-

μm thick Nb foil, glued to a Macor back plate and lapped. They were then coated with a 100-nm layer of gold to minimize patch field type interactions with the source.[11]

Two capacitor plates on opposite sides of the source are connected to form a Wheatstone-type capacitor bridge. As the source mass moves, the bridge output is recorded and fit to the prediction of a two-parameter model, with the source amplitude x_S and offset from the center x_0 as unknowns. Figure 6 shows an example, which yielded $x_S = 102\ \mu$m and $x_0 = -9\ \mu$m.

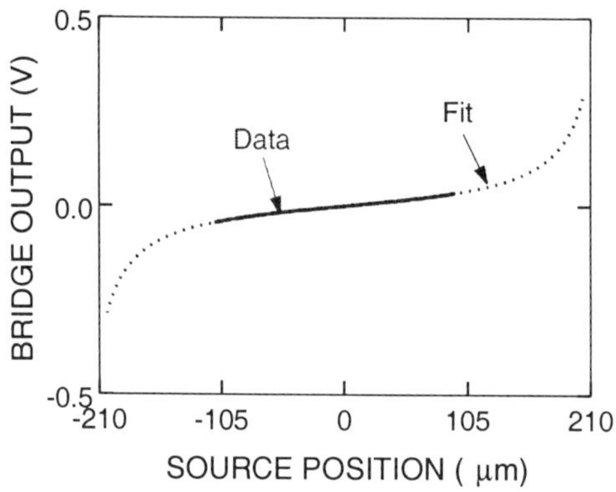

Fig. 6. Source position and amplitude determined from the capacitor bridge.

3. Procedure and Error Analysis

The experiment is mounted on a cryostat insert and enclosed in an aluminum (Al) vacuum chamber. A cryo-pump in the form of a charcoal getter is also installed in order to achieve the very low pressures required for the experiment. After evacuating the chamber, it is cooled to liquid He temperature (4.2 K).

Persistent currents are stored in the alignment circuits to align the detector with respect to the source mass. Currents are then stored in the sensing circuits to make them sensitive to motion of the test masses. The source mass is then driven at resonance in the pendulum mode, and its position and amplitude are read out from the capacitor bridge, as discussed above. The source amplitude is maintained and the two SQUID output channels, three capacitor bridge channels, and two tilt meter channels are recorded for a period of 10 – 12 hours for each data set.

In addition, a gravitational calibration signal in the form of two lead (Pb) bricks mounted on a turn table was used to directly calibrate the DM output. The CM output was calibrated by tilting the detector by known amounts and using earth's gravity to provide a common acceleration to the two test masses. Based on a preliminary analysis and observation, we have identified two dominant sources of error, which we label as a magnetic cross-talk error and a patch-field error.

3.1. *Magnetic cross-talk*

Even without any sensing current, the sensing circuits can be sensitive to test mass motion due to some residual magnetic flux trapped through the test masses as they cool below their superconducting transition temperature. This sensitivity can be reduced or eliminated by storing very small currents in the sensing circuit to counter the trapped flux. After the residual sensitivity to test mass motion was canceled, we observed that there still was a signal (fundamental and higher harmonics) in both SQUID outputs related to the source motion. A likely mechanism for this signal is due to magnetic flux, which is trapped through both the superconducting source mass and the shields, that couples directly into the sensing circuits, thus introducing a cross-talk between the source and the sensing circuits. Consequently, this cross-talk is also present when we store the full sensing current and thus constitutes a large error term. We found this cross-talk to vary with the relative position between the source and the detector, and also with the position of the test masses with respect to the shields.

This effect can be removed by making two measurements with the direction of the sensing currents reversed and subtracting the two results. Any real signal arising from the motion of the test masses *will* change sign with the reversed current but the cross-talk *will not*. So by differencing the two measurements, we can cancel out the cross-talk while doubling the acceleration signal.

3.2. *Patch-field forces*

When two metals are placed in contact with each other, a potential difference can appear between them. This difference depends on the difference in the work functions of the two metals and it originates from the difference between the electrochemical potentials of the metal. The electrochemical potential is the minimum energy needed to remove an electron from a metal surface. The difference in the electrochemical potential can arise either due to difference in the energies of the electrons in the material for different materials (known as 'contact potential'), or due to different surface properties of the material. For example, when the surface of a polycrystalline material is composed of patches with different crystallographic orientation. This is known as 'patch effect'. In addition to the crystallographic orientation, contaminants, adsorption layers, or temperature differences can affect the patch potential distribution and magnitude.[11,12] This type of coupling between the source mass and the detector can be an important source of error.

We found a rather large second and higher harmonic signal in the CM SQUID output related to the source motion. We attribute this to the patch field. Although gold was deposited on the surfaces of the source, shields, and capacitor plates, the size of the patches could be large compared to the gap due to surface contamination.[11] The observed force corresponds to an average patch voltage of ~ 0.2 V. The CM rejection is limited to a factor of 500; hence, a residual (unbalanced) CM signal is present in the DM SQUID output.

In the CM channel, the fundamental term seems to be about 10 times smaller than the $2f$ term and was found to vary greatly with the relative position of the source with respect to the detector. One possible mechanism to model this, is a force which produces a torque on the detector twice, in the same direction, for every one cycle of the source motion. For instance, if the source were rotated with respect to the detector, an attractive force due to patch fields near the edges closer to the detector would produce a torque in the same direction, twice for every source cycle, thus producing a $2f$ error.

Regardless of the exact mechanism of this effect, we found a novel way to cancel the second-harmonic term in the CM output by applying a DC bias voltage to one of the capacitor plates, as shown in Figure 7. This electrostatic force mimics the patch-field force, and due to the nonlinearity and asymmetry, it produces a second-order term, which can be tuned to *cancel out* the CM signal to better than 10%. This procedure is safe since we apply only a CM force. We found that the DM signal is reduced proportionately (after the cross-talk term is removed), as expected from the unvaried CM rejection.

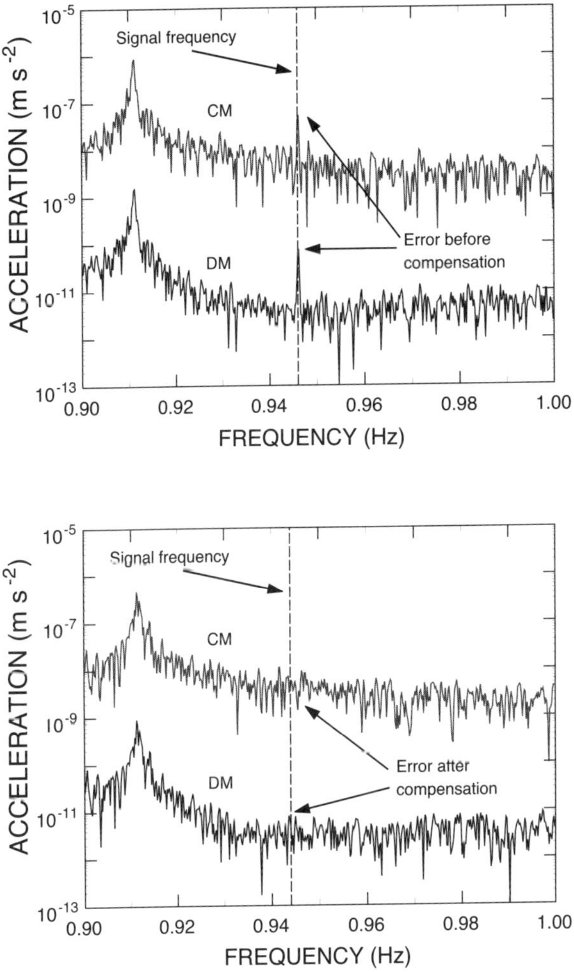

Fig. 7. Frequency response of the outputs of the sensing circuits before and after error compensation.

3.3. *Seismic noise*

If the test masses were perfectly aligned and centered with respect to each other, then the accelerometer would be insensitive to off-axis disturbances. In reality, misalignment between the test masses couples the ground (seismic) noise into the DM channel, thus reducing the sensitivity of the experiment. This currently limits the noise floor of the DM channel. In principle, this can be reduced by measuring the ground noise in the other two orthogonal directions and subtracting proportionally from the DM channel. We plan to add more coils and SQUIDs to measure the seismic noise in the future.

3.4. *Intrinsic noise of the detector*

The intrinsic power spectral density of a superconducting differential accelerometer can be written[13,14] as

$$S_a(f) = \frac{8}{m}\left[\frac{k_B T \omega_D}{Q_D} + \frac{\omega_D^2}{2\eta\beta} E_A(f)\right], \tag{2}$$

where m is the mass of each test mass, $\omega_D = 2\pi f_D$ and Q_D are the DM (angular) resonance frequency and quality factor, β is the electromechanical energy coupling coefficient, η is the electrical energy coupling coefficient of the SQUID, and $E_A(f)$ is the input energy resolution of the SQUID.

For our experiment, $T = 4.2$ K, $m = 8.7$ g, $f_D = 13$ Hz, $Q_D = 10^5$, $\eta = 0.25$, $\beta = 0.2$, and $E_A(f) = 1 \times 10^{-30}$ $(1 + 0.1\text{Hz}/f)$ J Hz^{-1} for the commercial dc SQUID used. This gives $S_a^{1/2}(f) = 1.2 \times 10^{-11}$ m s^{-2} Hz$^{-1/2}$ at $f = 0.94$ Hz.

4. Data Analysis

By combining the CM and DM calibration signals, the baseline of the differential accelerometer (distance between the centers of the test masses) is computed to be (3.65 ± 0.03) mm. Subtracting the source and the test mass thicknesses, the shield thicknesses and the estimated gap between the test masses and the shields, the average surface-to-surface distance between the source and each test mass is found to be (260 ± 40) μm. The source mass to shield gap is estimated to be (210 ± 40) μm.

At the time of writing this paper, we are accumulating the $1/r^2$ data to reduce the random noise. Here we present an analysis of a subset of our data.

In Figure 8, the observed residual DM signal is plotted against the residual CM signal for a source amplitude of 90 μm. Each data point represents the difference between two consecutive nights of data taken with the CM and DM currents reversed in direction. As we discussed in Section 3.2, by using this procedure, the cross-talk and the patch field errors have been canceled while the true acceleration signals have been added. The scatter in the data is partly due to imperfect cancelation of the CM force (patch

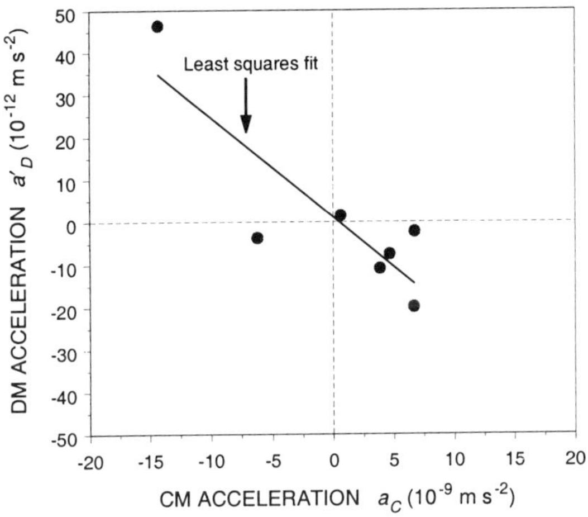

Fig. 8. Residual DM signal as a function of the residual CM signal and a least squares straight line fit.

field force) and the statistical noise in the experiment, dominated by the seismic noise.

Assuming the CM rejection remains constant throughout this data set, we can remove the unbalanced CM error further by fitting this data to a straight line:

$$a'_D = \Delta h_C \times a_C + a_D . \tag{3}$$

where Δh_C represents the residual CM sensitivity of the differential accelerometer and a_D is the true differential acceleration. The result is $a_D = (1.2 \times 10^{-12} \pm 7.6 \times 10^{-12})$ m s^{-2}.

By equating the result with the DM acceleration expected from the Yukawa term in Eq. 1, we find $\alpha < 2.3$ at $\lambda = 100$ μm and $\alpha < 1$ at $\lambda = 160$ μm for this set of data. This is a null result. At present, our result is limited mainly by the ability to resolve the DM signals due to insufficiently rejected seismic noise.

5. Improvements Planned

We plan to replace the Nb source mass with a source mass made out of tantalum (Ta). As the temperature of the new Ta source mass will be maintained above the superconducting transition temperature of Ta (4.46

K), it will not trap any magnetic flux through it and hence the cross-talk error should be significantly reduced.

We also plan to install additional accelerometers to measure motion of the detector in the other degrees of freedom. This will allow us to measure and subtract the seismic noise in our DM output thus enabling operation at the amplifier noise level. Furthermore, this will let us monitor any source induced detector motion in the other degrees of freedom, hence simplifying our error analysis considerably.

With these improvements, we hope to achieve a nearly amplifier noise limited sensitivity, enabling us to measure deviations from the inverse square law to the limit shown in Figure 1. To obtain this limit, the combination of the seismic noise and the other errors must be reduced to the level of the SQUID noise.

Acknowledgments

We have benefited from discussions with Fred Wellstood, Blas Cabrera, Aharon Kapitulnik, Francis Everitt, Sasha Buchman, and Jun Luo. We are indebted to Ron Norton for technical assistance and to Lvyuan Chen for analysis of source metrology error. This research was supported in part by the National Science Foundation under grants PHY0244966 and PHY0904614.

References

1. Arkani-Hamed, N., Dimopoulos, S., and Dvali, G., *Phys. Rev. D* **59**, 086004 (1999).
2. I. Antoniadis, N. Arkani-Hammed, S. Dimopoulos, and G. R. Dvali, *Phys. Lett. B* **36**, 257-263 (1998).
3. N. Arkani-Hammed, S. Dimopoulos, and G. R. Dvali, *Phys. Lett. B* **429**, 263-72 (1998).
4. Sundrum, R., Phys. Rev. D **69**, 044014 (2004).
5. L. Randall and R. Sundrum, *Phys. Rev. Lett.* **83**, 4690-469 (1999).
6. Tu L. C., *et al.*, *Phys. Rev. Lett.* **98**, 201101 (2007).
7. Kapner, D. J., *et al.*, *Phys. Rev. Lett.* **98**, 021101 (2007).
8. Long, J. C., et al., *Nature.* **421**, 922 (2003).
9. Chiaverini, J., *et al.*, *Phys. Rev. Lett.* **90**, 151101 (2003).
10. Paik, H. J., Prieto, V. A., and Moody, M. V., *Null Test of the Inverse Square Law at 100-Micrometer Distance*, in *Gravitation and Astrophysics: on the Occasion of the 90th Year of General Relativity*, eds. I. A. Nester, C. M. Chen and J. P. Hsu, Vol. Proceedings of ICGA-7, (World Scientific, Singapore, 2007), pp. 9-19.
11. N. A. Robertson, *et al.*, *Class. Quantum Grav.* **23**, 2665-2680 (2006).

12. C. C. Speake, *Class. Quantum Grav.* **13**, A291-A297 (1996).
13. Chan, H. A. and Paik, H. J., *Phys. Rev. D* **35**, 3551 (1987).
14. Chan, H. A., Moody, M. V., and Paik, H. J., *Phys. Rev. D* **35**, 3572 (1987).

TESTING RELATIVISTIC GRAVITY AND DETECTING GRAVITATIONAL WAVES IN SPACE*

WEI-TOU NI

Center for Gravitation and Cosmology, Purple Mountain Observatory, Chinese Academy of Sciences, No. 2, Beijing W. Rd., Nanjing, 210008 China

National Astronomical Observatories, Chinese Academy of Sciences, Beijing, 100012 China

e-mail: wtni@pmo.ac.cn

For testing gravity and detecting gravitational waves in space, deep-space laser ranging using drag-free spacecraft is a common method. Deep space provides a large arena and a long integration time. Laser technology provides measurement sensitivity, while drag-free technology ensures that gravitational phenomenon to be measured with least spurious noises. In this talk, we give an overview of motivations and methods of various space missions/proposals testing relativistic gravity and detecting gravitational waves, and refer to various references.

1. Introduction - Mapping the Gravitational Field in the Solar System

The solar-system gravitational field is determined by three factors: the dynamic distribution of matter in the solar system; the dynamic distribution of matter outside the solar system (galactic, cosmological, etc.) and gravitational waves propagating through the solar system. Different relativistic theories of gravity make different predictions of the solar-system gravitational field. Hence, precise measurements of the solar-system gravitational field test these relativistic theories, in addition to enabling gravitational wave observations, determination of the matter distribution in the solar-system and determination of the observable (testable) influence of our galaxy and cosmos. The tests and observations from various missions with various configurations could include: (i) a precise determination of the relativistic parameters with 3-5 orders of magnitude improvement over previous measurements; (ii) a 1-2 order of magnitude improvement in the measurement of $\dot{G}$; (iii) a precise determination of any anomalous, constant acceleration A_a directed towards the Sun; (iv) a measurement of solar angular momentum via the Lense-Thirring effect; (v) the

*This work is supported by the National Natural Science Foundation of China (Grant Nos. 10778710 and 10875171).

detection of solar g-mode oscillations via their changing gravity field, thus, providing a new eye to see inside the Sun; (vi) precise determination of the planetary orbit elements and masses; (vii) better determination of the orbits and masses of major asteroids; (viii) exploring the dynamic distribution of matter outside the solar system (galactic, cosmological, etc.) and testing large scale gravitational theories; (ix) detection and observation of gravitational waves from massive black holes and galactic compact binary stars in the low and middle frequency range 100 nHz to 10 Hz; and (x) exploring background gravitational-waves [1].

2. Methods of Measurement

For testing gravity and detecting gravitational waves in space, deep-space laser ranging using drag-free spacecraft is a common method. Deep space provides a large arena and a long integration time. Laser technology provides measurement sensitivity, while drag-free technology ensures that gravitational phenomenon to be measured with least spurious noises. Up to now, microwave technology is used for measuring the solar-system and testing solar-system gravity. However, this will be replaced by more precise laser technology. The precision has already been demonstrated by lunar laser ranging [2]. For last forty years, we have seen great advances in the dynamical testing of relativistic gravity. This is largely due to interplanetary radio ranging and lunar laser ranging [3]. Interplanetary radio ranging and tracking provided more stimuli and progresses at first. However with improved accuracy of 1 mm from 20-30 cm and long-accumulation of observation data, lunar laser ranging reaches similar accuracy in determining relativistic parameters as compared to interplanetary radio ranging.

There are various interplanetary optical missions proposed. As an example of interplanetary missions, we mention ASTROD; the baseline scheme of ASTROD is to have two spacecraft in separate solar orbits and one spacecraft near the Earth-Sun L1/L2 point carrying a payload of a proof mass, two telescopes, two 1-2 W lasers with spares, a clock and a drag-free system ranging coherently among one another using lasers. [1]

For technology development, we refer the readers to LISA proceedings and ASTROD proceedings.

3. Demonstration of Interplanetary Laser Ranging

Interplanetary laser ranging has been demonstrated by MESSENGER (MErcury Surface, Space ENvironment, GEochemistry, and Ranging) [4-6].

The MESSENGER spacecraft, launched on 3 August 2004, is carrying the Mercury Laser Altimeter (MLA) as part of its instrument suite on its 6.6-year voyage to Mercury. Between 24 May, 2005 and 31 May, 2005 in an experiment performed at about 24 million km before an Earth flyby, the MLA on board MESSENGER spacecraft performed a raster scan of Earth by firing its Q-switched Nd:YAG laser at an 8 Hz rate. Pulses were successfully received by the 1.2 m telescope aimed at the MESSENGER spacecraft in the NASA Goddard Geophysical and Astronomical Observatory at Gaddard Space Flight Center (GSFC) when the MLA raster scan passed over the Earth station. Simultaneously, a ground based Q-switched Nd:YAG laser at GSFC's 1.2 m telescope was aimed at the MESSENGER spacecraft. Pulses were successfully exchanged between the two terminals. From this two-way laser link, the range as a function of time at the spacecraft over 2.39×10^{10} m (~ 0.16 AU) was determined to ± 0.2 m (± 670 ps): a fractional accuracy of better than 1×10^{-11}.

A similar experiment was conducted by the same team to the Mars Orbiter Laser Altimeter (MOLA) on board the Mars Global Surveyor (MGS) spacecraft in orbit about Mars [6]. At that time, the MOLA laser was no longer operable after a successful topographic mapping mission at Mars. The experiment was one way (uplink) and the MOLA detector saw hundreds of pulses from 8.4 W Q-switched Nd:YAG laser at GSFC.

4. Missions for Testing Relativistic Gravity

In this section, we briefly review various missions for testing relativistic gravity and illustrate by looking into various ongoing / proposed experiments related to the determination of the PPN space curvature parameter γ. Some motivations for determining γ precisely to $10^{-5} - 10^{-9}$ are given in [7, 8].

First, we mention two recently completed experiments --- Cassini experiment and GP-B experiment. Cassini experiment is the most precise experiment measuring the PPN space curvature parameter γ up to now. In 2003, Bertotti, Iess and Tortora [9] reported a measurement of the frequency shift of radio photons due to relativistic Shapiro time-delay effect from the Cassini spacecraft as they passed near the Sun during the June 2002 solar conjunction. From this measurement, they determined γ to be 1.000021 ± 0.000023. GP-B experiment used quartz gyro at low-temperature to measure the Lense-Thirring precession and the geodetic precession in a polar orbit of the earth. The geodetic precession gives a measure of γ. The precision from their current analysis is around 0.003 [10].

Bepi-Colombo [11] is planned for a launch in 2013 to Mercury. A simulation predicts that the determination of γ can reach 2×10^{-6} [12].

GAIA (Global Astrometric Interferometer for Astrophysics) [13] is an astrometric mission concept aiming at the broadest possible astrophysical exploitation of optical interferometry using a modest baseline length (~3m). GAIA is planned to be launched in 2013. At the present study, GAIA aims at limit magnitude 21, with survey completeness to visual magnitude 19-20, and proposes to measure the angular positions of 35 million objects (to visual magnitude V=15) to 10 μas accuracy and those of 1.3 billion objects (to V=20) to 0.2 mas accuracy. The observing accuracy of V=10 objects is aimed at 4 μas. To increase the weight of measuring the relativistic light deflection parameter γ, GAIA is planned to do measurements at elongations greater than 35° (as compared to essentially 47° for Hipparcos) from the Sun. With all these, a simulation shows that GAIA could measure γ to $1\times10^{-5} - 2\times10^{-7}$ accuracy [14].

ASTROD I is a first step towards ASTROD. Its scheme is to have one spacecraft in a Venus-gravity-assisted solar orbit, ranging optically with ground stations with less ambitious, but still significant scientific goals in testing relativistic gravity. In the ranging experiments, the retardations (Shapiro time delays) of the electromagnetic waves are measured to give γ. In the astrometric experiments, the deflections of the electromagnetic waves are measured to give γ. These two kinds of experiments complement each other in determining γ. The Cosmic Vision ASTROD I (Single Spacecraft Astrodynamical Space Test of Relativity using Optical Devices) mission concept [15] is to use a drag-free spacecraft orbiting around the Sun using 2-way (both uplink and downlink) laser pulse ranging between Earth and spacecraft to measure γ and other relativistic parameters precisely. The γ parameter can be separated from the study of the Shapiro delay variation. The uncertainty on the Shapiro delay measurement depends on the uncertainties introduced by the atmosphere, timing systems of the ground and space segments, and the drag-free noise. A simulation shows that an uncertainty of 3×10^{-8} on the determination of γ is achievable.

LATOR (Laser Astrometric Test Of Relativity) [16] proposed to use laser interferometry between two micro-spacecraft in solar orbits, and a 100 m baseline multi-channel stellar optical interferometer placed on the ISS (International Space Station) to do spacecraft astrometry for a precise measurement of γ.

For ASTROD (Astrodynamical Space Test of Relativity) [1], 3 spacecraft, advanced drag-free systems, and mature laser interferometric ranging will be used and the resolution is subwavelength. The accuracy of measuring γ and other parameters will depend on the stability of the lasers and/or clocks. An uncertainty of 1×10^{-9} on the determination of γ is achievable in the time frame of 2025-2030.

Space mission proposals to the outer solar system for testing relativistic gravity and cosmology theories include Super-ASTROD [17], SAGAS [18] and Odyssey [19]. We refer the readers to the refereces mentioned.

5. Missions for Detecting Gravitational Waves

A complete classification of gravitational waves according to their frequencies is: (i) Ultra high frequency band (above 1 THz); (ii) Very high frequency band (100 kHz – 1 THz); (iii) High frequency band (10 Hz – 100 kHz); (iv) Middle frequency band (0.1 Hz – 10 Hz); (v) Low frequency band (100 nHz – 0.1 Hz); (vi) Very low frequency band (300 pHz – 100 nHz); (vii) Ultra low frequency band (10 fHz – 300 pHz); (viii) Hubble (extremely low) frequency band (1 aHz – 10 fHz); (ix) Infra-Hubble frequency band (below 1 aHz). The aims of gravitational-wave space missions are for detection of gravitational waves in the low frequency band (LISA, ASTROD, ASTROD-GW, and Super-ASTROD) and middle frequency band (DECIGO and Big Bang Observer). The space detectors are complimentary to the ground detectors which aim at detection of gravitational waves in the high frequency band.

LISA has nearly equilateral triangular spacecraft formation of armlength 5×10^6 km in orbit 20° behind earth.

ASTROD has a large variation in its triangular formation with armlength varying up to about 2 AU.

ASTROD-GW (ASTROD [Astrodynamical Space Test of Relativity using Optical Devices] optimized for Gravitation Wave detection) is an optimization of ASTROD to focus on the goal of detection of gravitational waves. The detection sensitivity is shifted 52 times toward larger wavelength compared to that of LISA. The scientific aim is focused for gravitational wave detection at low frequency. The mission orbits of the 3 spacecraft forming a nearly equilateral triangular array are chosen to be near the Sun-Earth Lagrange points L3, L4 and L5. The 3 spacecraft range interferometrically with one another with arm length about 260 million kilometers. After mission-orbit optimization, the changes of arm length are less than 0.0003 AU or, fractionally, less than $\pm 10^{-4}$ in ten years, and the Doppler velocities for the three spacecraft are less than ±4m/s. Both fit the LISA requirement and a number of technologies developed by LISA could be applied to ASTROD-GW. For the purpose of primordial GW detection, a 6-S/C formation for ASTROD-GW will be used for correlated detection of stochastic GWs.

The science goals for these missions are detection of Gravitational Waves (GWs) from (i) Supermassive Black Holes; (ii) Intermediate-Mass Black Holes; (iii) Extreme-Mass-Ratio Black Hole Inspirals; (iv) Galactic Compact Binaries; and (v) Primordial Gravitational Wave Sources, Strings, Boson Stars etc.

For direct detection of primordial (inflationary, relic) GWs in space, one may go to frequencies lower or higher than the LISA [20] bandwidth, where there are potentially less foreground astrophysical sources to mask detection. DECIGO [21] and Big Bang Observer [22-23] look for GWs in the higher frequency range while ASTROD [1], ASTROD-GW [24] and Super-ASTROD [17] look for GWs in the lower frequency range. In the following section, we address to the issue of detectability of primordial gravitational waves.

6. The Detectability of Primordial Gravitational Waves

Inflationary cosmology is successful in explaining a number of outstanding cosmological issues including the flatness, the horizon and the relic issues. More spectacular is the experimental confirmation of the structure as arose from the inflationary quantum fluctuations. However, the physics in the inflationary era is unclear. Polarization observations of Cosmic Microwave Background (CMB) missions may detect the tensor mode effects of inflationary gravitational waves (GWs) and give an energy scale of inflation. To probe the inflationary physics, direct observation of gravitational waves generated in the inflationary era is needed. The following figure shows the sen that the direct observation of these GWs with sensitivity down to Ω_{gw} ~ 10^{-20}-10^{-23} is possible using present projected technology development if foreground could be separated (Figure 1).

7. Outlook

Space missions using optical devices will be important in testing relativistic gravity, measuring solar-system parameters and detecting gravitational waves. Laser Astrodynamics in the solar system envisages ultra-precision tests of relativistic gravity, provision of a new eye to see into the solar interior, precise measurement of $\dot{G}$, monitoring the solar-system mass loss, and detection of low-frequency gravitational waves to probe the early Universe and study strong field black hole physics together with astrophysics of binaries. One spacecraft and multi-spacecraft mission concepts are in line for mission opportunities. In view of their importance both in fundamentals and in technology developments, mission concepts of this kind will be a focus in the near future.

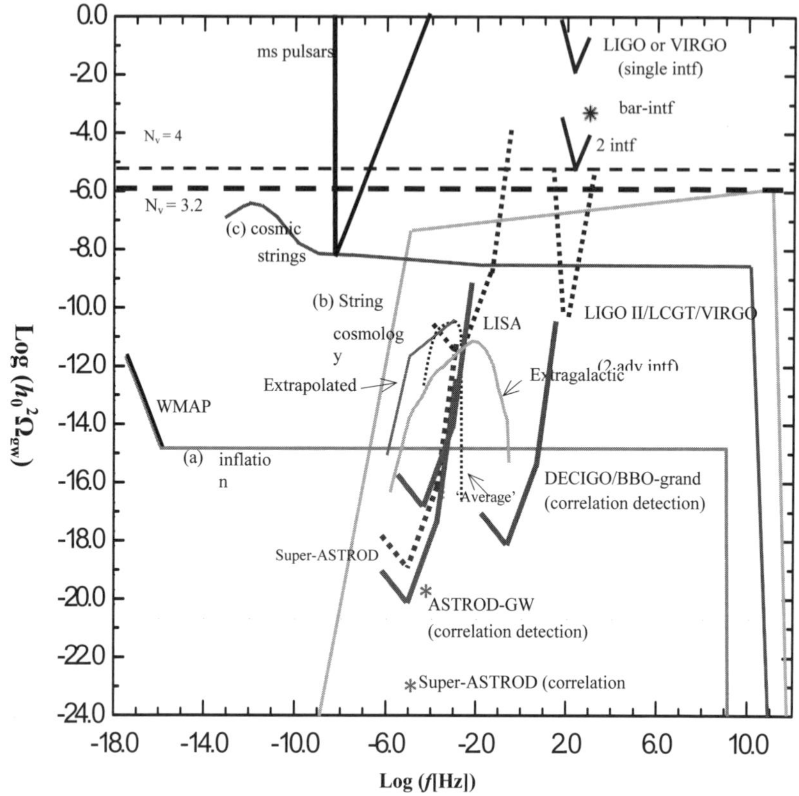

Figure 1. The stochastic backgrounds with bounds and GW detector sensitivities. (Adapted from Figure 3 and Figure 4 of Ref's 17 and 25 with sensitivity curve of DECIGO/BBO-grand added; the extragalactic foreground and the extrapolated foreground curves are from Ref. 27; see Ref's 17, and 25-28 for explanations).

References

1. W.-T. Ni, *Int. J. Mod. Phys. D* **17**, 921 (2008); and references therein.
2. T. W. Murphy, *et al*., *Publ. Astron. Soc. Pac.* **120**, 20 (2008); and references therein.
3. W.-T. Ni, *Int. J. Mod. Phys. D* **14**, 901 (2005); and references therein.
4. D.E. Smith, M.T. Zuber, X. Sun, G.A. Neumann, J.F. Cavanaugh, J.F. McGarry and T.W. Zagwodzki, *Science* **311**, 53 (2006).

5. X. Sun, G. A. Neumann, J. F. McGarry, T. W. Zagwodzki, J. F. Cavanaugh, J. J. Degnan, D. B. Coyle, D. R. Skillman, M. T. Zuber, and D. E. Smith, Laser Ranging between the Mercury Laser Altimeter and an Earth-based Laser Satellite Tracking Station over a 24 Million Kilometer Distance, presented at *OSA Annual Meeting, Tucson, Arizona, USA, October 16-20, 2005*.
6. J. J. Degnan, Laser Transponders for High Accuracy Interplanetary Laser Ranging and Time Transfer, pp. 231-242 in *Lasers, Clocks, and Drag-Free: Exploration of Relativistic Gravity in Space*, Eds. H. Dittus, C. Lämmerzahl and S. G. Turyshev (Springer, 2008).
7. T. Damour and K. Nordtvedt, Jr., *Phys. Rev. Lett.* **70**, 2217 (1993).
8. T. Damour, F. Piazza, and G. Veneziano, *Phys. Rev.* **D 66**, 046007 (2002).
9. B. Bertotti, L. Iess, and P. Tortora, *Nature* **425**, 374-376 (2003).
10. C. W. F. Everitt, private communication.
11. http://www.esa.int/esaSC/120391_index_0_m.html
12. A. Milani, D. Vokrouhlicky, D. Villani, C. Bonanno, and A. Rossi, *Phys. Rev.* **D 66** (8), 082001(21) (2004).
13. http://www.esa.int/esaSC/120377_index_0_m.html
14. A. Vecchiato, M. G. Lattanzi, B. Bucciarelli, *et al.*, *Astron. Astrophys.* **399** (1), 337-342, (2003).
15. T. Appourchaux *et al.*, *Exp. Astron.* **23** (2009) 5.
16. S. G. Turyshev, M. Shao, and K. Nordtvedt, *Class. Quantum Grav.* **21**, 2773 (2004).
17. W.-T. Ni, *Class. Quantum Grav.* **26,** 075021 (2009).
18. P. Wolf *et al.*, *Exp. Astron.* **23,** 651 (2009).
19. B. Christophe *et al.*, *Exp. Astron.* **23,** 529 (2009).
20. LISA Study Team, LISA (Laser Interferometer Space Antenna): A Cornerstone Mission for the Observation of Gravitational Waves, ESA System and Technology Study Report ESA-SCI 11 (2000).
21. S. Kawamura *et al.*, *Class. Quantum Grav.* **23,** S125 (2006).
22. J. Crowder and N. J. Cornish, *Phys. Rev. D* **72,** 083005 (2005); and references therein.
23. N. Seto, Phys. Rev. **D73**, 063001 (2006); and references therein.
24. W.-T. Ni, ASTROD optimized for gravitational-wave detection: ASTROD-GW – a pre-Phase A study proposal submitted to Chinese Academy of Sciences, February 26, 2009.
25. M. Maggiore, Stochastic backgrounds of gravitational waves (2000) [arXiv:gr-qc/0008027v1].
26. M. Maggiore, *Phys. Reports* **331,** 283 (2000).
27. A. J. Farmer and E. S. Phinney, *Mon. Not. Roy. Astron. Soc.* **346,** 1197 (2003).
28. W.-T. Ni, *Int. J. Mod. Phys. D* **18**, December issue (2009); arXiv:0905.2508 (2009).

CRYOGENIC ADVANCED GRAVITATIONAL WAVE DETECTOR (LCGT)

K. KURODA

ICRR, The University of Tokyo, Kashiwa, Chiba 277-8582, JAPAN
E-mail: kuroda@icrr.u-tokyo.ac.jp
http://www.icrr.u-tokyo.ac.jp

LCGT Collaboration*

LCGT shall be planned to be the large scale cryogenic gravitational wave telescope in order to firstly detect gravitational waves. After the detection, the detector will be served as an astronomical tool to observe the Universe. In collaborative observation by Advanced LIGO, GEO HF and Advanced Virgo projects, LCGT is desired with AIGO to contribute both the enterprise of detecting the gravitational wave events and the worldwide network for gravitational wave astronomy. This paper summarizes the status of LCGT.

Keywords: Gravitational wave, Laser interferometer, Cryogenics, Underground

1. Introduction

LCGT was planned during the construction of TAMA[1] and at the same time of GEO[2] construction to firstly detect gravitational wave from the coalescences of neutron star binaries, which occur within 200 Mpc in the most sensitive detector configuration. In order to fulfill the objective by the similar baseline length such as the initial LIGO[3] and Virgo[4] detectors, LCGT will be equipped with cryogenic mirror system to reduce thermal noise and be placed underground with advanced anti-vibration technique to suppress seismic noise at lower frequencies. Since its objective sensitivity is equivalent to those of the advanced designs of the LIGO and Virgo projects, LCGT is called as the second generation detector. However, its design is rather conservative except the adoption of cryogenics and the resonant sideband extraction scheme in a sense that its optical design is similar

*LCGT collaboration consists of 92 domestic researchers belonging to 17 universities or research institutes and 26 oversea members belonging to 14 organizations.

to that of the initial LIGO and Virgo. Therefore, in future, there is a room of sensitivity improvement beyond the sensitivity of the second generation detectors by optimizing its optical configuration.

Although the construction of the advanced LIGO is in progress, LCGT has possibility to largely contribute to the first detection of gravitational wave by earlier funding by Japanese government. After the first detection, LCGT will serve as an astronomical tool for the observation of the Universe through gravitational wave radiation under an appropriate worldwide network of observation. Among those projects such as LIGO, GEO, and Virgo, AIGO is also planned to be built in Western Australia.[5] Since at least three detectors are needed to determine the position of the gravitational wave source, both LCGT and AIGO are strongly desired to be realized in order to enhance the observation efficiency, which is expressed by a statement of the gravitational wave international committee (GWIC) under PaNAGIC (one of subcommittee of IUPAP).[6] This paper summarizes the scientific significance and the worldwide importance of LCGT project.

2. Gravitational wave sources and sensitivity

The objective of LCGT is to detect at least one gravitational wave event in a year. The coalescence of double neutron star (DNS) is the most important among other sources in the sense that its wave form is precisely predicted and its existence has certainly been confirmed. The wave form emitted at the time of coalescence is analytically calculated along with the time evolution until just before the merging moment. Around the moment of its merger, the wave form needs to be numerically solved and its numerical solution is close at hand.[7–11] These signal wave forms are used to build the matched filter that increases the signal to noise ratio for detecting weak signals buried in noises. Table 1 lists the DNSs that have been observed by astronomical observation in our Galaxy. There are nine pairs of DNS known so far, one of which lies in M15 galaxy. Two pairs of these DNS are waiting further observation to be identified as definite double neutron star. Fourth column in this Table represents the strength of relativistic nature of DNS, which does not necessarily mean the strong radiation of gravitational wave due to the partial dependence on the orbital eccentricity.

The coalescence rate is estimated by both the distribution of these binaries and their life time. After the discovery of the highly relativistic binary neutron star system,[12] a new young binary pulsar was detected.[13] The former discovery increased the coalescence rate from 10^{-6} to 10^{-5} a year for milky-way equivalent galaxy[14] and the latter pushed up by another factor

Table 1. Double neutron stars that are discovered in our Galaxy

DNS	Distance[kpc]	orbital period[d]	Advancement of periastron[deg·yr^{-1}]
B1913+16	7.3	0.323	4.3
B1534+12	1.1	0.421	1.76
J1518+4904	0.6	8.63	0.011
J1811-1736	0.5	18.8	0.009
J1829+2456	1.2	1.18	0.28
J0737-3039	0.6	0.102	16.8

of six. The merger rate is estimated for milky-way equivalent galaxy to be[15]

$$83.0^{+209.1}_{-66.1}(\text{C.I.}95\%)[\text{events}\cdot\text{Myr}^{-1}]$$

Since the distribution of galaxies is estimated as 0.01 per Mpc^3, LCGT that has sensitivity to detect the signal up to 250 Mpc at its optimum configuration is able to detect ≈ 6.4 events in a year. This value is obtained by averaging through the whole sky and also being averaged in the directional angles of the orbital axis of the neutron star binary system.

Since we cannot determine the source position by a single detector, we need to collaborate with other observatories that are placed each other far away in a global scale with similar sensitivity. With the advanced LIGO and the advanced Virgo, LCGT may contribute to achieve extremely high observation duty cycle for the whole sky coverage.[16] And if we can add AIGO to this network, the efficiency of the observation expands more.

The design sensitivity of LCGT is shown in Fig. 1 as the detection range versus mass of accompany star for the coalescence of binary neutron star and for its black hole quasi-normal mode ring down. We plan to adopt the broadband resonant-sideband-extraction (BRSE) scheme for the default design and the detuned resonant-sideband-extraction (DRSE) is considered to be its advanced optical configuration. The detection range in this figure is defined for signal to noise ratio of 10 assuming sources at optimal direction of the detector sensitivity. As for the nominal source of 1.4 $M_\odot$ pair, the default design gives 165 Mpc and the detuned design does 250 Mpc.

The expected sources of LCGT are plotted in the Fig. 2 for the sensitivity of the default design of BRSE scheme. The quasi-normal mode oscillation of the black hole will be observed by its ring down just after the formation of the black hole if the merged mass is something like 100 $M_\odot$. The signal from the ring down of 2.8 $M_\odot$ system with Kerr parameter of $a = 0.95$ lies outside of the sensitive frequency band of LCGT. If stellar core collapses happen at Galactic center, they can be well within the scope

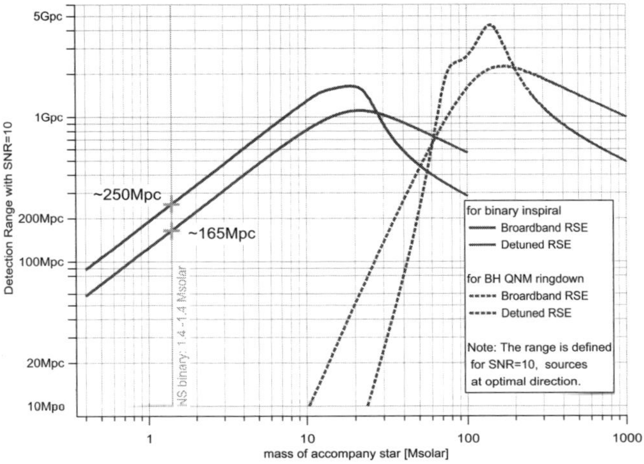

Fig. 1. The design sensitivity of LCGT. The lower red curve shows the default design of broadband resonant sideband extraction (BRSE) and the upper blue curve shows its advanced optical configuration of detuned resonant sideband extraction (DRSE). The vertical axis is the detection range and the horizontal axis is the mass of accompany star both for the coalescence of binary neutron star and for its black hole quasi-normal mode ring down. The range is defined for signal to noise ratio of 10 assuming sources at optimal direction of the detector sensitivity.

of the LCGT sensitivity for Dimmelmeier ' s wave forms (Fig. 2).[17]

As for the continuous waves from pulsars, LCGT can detect them by appropriate signal integration compensating Doppler effect by the Earth orbital motion around the Sun. Figure 3 estimates such possible candidates for integration times of 2 weeks, one year and ten years, respectively, in the default BRSE sensitivity.

There may be other possible gravitational wave sources in the universe other than the above known ones. The detection of gravitational waves from such unknown sources may open another new field of physics and astronomy.

3. Design of LCGT

The sensitivity of LCGT is attained by an interferometers located underground, using a three-kilometer length baseline, cooling mirrors at cryogenic temperature, and a high-power laser source employing 150 W output. The optical configuration is a power recycled Fabry-Perot-Michelson interferometer with the resonant-sideband-extraction (RSE) scheme (Fig. 4). The detailed design of the control system was tested for the RSE scheme.[18] Ta-

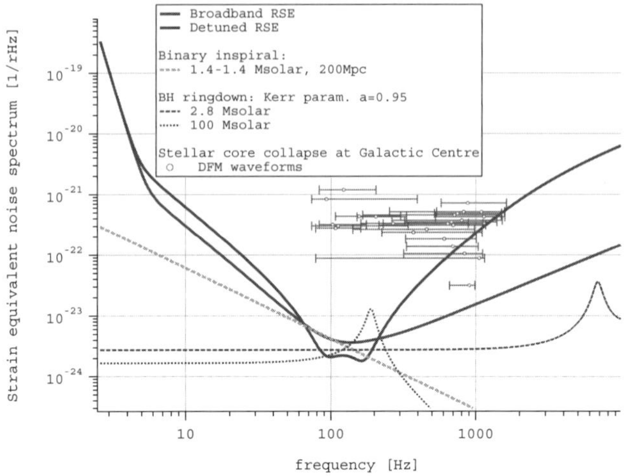

Fig. 2. Expected gravitational wave sources of LCGT in the default design of Broadband RSE scheme. The main source is the coalescence of binary neutron star. Also the quasi-normal mode oscillation of the black hole is observed by its ring down just after the formation of the black hole ($\sim$100 $M_\odot$). Stellar core collapse occurred at Galactic center may be detectable.

ble 2 lists the important parameters of LCGT, which were revised three times from the original design.[19]

3.1. *History of revision*

The first revision was made in 2002 adopting RSE scheme to reduce heat production in the input test mass with maintaining the similar optical power inside the main cavity.[20] In 2004, we increased the number of interferometers from one to two which are installed in the same vacuum tube to enhance the rejection ratio against false events.[21] This change corresponds to the emphasis on the first detection of the gravitational wave. Finally, we focused our attention to practical construction of the interferometer with similar sensitivity and revised its design.[22]

3.2. *Design policy and optical design parameters*

Ultimate sensitivity of a laser interferometer is determined by seismic noise at low frequencies (10-30 Hz) (which is reduced by improving the vibration isolation system), and it is limited by photon shot noise at higher frequencies (more than 300 Hz), which can be improved only by increasing the light

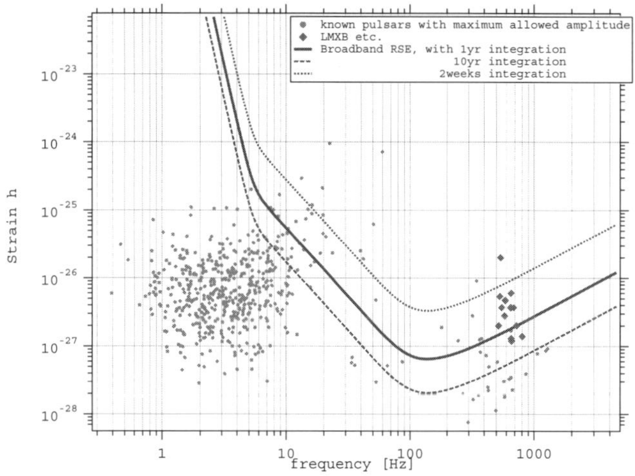

Fig. 3. Expected continuous wave sources of pulsars in the default design. Circular dots (color gold) represent the known pulsars with maximum allowed amplitudes and the symbol of diamond (color pink) represents the possible waves from low mass X-ray binaries. All curves assume the Broadband RSE sensitivity with integration of 2 weeks, 1 year and 10 year integration.

power in the main cavities. The sensitivity of middle frequencies (30-300 Hz) is limited by the fluctuation of the photon recoil force noise. Reaching this radiation pressure noise requires that thermal noise is well reduced both by decreasing the temperature and by decreasing the internal mechanical loss (*i.e.*, increasing the mechanical Q of vibration modes). The sources of thermal noise are mirror internal mechanical loss, mechanical loss of the optical coating and mechanical loss of the suspension fiber suspending the mirror. The reduction of thermal noise is attained by cooling both the mirror and the suspension system that suspends the mirror.

We adopt sapphire for the mirror substrate. Sapphire has relatively large optical loss inside the substrate. Since the optical absorption inside the input test mass produces heat in the mirror, which has to be extracted through fiber heat conduction, the power passes through inside the input test mass needs to be suppressed. Since high optical power is needed in the cavity, we increase the cavity finesse with keeping low power incidence through the input test mass. This results in a large finesse with low power recycling gain as shown in Table 2. Since high cavity finesse reduces the frequency band width, we apply RSE scheme in order to resonantly extract the gravitational wave signal, which equivalently expands the frequency

Table 2. LCGT design parameters

Item	Parameter
Baseline Length	3 km
Interferometer	One set
	Power recycled Fabry-Perot
	-Michelson with RSE
Optical Power	Laser:150 W; Finesse:1550
	Input power at BS: 825 W
	Cavity power 800 kW
Beam radius at End	3 cm
Main Mirror	Sapphire 30 kg, 20 K
	Diameter 25 cm
	Mechanical Q:10^8
Suspension	Frequency: 1 Hz; Q:1×10^8
pendulum	10 K
Vacuum	$\leq 10^{-7}$ Pa

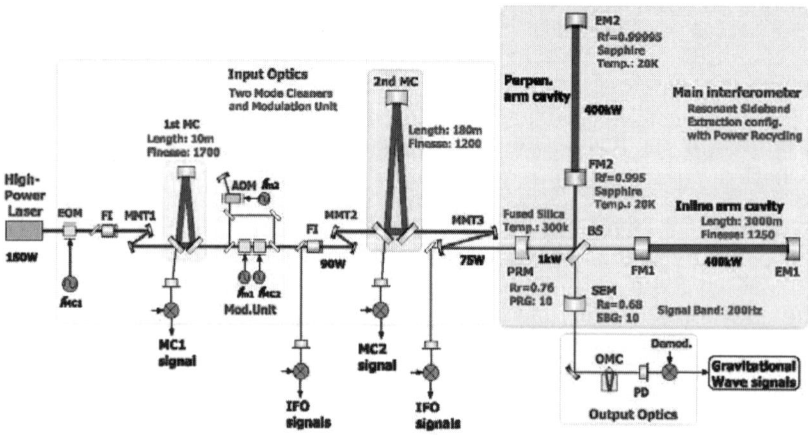

Fig. 4. Optical design of LCGT. The optical configuration is a power recycled Fabry-Perot-Michelson interferometer with the broad band resonant-sideband-extraction (BRSE) scheme.

bandwidth.

Based on the above default design of LCGT, we are studying to adopt advanced configurations; the detuned RSE and also removal of the 180 m length input mode cleaner.

3.3. *Tunnel construction*

The site is located in Gifu prefecture and at an elevation of 400 m in the heart of the high and massive mountain range, approximately 60 km inland from the Sea of Japan. The mountain consists of rocks of amphibolies and gneiss. The mining company (Kamioka Mining Co. Ltd) maintains the facility of infrastructure but mining activity is ceased at present. The designed scale of tunnel is 4 m in width and 4.5 m in height. The arm tunnels of the interferometer are placed well below the slope of the mountain by more than 200 m to keep the environment of low seismic noise. Construction time was estimated to be three years that were originally incompatible with the installation of the vacuum tube system required by the total schedule the construction but this time length is practically shortened down to two years that allow the earlier start of installing vacuum system.

3.4. *Vacuum system*

The whole vacuum system is placed in the tunnel. The main beam tube has 1 m inner diameter and is made of stainless steel of 304. The allocation of vacuum chambers is figured in Fig. 5. Only one vacuum chamber of the main mirror appears in the horizontal arm at right hand side, which is practically a vacuum cryostat with radiation tubes extending for both directions (roughly 10 m in length). Except arm tubes, there are two vacuum tubes for two mode cleaners. We produce 600 tubes, the unit of which is in 12 m long and has a flange for one side and a flange with short bellows for the other side. These tubes are introduced inside the tunnel and connected one by one. The flange connection expels the inefficient welding procedure inside the tunnel and the assembling time is expected to be shortened.

Since baking the tube underground is not an easy task due to limited space and closed air, we apply electro-chemical buffing (ECB) technique that was applied to the vacuum system of TAMA interferometer to cut out the baking procedure in situ.[23] In addition to this ECB technique, each tube product is baked in vacuum in prior to the transportation to the tunnel in order to assure the gas releasing rate of tube inside being maintained 10^{-9}Pa m^3 s^{-1} m^{-2} that is the value after 50 hours of pumping. Since the construction cost of the vacuum system is estimated to be approximately 40% of the total cost, we always have a strong pressure to reduce cost with keeping the quality of vacuum, which reflects on the maintenance cost of the vacuum system.

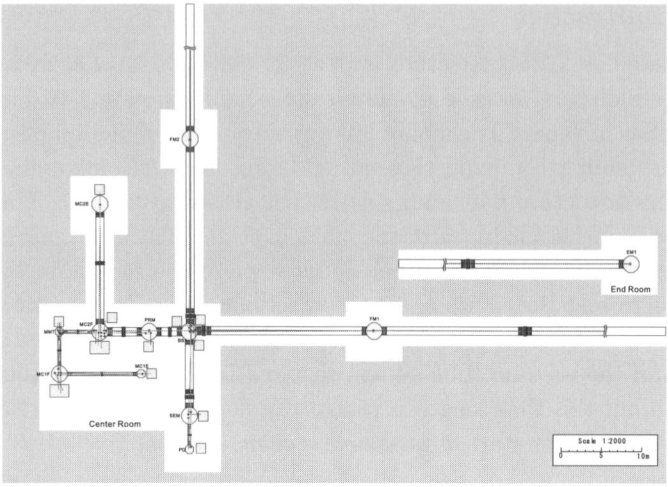

Fig. 5. The configuration of vacuum system with chambers and beam tubes. The main beam tube has 1 m inner diameter and is made of stainless steel of 304. The beam tube consists of 600 tubes of a unit tube in 12 m long. The unit tube has a flange for one side and a flange with short bellows for the other side. This flanged tube unit makes the construction safer and more speedy than in-situ welding.

3.5. *SPI*

A suspension point interferometer (SPI) is designed to cut mechanical vibration introduced through cryogenic heat conductors by refrigerators. Cavity test masses are suspended by controlled suspension points, the interval of which is determined by locking to the multiples of wavelength inside a Fabry-Perot cavity that is independent from the main cavity. A heat link attached to the suspension point makes one of the suspension points vibrate, which is compensated by the movement of the other suspension point by the feedback. The design of SPI closely relates with the design of the anti-vibration system in the next subsection.

3.6. *Anti-vibration system*

The design of the suspension system is the heart of this cryogenic system and had been already presented using the figure at the left hand side of Fig. 6.[24] The main mirror is suspended through the auxiliary mirror of the suspension point interferometer (SPI). We design three types of anti-vibration system as shown in this figure as A, B and C. Type A is a system of SAS (seismic attenuation system), which was originally developed in Virgo project and modified for TAMA under the research collaboration with Caltech. SAS consists of inverted pendulum and three stages of GAS (geometric anti-spring) filter. Type A is applied for two input test masses and two end test masses. The mirrors of beam splitter, power recycling, signal extraction, and second mode cleaner are suspended by the type B anti-vibration system that consists of SAS of two stages of GAS filter with inverted pendulum in room temperature. The most simple system C consisting of two stages of suspension on the stack isolation system houses mirrors of the first stage mode cleaner, matching telescope and photo detectors. The performance of SAS has been tested by installing to TAMA interferometer. The other anti-vibration systems are conventional and can be applied to LCGT without any troubles.

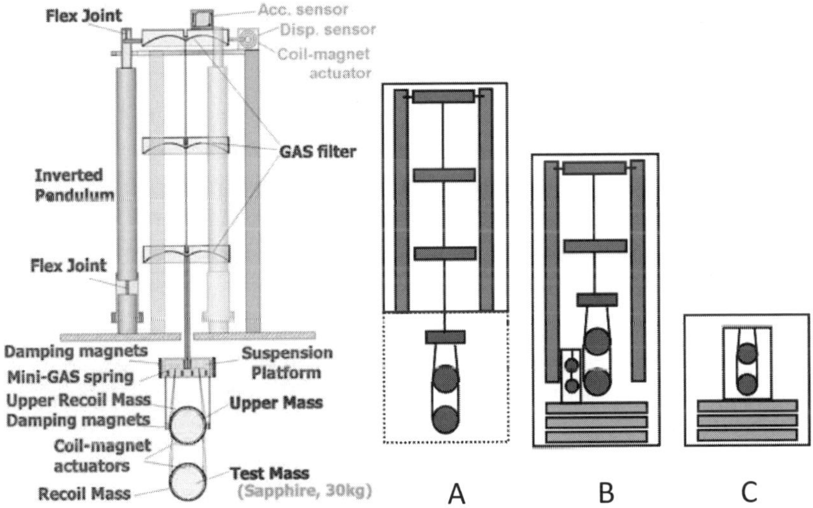

Fig. 6. The design of the suspension system is the heart of this cryogenic system. The main mirror is suspended through the auxiliary mirror of the suspension point interferometer (SPI). The SPI mirror is also suspended by a wire through the radiation shield hole, which connected to the last stage of GAS filter system that is supported by the inverted pendulum system. Three types of anti-vibration system are used for suspending mirrors to satisfy their functions.

4. R&D

The R&D of LCGT consists of independent scientific projects such as TAMA and CLIO. Apart from these projects, several research items are conducted by individual research groups.

4.1. *TAMA*

We developed a 300 m baseline interferometric gravitational wave detector, TAMA, at the Mitaka campus of the National Astronomical Observatory of Japan (NAOJ) and several observations had been conducted. TAMA project started in April, 1995, as a five-year project and it was extended by two years after 1999. TAMA is organized by researchers belonging to universities and national laboratories. We achieved nine data-taking runs that span from two to eight weeks, which ended in 2004. About TAMA project, we had presented its development of Seismic Attenuation System (SAS) newly installed for four main mirrors and tried to achieve its design sensitivity. However, we finally found that the sensitivity at low frequencies was limited by so-called up-conversion noise possibly arising from the suspension and actuation system and cannot be improved anymore without a drastic change of the suspension system utilizing more sophisticated actuator.

The achieved sensitivity of TAMA is shown with estimated main noise sources in Fig. 7. We recognized that there was a gap between the achieved sensitivity and the practically attainable one at low frequencies. However, we could demonstrate that the basic techniques for the interferometer operation was acquired by the fulfillment of the target noise curve at frequencies more than 800 Hz, where the optical system properly worked and the control system was appropriate. In order to improve the noise spectrum in the frequency range from 30 Hz to 800 Hz, we had installed the Seismic Attenuation System (SAS)[25] for four main mirrors and improved the performance of anti-vibration at low frequencies. By this installation the sensitivity at low frequencies was improved and we could find the up-conversion noise limit as shown by the blue curve in Fig. 7.

4.2. *CLIO*

After several basic experiments for cryogenic mirror system,[26] we have constructed a 100 m baseline length cryogenic interferometer (CLIO) placed underground at Kamioka. The objective of this project is to practically present the feasibility of cryogenic mirror system for LCGT.[27] Since the sensitivity of the interferometric gravitational wave detector is limited by thermal noise of the mirror and suspension system for frequencies of 30-300 Hz, the cryogenic mirror system possibly reduces the thermal noise according to the reduction of the temperature. This is correct only if the mechanical loss of the system does not decrease with temperature reduction.

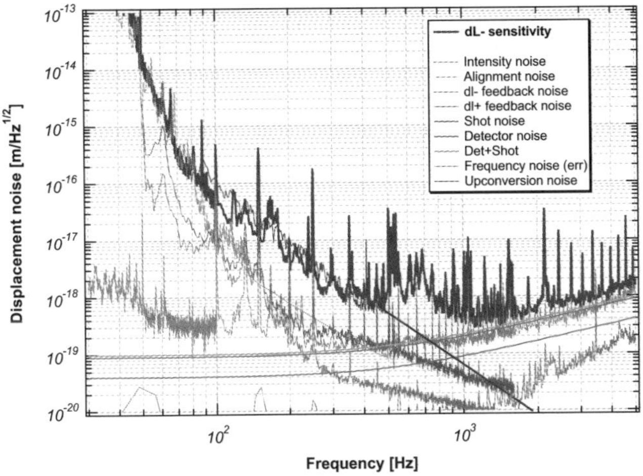

Fig. 7. Achieved sensitivity of TAMA. This curve was obtained after the installation of SAS systems for four main mirrors. We could demonstrate that the basic technique for the interferometer operation was acquired by the fulfillment of the target noise curve at frequencies more than 800 Hz, where the optical system properly worked and the control system was appropriate. The sensitivity in lower frequencies was limited due to up-conversion noise caused by relatively large seismic noise with unidentified non-linear mechanism, which was empirically tested.

That is, the power spectrum of thermal noise is proportional to T/Q, where T is the temperature and Q^{-1} represents the mechanical loss of the system. The most serious problem in cryogenics is that the increase of mechanical loss of fused silica at low temperature, which means that silica cannot be used at cryogenic mirror system. In place of silica, we adopt sapphire material for the mirror substrate and fibers that suspend the mirror. Since we cannot obtain a good quality mirror substrate by available technology at present, we need to cope with this problem, which means the high finesse with low power recycling cavity.

The construction of CLIO finished in 2006 and noise hunting to obtain thermally limited noise interferometer was conducted.[28] By the end of 2008, we attained the room temperature limited noise sensitivity as shown in Fig. 8. Adding minor improvements, cryogenic sensitivity is being sought. The lack of proper alignment actuator inside the dark cryostat and the lack of appropriate damping mechanism at cryogenic temperature make the commissioning consume much time at cryogenic temperature. However, these are not fatal problem for the achievement of the objective of CLIO.

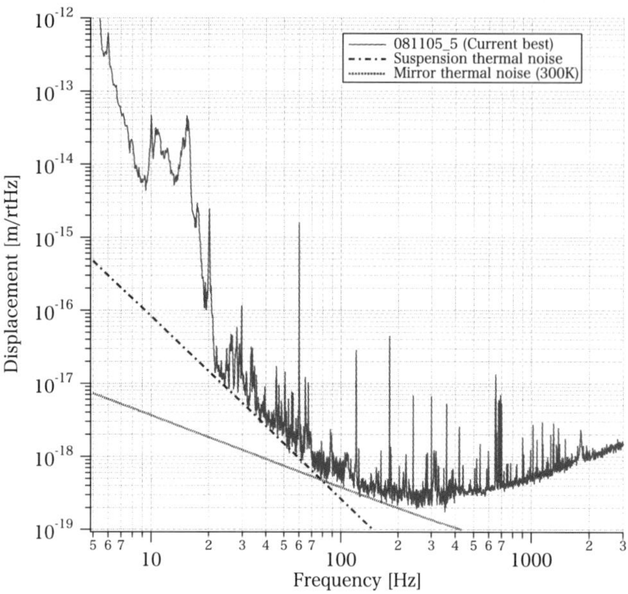

Fig. 8. Thermal noise limited sensitivity of CLIO at room temperature. Both suspension and mirror bulk thermal noises were measured by the CLIO system without cooling the mirror. Adding minor improvements, cryogenic sensitivity is being sought.

4.3. *Substrate quality measuring device*

The specification of the main mirror of LCGT is sapphire substrate, 25 cm in diameter and 12 cm in length ($\approx$30 kg). The optical absorption should be less than 20 ppm/cm and the fluctuation of optical refractive index needs to be less than $\approx$0.1 ppm to satisfy the optical design requirement. The optical loss of sapphire piece (purchased from Crystal Systems Inc. in 1997) at cryogenic temperature was measured by calorimetric method and gave $\approx$90 ppm/cm.[29] Five years after this measurement, we purchased other lots of sapphire from the same company. We applied three different measurements under research collaboration with UWA on these two samples: birefringence, scattering and optical loss, the measurement values of which were mapped and compared.[30] Since there was a clear correlation among these measurements, we can infer the other quality by knowing the result by only one of measurements. Just in case, the piece had the optical loss of $\approx$50 ppm/cm, which may reflect the progress of crystal production. However, we need to select better lot by one of measurements. In order to evaluate raw sample substrate pieces, we have developed an auto scanning measuring device of birefringence of sapphire and reconstructed for large pieces.[31] Figure 9 shows the measurement device of birefringence of

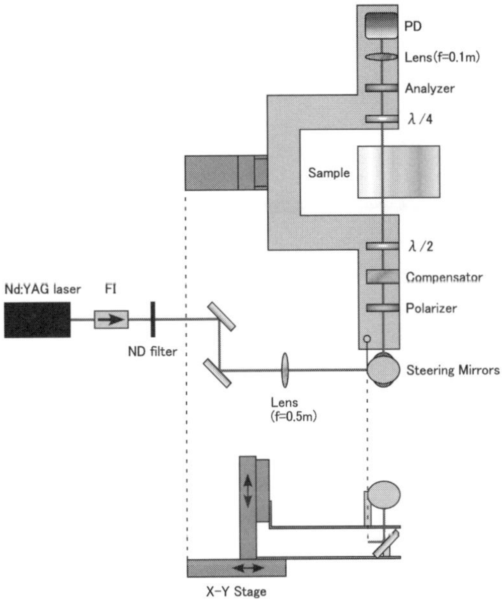

Fig. 9. Measurement device of birefringence of large sapphire pieces with automatic scanning mechanism. The upper figure shows horizontal view of the device and the lower corresponds to the vertical view. The large sample is fixed to the optical table, while the main part of optics is on the scanning X-Z stage. The laser source, the Faraday isolator and the ND filter are placed on the optical table.

large pieces with automatic scanning apparatus. We introduce the light of Nd:YAG laser, the wavelength of which is 1.064 nm, through the Faraday isolator with a ND filter into the optics arranged on the scanning X-Z stage. The measured sample is fixed on the optical table. Since the weight of the optics system is lighter than the sample, this scanning system causes less harmful vibration during the scanning. Without any birefringence of the sample, the light power on the photo detector is zero and some birefringence of the sample causes leakage of light, which is compensated by the compensator that shows the magnitude of the birefringence. By this device, we measured the 25 cm diameter sapphire piece (29 kg) produced for LIGO project, which was lent for this measurement by the courtesy of LIGO project. The result is shown in Fig. 10. A large un-homogeneity appears in the center, the value of which does not satisfy the specification of LCGT, which is worse roughly by a factor of several. Since the axial symmetric axis was crystal a-axis, the measurement data were corrected using the angle of the surface declination against the asymmetric axis with the known refractive index of the used laser wavelength.[32]

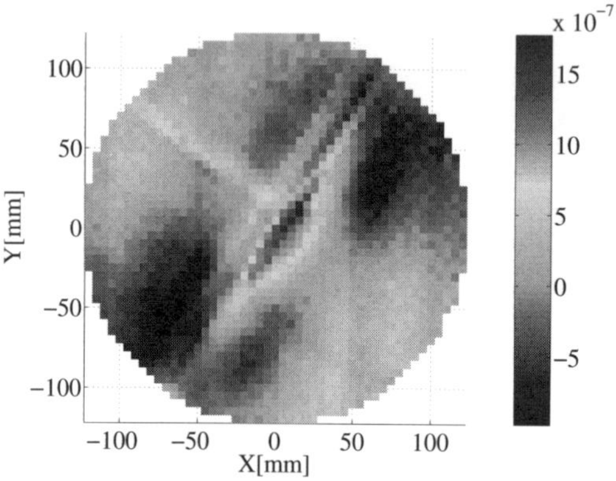

Fig. 10. The birefringence fluctuation map of a large sapphire sample, 25 cm in diameter, 29 kg in weight, which was lent by the courtesy of LIGO project. A large unhomogeneity appears in the center, the value of which does not satisfy the specification of LCGT that is worse roughly by a factor of several.

4.4. *Sensitivity improvement*

Figure 11 compares the sensitivity curve of LCGT with those of TAMA, CLIO, and LIGO (initial LIGO). The sensitivity at low frequencies of LCGT is attained by SAS. That of higher frequencies is attained by higher laser power. The mid-frequency region is improved by cryogenic mirror and suspension system, which is under being tested by CLIO. Although the improvement, especially two orders of magnitude at low frequencies, is adventurous, it is not impossible until the construction period of LCGT that takes four years after the beginning of the construction.

5. Schedule and Manpower organization

We submitted the budget request to the University of Tokyo for starting the construction in the financial year of 2010 and it was passed to the Ministry of Education, Sports, Culture, Science and Technology (MEXT) in June,

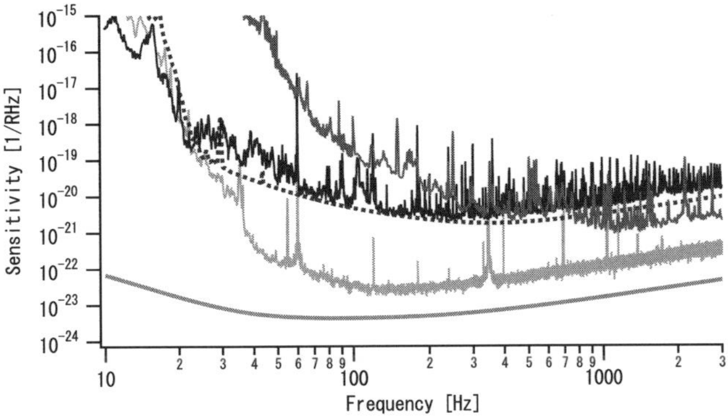

Fig. 11. LCGT sensitivity (bottom curve) compared with those of CLIO(October 2008, room temperature), CLIO (ultimate, room temperature), TAMA(October 2008 after SAS installation), LIGO(Science run 5) from the above to down at the right hand side of the figures. The horizontal axis is frequency [Hz] and the vertical axis represents sensitivity spectrum for gravitational waves [1/$\sqrt{}$Hz].

2009.[a] The schedule planed in LCGT is illustrated with other projects in Fig. 12, where LCGT is constructed in five years and starts observation after two years commissioning. The relation with other projects will affect the funding status of LCGT as described in the previous section.

LCGT is conducted under the association of researchers belonging to universities and research organizations in and outside of Japan. Especially, three organizations; ICRR of the University of Tokyo, National Astronomical Observatory, Japan (NAOJ), and High Energy Accelerator Research Organization (KEK), exchange the memorandum of understanding to promote gravitational wave research by LCGT, where ICRR hosts the project. The directors of these organizations are the main members of the LCGT board. LCGT collaboration consists of 92 domestic researchers belonging to 17 universities or research institutes and 26 oversea members belonging to 14 organizations. Total members are 118 researchers in summer, 2009. Task of domestic collaboration members is roughly determined and that of oversea members will be discussed after knowing the submission of our request from the government to the representative house, which usually happens in December. Figure 13 represents the project structure to cope with the design study and practical work for conducting LCGT project.

[a]However, this request was not passed to the Ministry of Finance, which does not mean the denial of the project. We only lost the chance of funding for FY 2010.

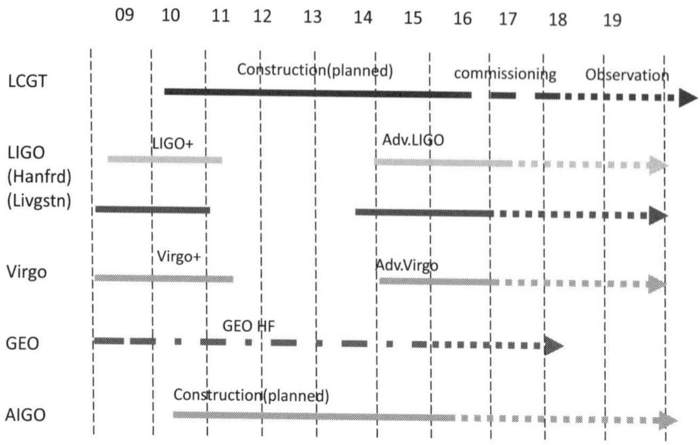

Fig. 12. Budget request starting the construction in April, 2010 was submitted to the Ministry of Education, Sports, Culture, Science and Technology (MEXT). We aim to construct in five years and begin observation after two years commissioning.

6. Project support from neighboring research areas

In 1993, gravitational wave telescope was nominated as one of future projects by the report of subcommittee of ICRR and early realization of gravity wave detectors was described in the Astronomical subcommittee of Science Council of Japan in 1994. At the time of TAMA project in 1994, MOU promoting gravitational wave research among directors of NAOJ (National Astronomical Observatory, Japan), KEK (High Energy Accelerator Organization), and ICRR, was exchanged and has been renewed every two years after that time. Space Science subcommittee of the academic council of MEXT nominated LCGT as one of projects in a fund waiting list with recommending the reinforcement of R&D in 2000. A special report of the Astronomical subcommittee of Science Council of Japan strongly requested the prompt funding of LCGT on behalf of the whole astronomical community in 2005. In 2007, ICRR was clearly positioned as the host institute for LCGT under revised MOU originally exchanged in 1994 and the future research plan committee of ICRR exclusively pushed LCGT in the same year. In 2008 GWIC strongly supported the funding of LCGT that made the observation start in 2015 possible. And also in the same year, Astronomy & Astrophysics subcommittee of Science Council of Japan raised the resolution to promote early LCGT funding. In 2009, combined subcommittee (IAU, Astronomy & Astrophysics) in Science Council of Japan made a resolution to endorse LCGT.

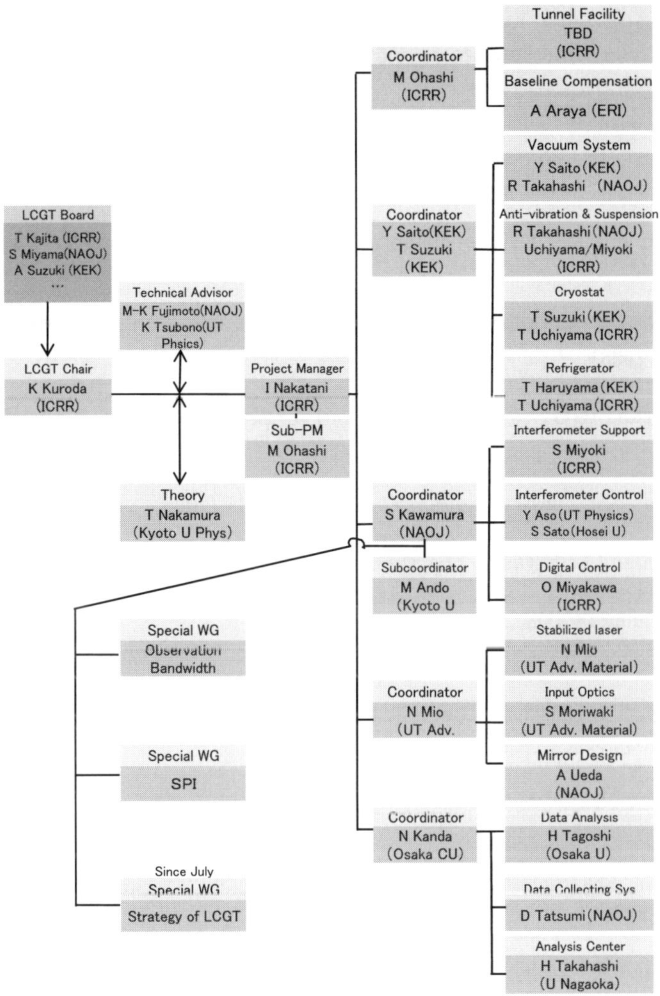

Fig. 13. Project structure in summer, 2009 to cope with the design study and practical work for conducting LCGT project.

7. Conclusion

Interferometer techniques (power recycling, Fabry-Perot Michelson interferometer, and control system) have been acquired by TAMA project and the feasibility of cryogenic mirror is under the process of evaluation by CLIO. If LCGT is funded soon, we reliably detect GW at the first time or at least early time in collaboration with other world-wide observatories such as Advanced LIGO, Advanced Virgo, GEO HF and AIGO. We are now at the corner where a long time-consuming R&D turns out to successful outcome of detecting gravitational waves.

References

1. R. Takahashi and The TAMA collaboration, *Class. Quantum Grav.* **21**, S403 (2004)

2. H. Grote and (for the LIGO Scientific Collaboration), *Class. Quantum Grav.* **25**, 114043 (2008)

3. D. Sigg (for the LIGO Science Collaboration), *Class. Quantum Grav.* **23**, S51 (2006)

4. F. Accernese *et al*, *Class. Quantum Grav.* **25**, 114045 (2008)

5. D. E. McClelland *et al*,*Class. Quantum Grav.* **23**, S41 (2006)

6. GWIC statements (2008), `http://gwic.ligo.org/statements/statements/0804Statements.html`

7. J. S. Read, C. Markakis, M. Shibata and K. Uryu, *Physical Review D* **79**, 124033 (2009)

8. T. Yamamoto, M. Shibata and K. Taniguchi, *Physical Review D* **78**, 064054 (2008)

9. K. Kikuchi, Y. Sekiguchi, M. Shibata and K. Taniguchi, *Physical Review D* **80**, 064037 (2009)

10. M. Shibata and K. Taniguchi, *Physical Review D* **73**, 064027 (2006)

11. K. Uryu *et al*, *Physical Review Letters* **97**, 171101 (2006)

12. M. Burgay *et al*, *Nature* **426**, 531 (2003)

13. D. R. Lorimer *et al*,*Astrophysical J.* **640**, 428 (2006)

14. C. Kim C, V. Kalogera and D. R. Lorimer, *Astrophysical J.* **584**, 985 (2003)

15. V. Kalogera *et al*, *Astrophysical J.* **601**, L179 (2004)

16. B. Schutz, *Capabilities of a Gravitational Wave Network in Fujihara seminar* (Hayama, Japan, 2009); `http://www.icrr.u-tokyo.ac.jp/~miyoki/`

fujihara_seminar_2009/Fujihara2009-Schutz.ppt

17. H. Dimmelmeier, J. Font, and E. Muller, *Astron. Astrophys.* **393**, 523 (2002)

18. F. Kawazoe *et al, presented at 7th Edoardo Amaldi Conference on Gravitational Waves* (Sydney, Australia, 2007)

19. K. Kuroda and LCGT collaboration, *Int. J. Mod. Phys. D* **8**, 557 (1999)

20. K. Kuroda *et al, Class. Quantum Grav.* **20**, S871 (2003)

21. K. Kuroda *et al, Class. Quantum Grav.* **23**, S215 (2006): K. Kuroda *et al, Prog. Theor. Phys. Suppl.* **163**, 54 (2006)

22. K. Kuroda *et al, presented at 7th Edoardo Amaldi Conference on Gravitational Waves* (Sydney, Australia, 2007)

23. Y. Saito *et al, Vacuum* **53**, 353 (1999)

24. T. Uchiyama *et al, Class. Quantum Grav.* **21**, S1161 (2004)

25. R. Takahashi *et al, Class. Quantum Grav.* **25**, 114036 (2007)

26. T. Uchiyama *et al, Physics Letter A* **242**, 211 (1998):T. Uchiyama *et al, Physics Letter A* **261**, 5 (1999): S. Miyoki *et al, Cryogenics* **41**, 415 (2001)

27. S. Miyoki *et al, Class. Quantum Grav.* **21**, S1173 (2004)

28. K. Agatsuma *et al, Proceedings of 8th Edoardo Amaldi Conference on Gravitational Waves* (New York, 2009)

29. T. Tomaru *et al, Physics Letter A* **283**, 80 (2001)

30. Z. Yan *et al, Appl. Opt.* **45**, 2631 (2006)

31. M. Tokunari *et al, J. Physics: Conference Series* **32**, 423 (2006)

32. M. Tokunari, *Doctor thesis of the University of Tokyo*(Tokyo, Japan, 2007) (in Japanese)

GROUND-BASED STUDY OF AN INERTIAL SENSOR WITH AN ELECTROSTATIC-CONTROLLED TORSION PENDULUM

HAI-BO TU, YAN-ZHENG BAI, LIN CAI, LI LIU, JUN LUO and ZE-BING ZHOU*

School of Physics, Huazhong University of Science and Technology, Wuhan 430074, P.R. China

**E-mail: E-mail: zhouzb@mail.hust.edu.cn*

An electrostatic-controlled torsion pendulum was constructed to test the in-flight performance of an electrostatic inertial sensor on ground. An electrostatic-controlled torque resolution of 6×10^{-13} N m Hz$^{-1/2}$ and a force resolution of 3.6×10^{-11} N Hz$^{-1/2}$ from 2 mHz to 0.1 Hz were achieved.

Keywords: Inertial sensor, electrostatic suspension accelerometer, Electrostatic-controlled torsion pendulum.

1. Introduction

High precision inertial sensors play a paramount role in gravitational experiments in space [1-6]. These inertial sensors have two operation modes. In one mode, the sensors act as free-falling references to guide the micro-Newton thrusters to compensate the non-gravitational forces acting on the spacecraft, namely the drag-free control, such as those in the LISA [7-9] missions. In another mode, the sensors act as accelerometers to measure the accelerations of the spacecraft or the expected forces, such as those in the CHAMP [10] and GRACE [11-12] missions. the sensors can also be operated in the combined modes, where the sensors are used to guide the thrusters to compensate the drag forces of the spacecraft along the velocity direction. In this case the sensors can measure the residual accelerations of the spacecraft with a high resolution, such as those in the GOCE [13-14] and MicroSCOPE [15-16] missions, which is called as the drag compensation.

For high precision inertial sensors, the performance verification on ground is mainly limited by 1 g Earth's gravity acceleration. One scheme is to use high voltage to levitate the proof mass (PM), with which ONERA has succeeded to develop the STAR, SuperSTAR, and GRADIO accelerometers

[17-18]. However, the ground-based measurement was limited at the level of 10^{-9} m s^{-2} Hz$^{-1/2}$ due to the very strong coupling from the large electric field and the seismic noises. Another scheme is to suspend the PM by a thin fiber to act as a torsion pendulum or balance, and a torque resolution better than 10^{-14} N m Hz$^{-1/2}$ has been achieved by several groups [19-21]. In addition, a drop tower can offer a short micro-gravity environment to test the performance of the inertial sensor [22].

The torsion pendulum or balance can be used not only to investigate the residual disturbances of the PM [23-25], but also to simulate the operation of the inertial sensor in flight based on the weak fiber suspension [26-27]. For the second application, the gaps between the electrodes and the PM are in general smaller than 1 mm, and the parasitic negative stiffness k_e induced by the capacitance transducer is much larger than that of the fiber k_f. It means that k_f can be negligible in this case and the torsion pendulum is unstable without the feedback control. It is beneficial to simulate the in-orbit operation conditions, which allow to test the performance of the electrostatic actuator, to investigate the cross-coupling between the non-sensitive and sensitive motions of the PM for the LISA mission, to study the drag-free control on ground and so on.

An electrostatic-controlled torsion pendulum and an controlled torsion balance were developed to investigate the performance of the inertial sensor in our laboratory, and the progress is outlined in Section 3. A multi-stage electrostatic-controlled torsion pendulum is being designed to investigate the performance of two or more degrees of freedom (DOFs) and the cross-couplings of the inertial sensor on ground, as outlined in Section 4.

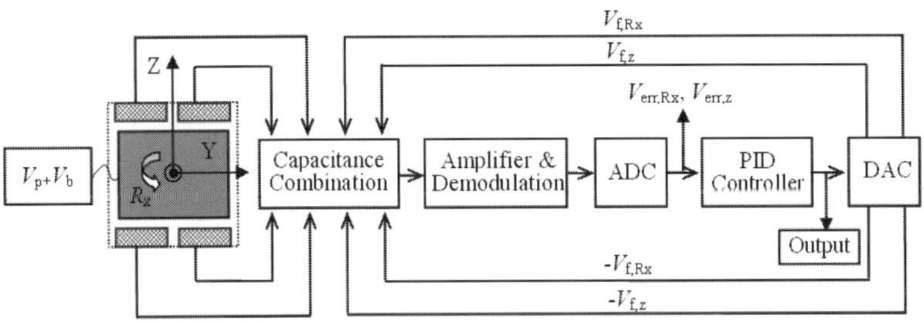

Fig. 1. Schematic diagram of the combined accelerations measurement for Z and R_x DOFs in the acceleration measurement mode. Two pairs capacitive electrodes combined with a servo circuitry act as a two-dimensional displacement transducer to monitor the relative motions of the PM with respective to the frame along Z axis and rotating about X axis, and simultaneously electrostatic-control the PM motionless at the frame center.

2. Operational principle of the electrostatic inertial sensor

In general, twelve capacitive electrodes surrounding a cubic PM combined with a circuitry can simultaneously realize the six-DOF acceleration measurement [28]. The schematic diagram of the combined measurement for a translational and a rotational DOFs is shown in Figure 1 when the inertial sensor operates in the acceleration measurement mode. The probe consists of two pairs of electrodes and a PM locked in the center of the electrodes. The electronic circuitry consists of an excitation modulation unit, a capacitance combination unit, an amplifier and demodulation unit, and a proportional-integral-differential (PID) controller. A DC bias voltage V_b is added on the PM in order to improve the linearity of the electrostatic actuator, and an AC voltage V_p at a frequency of 100 kHz is used to excite the capacitance transducer to suppress the low frequency noises of the circuitry.

As an example, Figure 2 shows the operational principle of a single translational DOF, where H_m, H_s, H_c, H_a are the transfer functions of the probe, the capacitance transducer, the PID controller and the electrostatic actuator, respectively. a_{in} is the input acceleration, $a_{n,d}$ is the direct acceleration disturbance of the PM induced by the environment, x_n is the equivalent displacement noise of the capacitance transducer, V_n is the voltage noise acting on the actuator [29]. The output of the sensor can be expressed as

$$V_f = \frac{1}{H_a} \cdot \frac{H_{open}}{1 + H_{open}}(-a_{in} + a_{n,total}) \tag{1}$$

with $H_{open} = H_m H_s H_c H_a$, $a_{n,total}$=$a_{n,d}$+$(\omega^2 + \omega_e^2)x_n$+$H_a V_n$, ω_e^2=-k_e/m, where m is the mass of the PM. It is obvious that, to achieve a high precision, the PM should be well isolated from any residual disturbances, the parasitic stiffness k_e should be as small as possible, and high-resolution capacitive transducers and ultra-low noise actuators are required.

3. Single degree-of-freedom measurement with an electrostatic-controlled torsion pendulum

An electrostatic-controlled torsion pendulum, as shown in Figure 3, was constructed for on ground performance testing of a rotational DOF of the inertial sensor. A titanium PM of about 4 cm × 4 cm × 1 cm combined with a mirror was suspended by a 25 μm tungsten fiber with a length of about 40 cm. The total mass of the pendulum is about 78 g and the moment of inertia is about 1.995×10^{-5} kg m^2. A 5 μm Au wire with a length of about

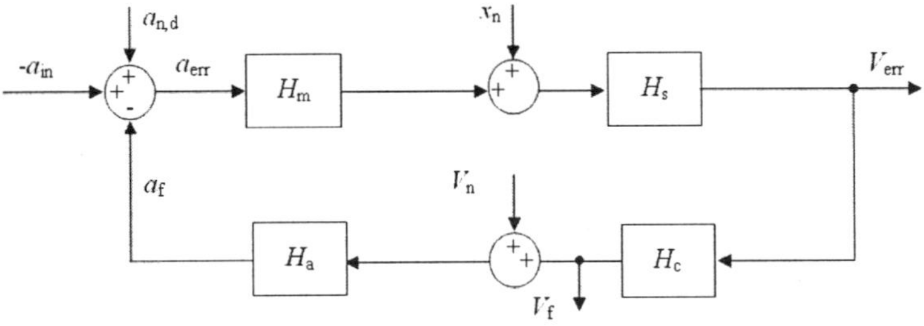

Fig. 2. The diagram of the transfer function of the inertial sensor

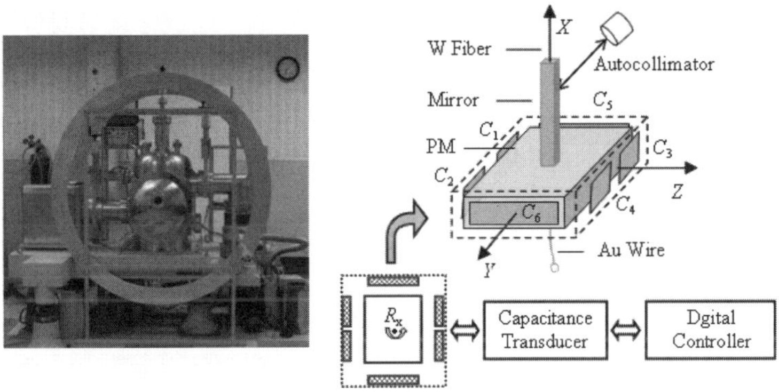

Fig. 3. Left: Photo of the experimental setup; Right: Schematic diagram of the electrostatic-controlled torsion pendulum. It consists of a probe, a capacitance transducer and a digital controller. Three pairs of electrodes are used to monitor both the two horizontal translations and the twist motion of the PM, and only the twist motion is controlled w.r.t. the electrode frame.

4 cm was used to discharge the PM, and its elastic coefficient is estimated to be about 4.2×10^{-11} N m rad^{-1} by comparing the free oscillating periods of the torsion pendulum with and without the discharge wire. The elastic coefficient of the charged gold wire is about 300 times smaller than that of the suspension tungsten fiber. The free oscillation period T_0 of the pendulum with the discharge wire is (248.4 ± 0.9) s, and the quality factor Q is about 2800 in a vacuum of about 2×10^{-5} Pa. The total elastic coefficient k_f is $(1.28 \pm 0.01) \times 10^{-8}$ N m rad^{-1}.

Three pairs of electrodes were used not only to measure the Y, Z translations and twist motions of the PM, but also to control the twist motion of the PM. The equilibrium capacitance gaps d_0 are about 150 μm. C_1

$\sim C_4$ and $C_5 \sim C_6$ as shown in Figure 3 are about 4.9 pF and 13.3 pF, respectively. We have also developed a capacitance transducer with a resolution of 2×10^{-6} pF Hz$^{-1/2}$ at 0.1 Hz based on an AC capacitance and transformer bridge, combined with a digital control loop based on a field-programmable-gate-array (FPGA)[30].

In order to investigate disturbances induced by the magnetic field, a pair of large Helmholtz coils with diameters of about 1.6 m was symmetrically set outside the vacuum chamber along Y axis, as shown in Figure 3. Another two small coils were mounted near the electrodes inside the chamber along Z axis, as shown in Figure 4. Both of the small coils are symmetrically departed about 20 mm along Y axis and 61.5 mm along Z axis away from the center of the PM to distinguish the torque from the susceptibility and residual dipole moment of the PM. Four heaters and four temperature sensors were set close to each electrode along Z axis to investigate the temperature effects. An aluminum shield box was used to isolate the PM from the disturbances of residual gas and heat flow. Based on the experimental design, the magnetic parameters such as the susceptibility and the residual dipole moment of the PM and the temperature effect were measured [31].

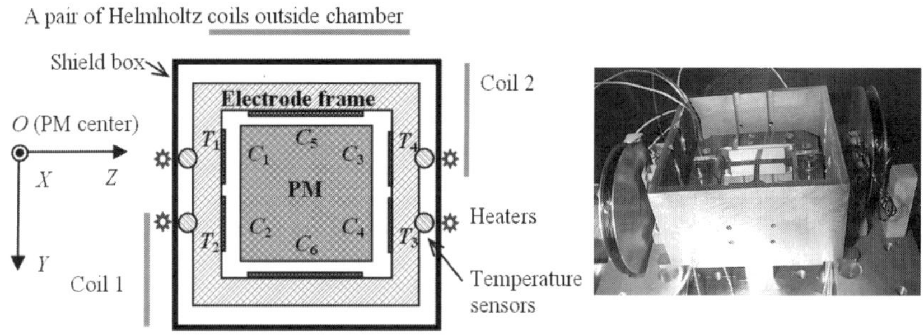

Fig. 4. Left: Schematic diagram of the residual disturbances experiment. Right: Photo of the layout in chamber.

A torque originating from the interaction of the remnant dipole moment $\boldsymbol{M}_r$ of the PM with an applied magnetic field $\boldsymbol{B}$ induced by the Helmholtz coils was used to calibrate the torque sensitivity of the electrostatic actuator. The magnetic torque is firstly measured based on the elastic torque of the suspending fiber in the case of the free oscillation pendulum, and then the sensitivity of the electrostatic actuator can be determined by the same magnetic torque under the control mode. The torque sensitivity was determined as $S_{Nm} = (1.73 \pm 0.31) \times 10^8$ V (N m)$^{-1}$, which agreed well with the theoretical estimation given by $d_0/(V_b \sum_{i=1}^4 C_i)$.

For the rotational DOF, the parasitic stiffness k_e can be theoretically expressed as

$$k_e \approx -\frac{(3b^2 + 3ad_0 + h^2)V^2}{12d_0^2} \sum_{i=1}^{6} C_i \tag{2}$$

with $V^2 = V_f^2 + V_b^2 + V_{p0}^2/2$, where a is the side length of the PM, b is the distance between the two centers of the adjacent electrodes, h is the width of the electrode, V_{p0} is the amplitude of the AC modulation voltage. The twist position of the PM relative to the frame was periodically set by the digital controller. Thanks to the electrostatic control, the PM was controlled to follow up with the setting position. The electrostatic-controlled torque is equal to the torque induced by k_e if neglecting the effect of the suspension fiber, which is given by

$$\Delta V/S_{Nm} = k_e \times \Delta\theta, \tag{3}$$

and k_e was obtained as $(5.6 \pm 1.4) \times 10^{-6}$ N m rad^{-1}, which agrees with the theoretical value given by Equation (2) and $k_c/k_f \backsim 438$ was deduced.

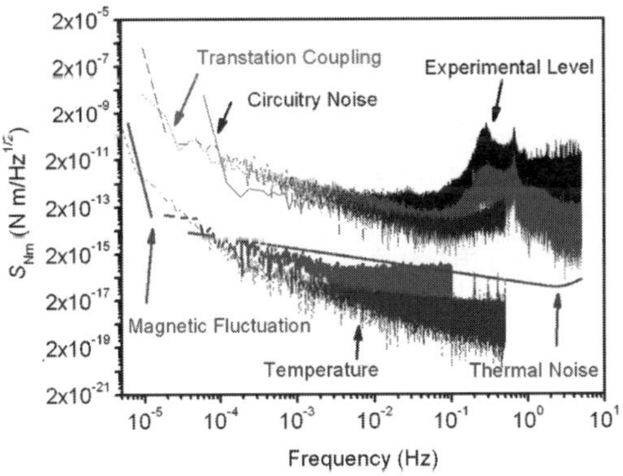

Fig. 5. Torque noise of the controlled torsion pendulum.

The equivalent torque noise of the controlled pendulum comes to 6×10^{-13} N m Hz$^{-1/2}$ in the frequency range from 2 mHz to 0.1 Hz, as shown in Figure 5, which is mainly limited by the coupling disturbance from the translational seismic noises. The noise level due to the capacitance transducer is 3 times lower than the experimental level. For the ground-based testing of the inertial sensor, the effects of the temperature and magnetic field fluctuations as well as the thermal noise of the mechanical pendulum were much lower than the present experimental level [31].

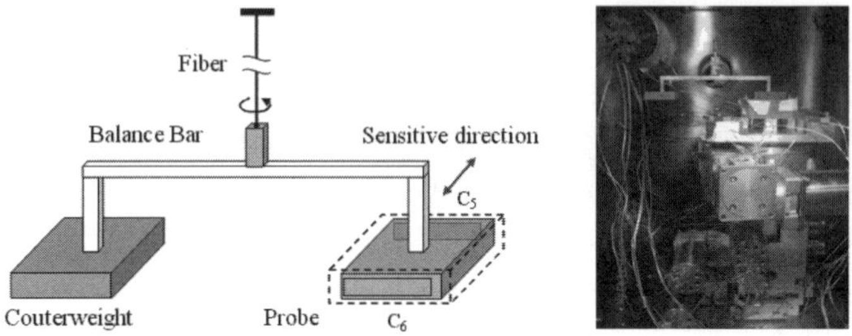

Fig. 6. Left: Schematic diagram of the electrostatic-controlled torsion balance. The PM can move along the sensitive direction flexibly due to the weak stiffness of the suspension fiber, which allows to simulate the operation of a translational DOF of the inertial sensor in flight. Right: Photo of the layout in chamber.

In addition, an electrostatic-controlled torsion balance was constructed to test the performance of a translational DOF of the inertial sensor on ground, as shown in Figure 6. A PM and a counterweight were connected with a bar to act as a balance. The nearly free twist motion of the balance with a small rotation angle is equivalent to the translation of the PM with respective to the electrodes. Therefore, a translation motion of the PM is flexibly due to the weak stiffness of the suspension fiber, which allows to simulate the operation of a translation DOF of the inertial sensor in flight. The moment of inertial of the whole torsion balance is about 7.8×10^{-4} kg m^2, and the free-oscillation period was measured to be about 373.6 s. Then the stiffness of the fiber was calculated to be 2.2×10^{-7} N m/rad, which also could be negligible compared with the equivalent stiffness induced by the translational parasitic stiffness of the probe. The current force noise of the controlled torsion balance comes to 3.6×10^{-11} N $Hz^{-1/2}$ from 2 mHz to 0.1 Hz.

4. Design of a multi-stage electrostatic-controlled torsion pendulum for further cross-couplings measurement

As further development of the controlled torsion pendulum scheme, a 2-stage or 3-stage electrostatic-controlled torsion pendulum is being considered to test the cross-couplings of the inertial sensor.

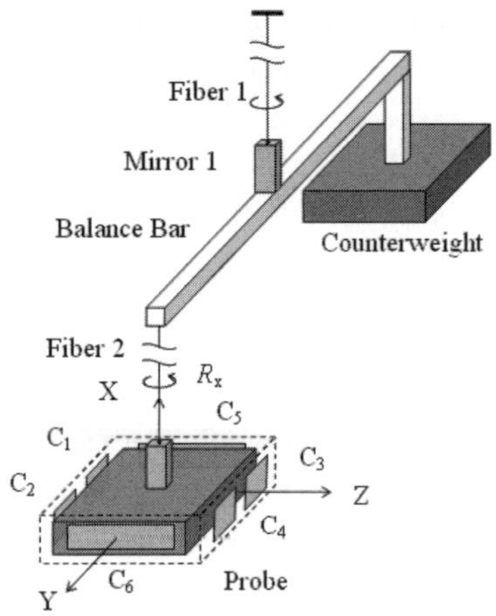

Fig. 7. Schematic diagram of the two-stage electrostatic-controlled torsion pendulum. Two fibers suspension allow to test the performances of translational motion along the Z axis and rotational motion along X axis of the PM on ground.

The design of a two-stage electrostatic-controlled torsion pendulum is shown in Figure 7, which can be considered as the combination of the controlled torsion pendulum and torsion balance, as shown in Figure 3 and Figure 6, respectively. A bar is suspended by a fiber, and the PM is suspended by the second fiber. The design should ensure that the stiffnesses of both fibers are negligible compared with the parasitic stiffnesses. Therefore, this pendulum allows to test the in-flight performance of the electrostatic-controlled inertial sensor such as a translational and a rotational motions and their coupling, and also provide a chance to test the coupling effect between a controlled and free translational motions, for example, Z and Y axes as shown in Figure 7, which is important for the LISA mission.

A three-stage electrostatic-controlled torsion pendulum is also being considered to simultaneously investigate two translational DOFs and a rotational DOF, in which the PM will be suspended by three fibers, and two balance bars are orthogonal to each other.

In these schemes, one or two translational DOFs of the PM will have much lower mechanical coupling stiffness, and be electrostatic-controlled simultaneously, so that the translation-twist couplings can be significantly suppressed. Thanks to the control scheme, the mechanical components of the pendulum have much smaller motion amplitude, therefore, the coupling effect from environmental gravity gradient is much lower than that of the free-oscillation torsion pendulum scheme. The thermal noise limit is estimated as low as 3×10^{-15} N m at 1 mHz with a quality factor of about 3000, which could be improved by a higher quality factor torsional pendulum constructed such as with a fused-silicon fiber suspension [32]. However, the ground-based investigation would be limited by the capacitance transducer and the seismic noises, which should be studied in the future.

5. Discussions

The electrostatic-controlled torsion pendulum can be used to simulate the operation of the inertial sensor with feedback control in flight, and thus to investigate the performance under close-loop state on ground because of its weak mechanical stiffness in its sensitive DOFs.

As further experiments, a two-stage or three-stage controlled torsion pendulum will be constructed to test the multi-DOF control of the PM, and also investigate their cross-coupling.

Acknowledgments

This work was supported by the National Natural Science Foundation of China under grants 10675047.

References

1. P. Touboul, B. Foulon and E. Willemenot, *Acta Astronautica*, **45**, 605 (1998).
2. V. Iafolla, A. Mandiello and S. Nozzoli, *Adv. Sp. Res.*, **25**, 1241 (2000).
3. P. Touboul, B. Foulon, M. Rodrigues and J. P. Marque, *Aerosp. Sci. Technol.*, **8**, 431 (2004).
4. J. Mester, S. Buchman, A. M. Cruise et al., *Nuclear Physics B*, **134**, 147 (2004).
5. V. Josselin, M. Rodrigues and P. Touboul, *Acta Astronautica*, **49**, 95 (2001).
6. B. F. Schutz, *Class. Quantum Grav.*, *Class. Quantum Grav.*, **26**, 094020 (2009).
7. R. Dolesi, D. Bortoluzzi, P. Bosetti et al., *Class. Quantum Grav.*, **20**, S99 (2003).
8. F. Nappo, P. Sarra, S. Vitale, R. Dolesi and W. J. Weber, *54th International Astronautical Congress*, (Bremen, Germany, 2003).
9. M. Rodrigues and P. Touboul, *Adv. Sp. Res.*, **32**, 1251 (2003).
10. P. N. A. M. Visser and J. van den Ijssel, *Adv. Sp. Res.*, **31**, 1905 (2003).
11. T. V. Helleputte, E. Doornbos and P. Visser, *Adv. Sp. Res.*, **43**, 1897 (2009).
12. J. Flury, S. Bettadpur and B. D. Tapley, *Adv. Sp. Res.*, **42**, 1414 (2008).
13. J. Flury, S. Bettadpur and B. D. Tapley, *Adv. Sp. Res.*, **54**, 167 (2003).
14. P.N.A.M. Visser, *Adv. Sp. Res.*, **39**, 1630 (2007).

15. P. Touboul and M. Rodrigues, *Class. Quantum Grav.*, **18**, 2487 (2001).
16. D. Hudson, R. Chhun and P. Touboul, *Adv. Sp. Res.*, **39**, 307 (2007).
17. E. Willemenot and P. Touboul, *Rev. Sci. Instrum.*, **7**, 302 (2000).
18. A. Bernard and P. Touboul, *Workshop ESA/NASA on the Solid-Earth Mission ARISTOTELES*, (Anacapri, Italy, 1991).
19. S. Schlamminger, K. Y. Choi, T. A.Wagner, J. H. Gundlach and E. G. Adelberger, *Phys. Rev. Lett.*, **100**, 041101 (2008).
20. L. C. Tu, S. G. Guan, J. Luo, C. G. Shao and L. X. Liu, *Phys. Rev. Lett.*, **98**, 201101 (2007).
21. L. Carbone, G. Ciani, R. Dolesi, M. Hueller, D. Tombolato, S. Vitale, W. J. Veber and A. Cavalleri, *Phys. Rev. D*, **75**, 042001 (2007).
22. H. Dittus, W. Vodel, R. Greger et al., *Adv. Sp. Res.*, **32**, 1301 (2003).
23. A. Cavalleri, G. Ciani, R. Dolesi et al., *Class. Quantum Grav.*, **26**, 094017 (2009).
24. L. Carbone, A. Cavalleri, R. Dolesi et al., *Phys. Rev. Lett.*, **91**, 151103 (2003).
25. L. Carbone, A. Cavalleri, R. Dolesi et al., *Class. Quantum Grav.*, **21**, S611 (2004).
26. Z. B. Zhou, S. W. Gao and J. Luo, *Class. Quantum Grav.*, **22**, S537 (2005).
27. Z. B. Zhou, S. B. Qu, H. B. Tu, Y. Z. Bai, S. C. Wu, Q. Y. Wan an J. Luo, *Int. J. Mod. Phys. D.*, **17**, 985 (2008).
28. H. B. Tu, S. C. Wu, Y. Z. Bai et al., *J. Jpn, Soc. Microgravity Appl.*, **24**, (2007).
29. C. C. Speake, *Class. Quantum Grav.*, **13**, A291 (1996).
30. Y. Z. Bai, Z. B. Zhou, H. B. Tu, et al., *Front. Phys. China*, **4**, 205 (2009).
31. H. B. Tu, Z. Y. Bai, B. Z. Zhou and J. Luo, *Journal of Physics: Conference Series* , **154**, 012036 (2009).
32. S. E. Pollack, S. Schlamminger and J. H. Gundlach, *AIP Conference Proceedings*, **873**, 158 (2006).

ORBIT DESIGN AND OPTIMIZATION FOR THE GRAVITATIONAL WAVE DETECTION OF LISA

Y. XIA*, G. LI, and Y. LUO

Purple Mountain Observatory,
Nanjing, 210008, China
**E-mail: yxia@pmo.ac.cn*

Z. YI

Department of Astronomy, Nanjing University,
Nanjing, 210008, China

G. HEINZEL, A. RÜDIGER

Max Planck Institute for Gravitational Physics,
Hannover, D-30167, Germany

The Laser Interferometer Space Antenna (LISA) is a joint ESA-NASA mission for detecting low-frequency gravitational waves, by using accurate distance measurements with laser interferometry between three spacecrafts, which will be launched around 2018 and one year later reach their operational orbits around the Sun. In order to operate successfully, it is crucial for the constellation of the three spacecrafts to have extremely high stability. Based on the study of operational orbits for a 2015 launch, we designed the operational orbits of beginning epoch on 2019-03-01, analyzed the acceptable error range of the injection, and introduced the method of orbit design and optimization.

Keywords: co-orbital restricted problem; orbit design; orbit optimization.

1. Introduction

The Laser Interferometer Space Antenna (LISA) is a joint ESA-NASA mission for detecting low-frequency gravitational waves in the frequency range from 0.1 mHz to 1 Hz. Three spacecrafts will be launched around 2018, and reach their operational orbits around the Sun after 14 months. Observation will last for 5 to 10 years.

As shown in Figure 1, the three spacecrafts form an equilateral triangle with an arm-length (side length) of around 5×10^6 km. The center of the

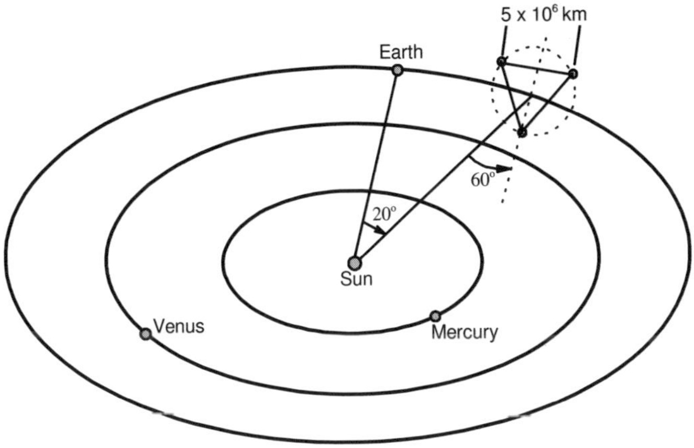

Fig. 1. LISA constellation moving around the Sun

constellation moves on an Earth-like orbit, trailing 20° behind the Earth. The angle between the direction from the constellation center to the Sun and the constellation plane is about 60°. Distance variations between the spacecraft will be measured by laser interferometry to detect gravitational waves. In order for LISA to operate successfully, it is crucial that the constellation of the three spacecrafts have extremely high stability. Several parameters of the constellation should be controlled so as to vary in only a limited range, and the detailed requirements are shown in Table 1.

Table 1. The variation ranges of the parameters of the LISA constellation

Parameter	Average value	Requirement
Arm-length l	5×10^6 km	$\Delta l = \pm 5 \times 10^4$ km
Internal angle β	60°	$\Delta\alpha = \pm 1.5°$
Relative velocity v_r	0	$v_r < 15$ m/s
Trailing angle to Earth θ	about 20°	as small as possible

Thus, the problem of optimizing the orbits of the LISA constellation can be summarized as follows: finding a set of orbital elements a_i, e_i, ω_i, Ω_i, I_i, M_{0i}, (i=1, 2, 3) to make the following cost function take its minimum:

$$Q(a_i, e_i, \omega_i, \Omega_i, I_i, M_{0i}) = w_1 \Delta l^2 + w_2 \Delta\beta^2 + w_3 \Delta {v_r}^2 + w_4 \Delta\theta^2 \quad (1)$$

with appropriately chosen weights w_j.

Li Guangyu, Yi Zhaohua et al.[1,2] optimized the operational orbits for an initial epoch of 2015-01-01 successively and presented the co-orbital restricted problem as an instruction of the orbit design for LISA. Now the mission of LISA is put off to about 2018, we optimized the operational orbits for an initial epoch on 2019-03-01. Furthermore, we presented the acceptable error range of the injection in this paper.

2. Selection of the Starting Orbits for optimization

The orbit of one spacecraft is described by the six starting orbit elements. The problem will be discussed in the following subspace of the orbital space of six dimensions

$$\chi\left\{x_{i0} - g_i \leq x_i \leq x_{i0} + g_i, i = 1, 2, \cdots, 6\right\} \tag{2}$$

The problem of orbit optimization becomes to find the minimum value of a cost function $Q(x)$ in the subspace of the orbital space χ.

Selection of the starting orbits for optimization is to solve the x_{i0} in equation (2). The starting orbits are such a set of orbits that determine the starting point of the optimization trajectory in the orbit space, along which we can get to the aimed point and obtain the optimization orbits by given optimization algorithms. These orbits have been discussed in detail in many papers.[1,3–6]

Since the spacecraft will be launched in about 2018 and enter their operational orbits 14 months later, we choose the initial epoch of the operational orbits as 2019-03-01. Then with the method in the above references, we get the starting orbits as shown in Table 2.

Table 2. The elements of the starting orbits for optimization (Epoch 2458543.5, 2019-03-01)

	a	e	ω	Ω	I	M_0
SC1	0.9992 AU	0.009648	270.0°	347.061°	0.9529°	238.0°
SC2	0.9992 AU	0.009648	270.0°	107.061°	0.9529°	118.0°
SC3	0.9992 AU	0.009648	270.0°	227.061°	0.9529°	358.0°

3. Tabu Search algorithm

For the optimization of the operational orbits, the cost function has 18 independent variables, and there are many local minima in the orbital space, which often makes the local optimization algorithm get stuck in a certain local point and never search for a global optimization. This can be solved by a global optimization algorithm, the Hybrid Reactive Tabu Search algorithm.[1,7]

The algorithm divides the continuous orbital space into many boxes, and these boxes are identified with different 2-bit binary numbers. A 2-dimensional example is shown in Figure 2. By inverting any bit of the identifier of a box, the box obtained is called a neighbor of the primary box. Calculate the cost function of different points of one box, and the minimum of the values of the cost function is regarded as the evaluation of the box. The search always transfers from the current box to its available and best evaluated neighbor.

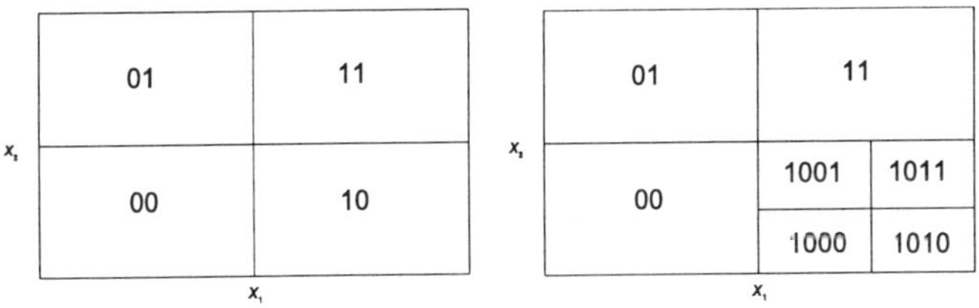

Fig. 2. The identifiers of the boxes in 2-dimensional space

In order to avoid getting stuck in a certain local point and search for a global optimization point, set a tabu period for every box. During the tabu period, the box is forbidden to search, and is ignored in the transfer between the neighborhoods, so as to get off a certain box and to return to a global search.

4. The result of orbit design and optimization

Optimizing the starting orbits in Table 2 with the Tabu Search algorithm, the result is shown in Table 3. The comparison of the range of the variation of constellation parameters before and after optimization is shown in Table 4, it can be seen the optimized orbits meet the requirements of the stability of the constellation, but the trailing angle is somewhat bigger, and we will do further research about it.

Considered the injection error in practice, we tested the acceptable error

Table 3. The elements of the optimized orbits (epoch 2458543.5, 2019-03-01)

	a (AU)	e	ω	Ω	I	M_0
SC1	0.99939441	0.0095250198	269.59762°	344.81853°	0.95341394°	238.75688°
SC2	0.99940953	0.0096189996	270.19189°	104.53895°	0.95637030°	118.45536°
SC3	0.99940926	0.0096209991	269.35168°	224.70675°	0.95009466°	359.12560°

Table 4. The change of parameters during the optimization

Parameter	Requirement	Before optimization	After optimization
Side length variation Δl	$\pm 5 \times 10^4$ km	$\pm 25.3 \times 10^4$ km	$\pm 4.42 \times 10^4$ km
Inner angle variation $\Delta\beta$	$\pm 1.5°$	$\pm 3.16°$	$\pm 0.74°$
Relative velocity v_r	< 15 m/s	< 53.52 m/s	< 9.93 m/s
Trailing angle θ	As small as possible	$< 26.99°$	$< 28.86°$

range of the orbit elements in Table 3, and the result is shown in Table 5. In this range, the requirement of the LISA constellation stability can also be fulfilled.

Table 5. The acceptable error range of the elements of the optimized orbits

	a (AU)	e	ω	Ω	I	M_0
SC1	1.6×10^{-7}	1.3×10^{-5}	1.0×10^{-4}°	5.4×10^{-5}°	4.3×10^{-4}°	4.2×10^{-4}°
SC2	1.8×10^{-7}	1.4×10^{-5}	1.0×10^{-4}°	6.0×10^{-5}°	5.8×10^{-4}°	2.4×10^{-4}°
SC3	1.2×10^{-7}	1.2×10^{-5}	1.2×10^{-4}°	6.0×10^{-5}°	4.8×10^{-4}°	2.4×10^{-4}°

Acknowledgments

Thank the National Science Foundation (Grant No.10573037, 10503013) and the Foundation of Minor Planets of Purple Mountain Observatory for the financial support.

References

1. Li, G., Yi Z., Heinzel, G., Rüdiger, A., Jennrich, O., Wang, L., Xia, Y., Zeng, F., & Zhao, H., Methods for Orbit Optimization for the LISA Gravitational Wave Observatory, Int. J. Mod. Phys. D, 17, 1021-1042, 2008.
2. Yi, Z., Li, G., Heinzel, G., Rüdiger, A., Jennrich, O., Zeng, F., & Zhao, H., Coorbital Restricted Problem and its Application in the Design of the Orbits of the LISA Spacecraft, Int. J. Mod. Phys. D, 17, 1005-1020, 2008.
3. Folkner, W. M., Hechler, F., Sweetser, T. H., Vincent, M. A., & Bender, P. L., LISA Orbit Selection and Stability, Class. Quantum Grav., 14, 1405-1410, 1997.
4. Dhurandhar, S. V., Nayak, K. R., Koshti, S., & Vinet, J. Y., Fundamentals of the LISA Stable Flight Formation, Class. Quantum Grav., 22, 481-487, 2005.

5. Sweetser, T. H., An end-to-end trajectory description of the LISA mission, Class. Quantum Grav., 22, S429-S439, 2005.
6. Nayak, K. R., Koshti, S., Dhurandhar, S. V., & Vient, J. Y., On the Minimum Flexing of LISA's Arms, Class. Quantum Grav., 23, 1763-1778, 2006.
7. Battiti, R., & Techiolli, G., The continuous Reactive Tabu Search: Blending Combinatorial Optimization and Stochastic Search for Global Optimization, Annals of Operations Research, 63(2), 151-188, 1994.

ANGULAR RESOLUTION OF MULTI-LISA CONSTELLATIONS

YAN WANG*

Department of Astronomy, Nanjing University, Nanjing, 210093 China;
Albert Einstein Institute, Callinstr. 38, 30167 Hannover, Germany.
**E-mail: wangyan7677000@gmail.com*

XUE-FEI GONG

Institute of High Energy Physics, Chinese Academy of Sciences, Beijing, 100049, China;
Theoretical Physics Center for Science Facilities, Chinese Academy of Sciences, Beijing, 100049, China;
Academy of Mathematics and System Sciences, Chinese Academy of Sciences, Beijing, 100190, China.

In this article, we present a detailed derivation of the angular resolution of arbitrary sets of LISA (Laser Interferometer Space Antenna) constellations with a toy model for gravitational wave signals, and further generalized to more complicated cases with slowly varying gravitational wave signals of well-defined frequency at any time instant. For future space-borne LISA-like gravitational wave detectors, our results may serve as a conservative quick estimate of the detector's angular resolution and hopefully moreover a reference for the configuration designs.

Keywords: gravitational wave detection, LISA, angular resolution

1. Introduction

In the past two decades, with the flourishing efforts in the development of interferometric gravitational wave detectors working both on ground and in space covering a broad band in gravitational wave frequency, the goal of gravitational wave hunting has experienced a conceptual change from bare detection to gravitational wave (GW) astronomy, which tries to serve as a counterpart of traditional electro-magnetic astronomy to accomplish the so called multi-messenger astronomy. Partly this is due to the data analysis scheme employed in gravitational wave detections, and more importantly the unique ability of gravitational wave observations in revealing the secret processes blind to optical and radio astronomy. To this end, among the desires in obtaining astrophysical information encoded in gravitational waves, crucial that one can determine sufficient accurately the celestial positions and luminosity distances of the sources. Given those parameters nicely measured, not only astrophysical information of the gravitational wave source like host galaxy for massive black holes and stellar environment of the source can be extracted, one can also carry out more valuable sciences like for instance precision cosmology (together with electro-magnetic observations) and hunting for electromagnetic follow-ups in the circumstances of violent gravitational wave releases.[1]

Angular resolution is particularly important for space-borne gravitational wave detectors due to the variety of GW sources they would detect hence the rich astrophysics involved and the characteristics of the detector operation and the signal type. The most mature space-borne detector LISA (Laser Interferometry Space Antenna) is a joint ESA-NASA mission aiming at detecting various kinds of gravitation wave sources (including massive black hole mergers, extreme mass ratio inspirals and white dwarf binaries etc.) in a low $10^{-4} \sim 1Hz$ frequency band.[2] Based on the same concept, several LISA follow-on missions have been proposed, including multi-LISA configurations, advanced LISA with shorter armlengths (ALIA) and the ambitious Big Bang Observer (BBO).[3] These proposed missions work at the slightly higher deci-Hertz windows and aim at gravitational waves from intermediate mass black hole mergers, cosmological compact binaries, supernovae and even the primordial gravitational waves generated at the very beginning of the universe. For all these proposals, important issues related to the improvement of positioning of the sources include for instance that for one specific mission, how to design the detector configuration when more than one triangle LISA like configurations are involved and how much more science could it bring compared to the nowadays one triangle design of LISA, and moreover if for one day we can have different missions operating overlap in time in neighboring frequency bands, how much we can benefit from the coincident detections in a precision measurement point of view.

We understand that space-borne gravitational wave detectors like LISA decode information about the source position in the measured signal through mainly three ways, the periodic Doppler shift imposed on the signal by the orbit motion around the Sun and time-varying orientation, the relative amplitudes and phases of the two polarization components, and the difference of the time-of-arrivals (in the multi-LISA case).[4–6] So detailed positioning accuracy for a specific source depends on very detailed characters of the gravitational wave it emits, the waveform, the relative position and orientation of the source to the detector, the frequency it shows up and therefore the response of the detector at there etc. But at the moment in this work, rather than angular resolution for specific LISA sources,[6–9] we focus on the influence of multi-detector configurations on the improvement of angular resolution. Due to the different question we are trying to answer here, what we need to carry out in a very first understanding is just a rough estimate in which we do not use specific knowledge of the sources but instead make quite general assumptions. By doing this we hope the work could serve as a conservative guide to a quick answer of the addressed question.

The rest of the work is organized as follows: in section 2, with certain assumptions and reasonable approximations, we derive analytically the conservative multi-LISA angular resolution for very narrow band oscillating signals and compare the results to the angular information obtained by only one triangle configuration; in section 3 we then generalize our results to more general circumstances with slowly varying amplitude and frequency gravitational wave signals and this ends up the article.

2. Derivation of angular resolution of multi-LISA constellation in the simplest signal model

To make the derivation as simple as possible while keeping the validity and reality of the results, we neglect all the parameters other than the angular positions. This is roughly equivalent to using only one polarization of the two polarization modes (or neglecting the relative phases and amplitudes between the two polarization modes) of a monochromatic gravitational wave source and neglecting the effect imposed by the time-varying orientation of LISA. In reality, these two kinds of effects can be used to further improve the angular resolution to some extent. What we are making use of to decode the information of the angular position are the Doppler shift caused by the detectors' orbital motions and the differences of the time of arrivals between different LISA constellations. In this way, we can model the gravitational wave signal observed by the i-th LISA constellation as[10]

$$h(t) = \mathcal{A}\cos(2\pi f\xi) = \mathcal{A}\cos(2\pi f(t + R\sin\theta\cos(2\pi f_{orb}t + \kappa_i - \varphi))) \quad (1)$$

$$\xi = t + R\sin\theta\cos(2\pi f_{orb}t + \kappa_i - \varphi). \quad (2)$$

Here f is the constant GW frequency of the astrophysical source during the observation, θ (co-latitude) and φ are the celestial position of the GW source in ecliptic coordinates, R is the radial distance to the centroid of the i-th LISA constellations, $f_{orb} = 1/1year$ is the LISA orbital frequency, and finally κ_i the initial ecliptic longitude of the centroid of the i-th constellation.

Given the one-sided noise spectral density $S_n(f)$, we can define an inner product of two functions of time $x(t), y(t)$[11,12]

$$\langle x(t)|y(t)\rangle = 4\mathcal{R}\left(\int_0^{+\infty} \frac{\widetilde{x}^*(f)\widetilde{y}(f)}{S_n(f)}df\right) \tag{3}$$

where tilde and star denote the Fourier transform of the corresponding time series and its complex conjugate respectively. With the introduction of this inner product, we can easily calculate the so called Fisher Information Matrix (FIM) for the angular parameters θ and φ,

$$F_{\mu\nu} = \langle h_{,\mu}(t)|h_{,\nu}(t)\rangle \tag{4}$$

$$= \begin{bmatrix} F_{\theta\theta} & F_{\theta\varphi} \\ F_{\varphi\theta} & F_{\varphi\varphi} \end{bmatrix} \equiv \begin{bmatrix} A & C \\ C & B \end{bmatrix} \tag{5}$$

where $h_{,\mu}(t) = \partial h(t)/\partial x^\mu$ and $x^\mu = (\theta, \varphi)$, with $\mu = 1, 2$. The uncertainties in the estimation of the angular parameters through the optimal filtering of gravitational waves with well understood waveforms buried in Gaussian noise are given by the inverse of the FIM,

$$(F^{-1})^{\mu\nu} = \begin{pmatrix} <\Delta\theta^2> & <\Delta\theta\Delta\varphi> \\ <\Delta\varphi\Delta\theta> & <\Delta\varphi^2> \end{pmatrix}. \tag{6}$$

As usual, the error solid angle is defined to be twice the area of the $1-\sigma$ error ellipse[13]

$$\Delta\Omega = 2\pi\sin\theta\sqrt{<\Delta\theta^2><\Delta\varphi^2> - <\Delta\theta\Delta\varphi>^2} \tag{7}$$

$$= \frac{2\pi\sin\theta}{\sqrt{AB - C^2}}. \tag{8}$$

According to our assumption that $h(t)$ is a very narrow band signal at frequency f_0, $S_n(f)$ nearly remains constant around $S_n(f_0)$ in the relevant signal frequency range. Thus the inner product can be simplified and calculated analytically,

$$\langle h_{,\mu}(t)|h_{,\nu}(t)\rangle = 4\mathcal{R}\left(\int_0^{+\infty}\frac{\widetilde{h_{,\mu}}^*(f)\widetilde{h_{,\nu}}(f)}{S_n(f)}df\right) \tag{9}$$

$$= \frac{4}{S_n(f_0)}\mathcal{R}\left(\int_0^{+\infty}\widetilde{h_{,\mu}}^*(f)\widetilde{h_{,\nu}}(f)df\right) \tag{10}$$

$$= \frac{2}{S_n(f_0)}\mathcal{R}\left(\int_{-T/2}^{T/2}h_{,\mu}(t)h_{,\nu}(t)dt\right) \tag{11}$$

where T is the total observation time. Each component of the FIM can be calculated successively

$$A = F_{\theta\theta} = \frac{2}{S_n(f_0)} \times \sum_{i=1}^{n}\mathcal{R}\left(\int_{-T/2}^{T/2}(2\pi f_0 R\mathcal{A})^2\sin^2(2\pi f_0\xi)\cos^2(2\pi f_{orb}t + \kappa_i - \varphi)\cos^2\theta dt\right) \tag{12}$$

For any kinds of LISA sources we always have $f_0 \gg f_{orb}$, therefore the term $\sin^2(2\pi f_0\xi)$ in the above equation experiences many oscillations during a typical observation time T, and hence can be replaced by 1/2 to a very good precision. This yields

$$A = \frac{2cos^2\theta(\pi f_0 R\mathcal{A})^2}{S_n(f_0)} \times \sum_{i=1}^{n}\mathcal{R}\left(\int_{-T/2}^{T/2}2cos^2(2\pi f_{orb}t + \kappa_i - \varphi)dt\right) \tag{13}$$

$$= \frac{2cos^2\theta(\pi f_0 R\mathcal{A})^2}{S_n(f_0)}T \times \left[n + sinc(2\pi f_{orb}T)\sum_{i=1}^{n}cos(2\kappa_i - 2\varphi)\right] \tag{14}$$

$$B = \frac{2sin^2\theta(\pi f_0 R\mathcal{A})^2}{S_n(f_0)}T \times \left[n - sinc(2\pi f_{orb}T)\sum_{i=1}^{n}cos(2\kappa_i - 2\varphi)\right] \tag{15}$$

$$C = \frac{sin(2\theta)(\pi f_0 R\mathcal{A})^2}{S_n(f_0)}T \times sinc(2\pi f_{orb}T)\sum_{i=1}^{n}sin(2\kappa_i - 2\varphi) \tag{16}$$

where $sinc(x) \equiv sin(x)/x$. Now it is straightforward to calculate the angular resolution for n-pairs of LISA constellations,

$$\triangle\Omega_n = \frac{\pi S_n(f_0)}{n(\pi f_0 R\mathcal{A})^2|cos\theta|T} \times \frac{1}{\sqrt{1 - sinc(2\pi f_{orb}T)\left[\frac{1}{n} + \frac{2}{n^2}\sum_{i<j}cos2(\kappa_i - \kappa_j)\right]}} \tag{17}$$

As a consistency check, we take $n = 1$, the above formula reduces to the single LISA case,

$$\Delta\Omega_1 = \frac{\pi S_n(f_0)}{(\pi f_0 R\mathcal{A})^2 |cos\theta| T} \times \frac{1}{\sqrt{1 - sinc(2\pi f_{orb} T)}}. \tag{18}$$

For signals with short observation times (for instance burst-like signals), it can be seen immediately that

$$\Delta\Omega_1 \approx \frac{\pi S_n(f_0)}{(\pi f_0 R\mathcal{A})^2 |cos\theta| T} \times \frac{\sqrt{6}}{2\pi f_{orb} T} \propto 1/T^2 \tag{19}$$

$$\Delta\Omega_n \approx \frac{\pi S_n(f_0)}{n(\pi f_0 R\mathcal{A})^2 |cos\theta| T} \times \frac{1}{\sqrt{1 - \left[\frac{1}{n} + \frac{2}{n^2}\sum_{i<j} cos2(\kappa_i - \kappa_j)\right]}}$$

$$\propto 1/T \tag{20}$$

$$\tag{21}$$

which means that for gravitational wave signals with short durations, the angular resolution of multi-LISA is better than that of a single LISA.

Figure.1 is a comparison of angular resolutions of a single LISA detector with various multi-LISA configurations for the simplest periodic signals. For the two-triangle configurations, two different placing of the detectors are considered and the difference between them is remarkable. The design with two identical LISA orbiting face to face (i.e. with an π delay in orbital phase) could not gain much in the ability of angular resolving as expected, while putting the two triangles in a $\pi/2$ delay fashion will help to improve the angular resolution significantly even to a level that comparable to a evenly separated three-triangle design.

3. Generalization to more complicated cases

In this section, we consider the more realistic case: a gravitational wave signal with slowly varying amplitude and frequency,

$$\begin{aligned} h(t) &= \mathcal{A}(t) cos(2\pi f(t)\xi) \\ &= \mathcal{A} cos(2\pi f(t)(t + Rsin\theta cos(2\pi f_{orb} t + \kappa_i - \varphi))) \end{aligned} \tag{22}$$

$$\xi = t + Rsin\theta cos(2\pi f_{orb} t + \kappa_i - \varphi) \tag{23}$$

In this circumstance, we can split the total observation time T into successive short durations T_k, while making sure that $f(t)$ and $\mathcal{A}(t)$ do not change much in each short duration. Then to a very good approximation, $f(t)$ and $\mathcal{A}(t)$ can be taken as the average values f_k and $\mathcal{A}_k$ in each period and the

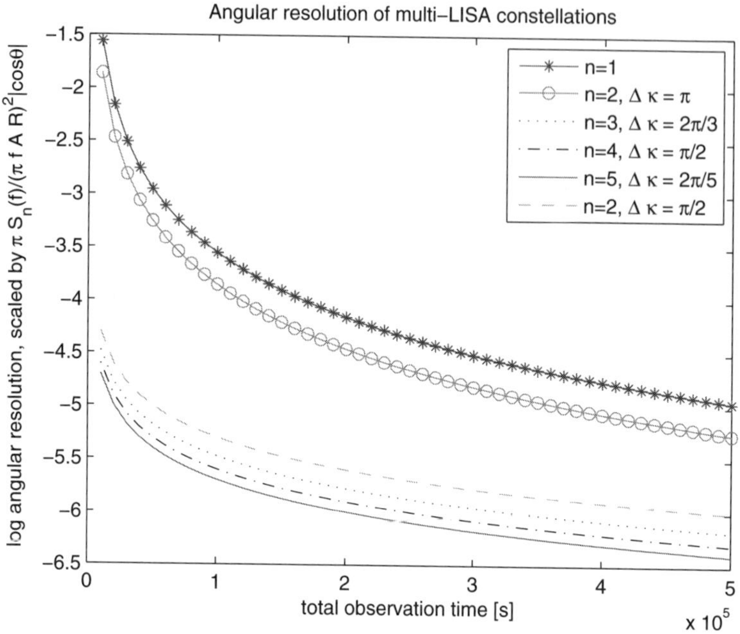

Fig. 1. a comparison of angular resolutions of a single LISA detector with various multi-LISA configurations for the simplest periodic signals. The Y axis is log_{10} scaled angular resolution.

components of the FIM in period $T_k (k = 0, 1, ..., M)$ can be calculated as before

$$A_k = \frac{2cos^2\theta(\pi f_k R A_k)^2}{S_n(f_k)} T_k \times \left[n + sinc(2\pi f_{orb} T_k) \sum_{i=1}^{n} cos2 \left(2\pi f_{orb} \sum_{j=0}^{k-1} T_j + \kappa_i - \varphi \right) \right] \quad (24)$$

$$B_k = \frac{2sin^2\theta(\pi f_k R A_k)^2}{S_n(f_k)} T_k \times \left[n - sinc(2\pi f_{orb} T_k) \sum_{i=1}^{n} cos2 \left(2\pi f_{orb} \sum_{j=0}^{k-1} T_j + \kappa_i - \varphi \right) \right] \quad (25)$$

$$C_k = \frac{sin(2\theta)(\pi f_k R A_k)^2}{S_n(f_k)} T_k$$
$$\times sinc(2\pi f_{orb} T_k) \sum_{i=1}^{n} sin2\left(2\pi f_{orb} \sum_{j=0}^{k-1} T_j + \kappa_i - \varphi\right) \quad (26)$$

According to the additive nature of the FIMs, we can add up these values to obtain the components of the FIM for the full observation, $A = \sum_{k=0}^{M} A_k, B = \sum_{k=0}^{M} B_k, C = \sum_{k=0}^{M} C_k$. The error solid angle can then be calculated in the same way as before.

Acknowledgments

The authors would like to thank Yun-Kau Lau and Zi-Ren Luo for useful discussions. Y. Wang is grateful for the hospitality of Yun-Kau Lau at Academy of Mathematics and System Sciences, CAS, Beijing, where part of the work was done. This work is supported by NSF, China under contract number 10773005.

References

1. B. F. Schutz, Nature (London) 323, 675 (1986); B. F. Schutz, Class. Quantum Grav. 6, 1761 (1989).
2. P. Bender et al., LISA prephase A report (1998).
3. P. Bender, Class. Quantum Grav. 21, S1203 (2004).
4. N. J. Cornish and L. J. Rubbo, Phys. Rev. D 67, 022001 (2003).
5. L. J. Rubbo, N. J. Cornish and O. Poujade, Phys. Rev. D 69, 082003 (2004).
6. C. Cutler, Phys. Rev. D 57, 7089 (1998).
7. M. Peterseim, O. Jennrich, and K. Danzmann, Class. Quantum Grav. 13, 279 (1996).
8. T. A. Moore and R. W. Hellings, Phys. Rev. D 65, 062001 (2002).
9. E. K. Porter and N. J. Cornish, Phys. Rev. D 78, 064005 (2008).
10. J. Crowder and N. J. Cornish, Phys. Rev. D 72, 083005 (2005).
11. C.W. Helstrom, Statistical Theory of Signal Detection, 2nd ed. (Pergamon Press, London, 1968)
12. A. Wainstein and V.D. Zubakov, Extraction of Signals from Noise (Prentice-Hall, Englewood Cliffs, 1962).
13. L. Wen, X. Fan and Y. Chen, J. Phys.: Conf. Ser. 122 012038 (2008).

DEVELOPMENT OF A DMT MONITOR FOR STATISTICAL TRACKING OF GRAVITATIONAL-WAVE BURST TRIGGERS GENERATED FROM THE OMEGA PIPELINE

JUN-WEI LI

Research Institute of Information Technology, Tsinghua University
Beijing 100084, P. R. China

JUN-WEI CAO

Research Institute of Information Technology
Tsinghua National Laboratory for Information Science and Technology
Tsinghua University, Beijing 100084, P. R. China

One challenge in large-scale scientific data analysis is to monitor data in real-time in a distributed environment. For the LIGO (Laser Interferometer Gravitational-wave Observatory) project, a dedicated suit of data monitoring tools (DMT) has been developed, yielding good extensibility to new data type and high flexibility to a distributed environment. Several services are provided, including visualization of data information in various forms and file output of monitoring results.

In this work, a DMT monitor, OmegaMon, is developed for tracking statistics of gravitational-wave (GW) burst triggers that are generated from a specific GW burst data analysis pipeline, the Omega Pipeline. Such results can provide diagnostic information as reference of trigger post-processing and interferometer maintenance.

1. Introduction

Many scientific applications require manipulation of large amount of data in real-time, e.g. astronomical and astrophysical experiments. The Very Long Baseline Array (VLBA), a system of ten radio telescopes, has a wideband, high-density recording system which generates data at a sustained rate of 128Mb/s, and peak rate of up to 512Mb/s[1]. The Sloan Digital Sky Survey (SDSS)'s motivation is to digitally map about a quarter of the whole sky. The data rate from its 120 million pixels camera is 8 Mb/s[2].

Another common feature for most of scientific applications is a distributed system and environment. For example, in VLBA there is an array of ten 25-m-diameter high-performance antennas scattered across the US. Another example is the High Energy Stereoscopic System (HESS) which consists of four imaging Cherenkov telescopes[3].

In order to maintain the integrity of the whole system and to obtain data quality information as the reference for subsequent data analysis, an efficient set of tools which can monitor critical system activities and important data statistics is required. Due to large amount of data and the distributed environment, demands for monitoring become greater.

LIGO (Laser Interferometer Gravitational-wave Observatory) initiated the Data Monitor Tool (DMT) project, aiming at defining tools and environments necessary to support continuous data monitoring of LIGO interferometers[4]. In this work, the DMT is applied for gravitational-wave burst (GWB) data analysis. In particular, a DMT monitor, OmegaMon, is developed for tracking statistics of gravitational-wave (GW) burst triggers that are generated from a specific GW burst data analysis pipeline, the Omega Pipeline.

The paper is organized as follows: the overview of LIGO and DMT is given in Sections 2 and 3, respectively. We introduce OmegaMon implementation in Section 4. Finally, we summarize our conclusions and future works in Section 5.

2. Overview of LIGO

Albert Einstein predicted the existence of gravitational waves in 1916 as part of the theory of general relativity in which he described space and time as different aspects of reality in which matter and energy are ultimately the same[5]. And the existence of gravitational wave is the most important prediction in general relativity theory.

Gravitational wave is a fluctuation in the curvature of space-time which propagates as wave. Until now, gravitational wave hasn't been directly detected, but observations of the orbital decay of the first binary pulsar PSR B1913 + 16 have provided significant indirect evidence for their existence since the late eighties[6].

To directly detect gravitational wave, LIGO was built by California Institute of Technology (Caltech) and the Massachusetts Institute of Technology (MIT). LIGO uses Michelson laser interferometers to detect gravitational waves by measuring the interference of two laser beams whose length is changed when the gravitational wave passes by[7]. In other words, the laser interferometer is similar as a microphone that converts gravitational waves into electrical signals. Currently LIGO has two observatory sites in the US with a total of three laser interferometers. The LIGO Hanford Observatory (LHO) in Washington State has two laser interferometers called H1 and H2. The LIGO Livingston Observatory (LLO) in Louisiana has a single interferometer called L1[8]. In the fifth science run of LIGO, the total data acquisition rate of three interferometers exceeded 16Mb/s[9].

In addition to the two observatory sites in the US, LIGO also collaborates with the British-German GEO[10] 600m detector located near Hannover, Germany; and shares data with the Italian/French VIRGO[11] detector. VIRGO's data acquisition rate is at about 20Mb/s, which means about 690GB per day written on disk[12]. GEO's data acquisition system is required to be capable of handling a data rate of up to 1Mb/s[13].

Summing up the above, all these gravitational wave detecting projects have large amount of real-time data streams to monitor. LIGO's DMT project commendably performs the monitoring on large volumes of data.

3. Overview of DMT

3.1. *DMT architecture*

DMT consists of three main components:

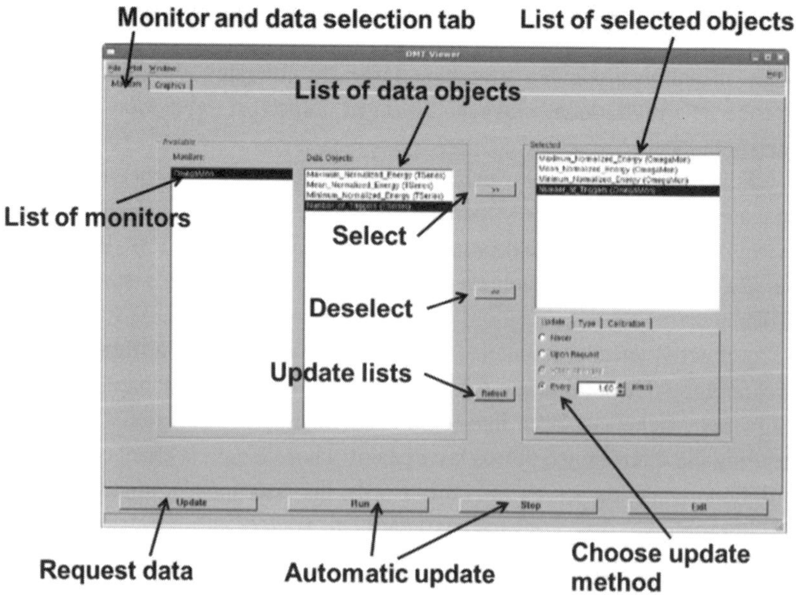

Figure 1. The DMT viewer GUI which mainly consists of list textboxes of available monitors, available and selected data objects.

- The monitors in various types. Each monitor has its own one or several specific data types to monitor. A data type could be a shared memory buffer or simple text file on disk. For each data type, a monitor could generate several classes of monitoring results. A class of corresponding monitoring result is called a class of data object.
- The name server. It is responsible for handling monitor registration and monitor information query.
- The DMT viewer. It is a tool for visualizing monitors' monitoring results, as shown in Figure 1.

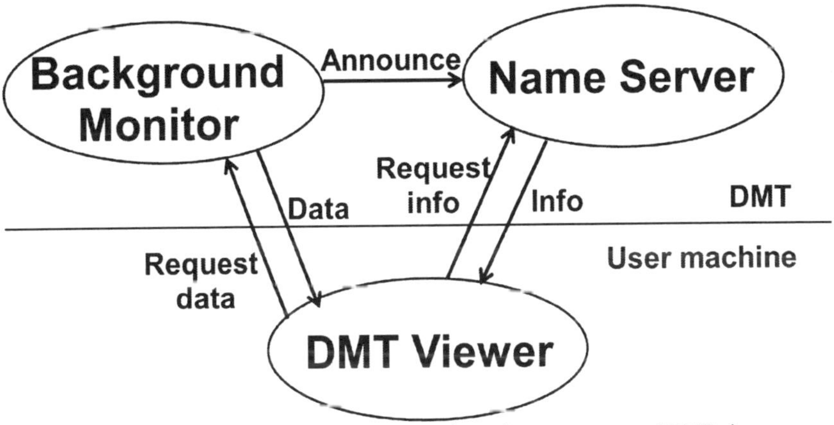

Figure 2. Three main components of DMT: monitors, the name server and DMT viewer.

Figure 2[14] shows the overview of the DMT architecture. In general, monitors run at background without displaying any graphics by itself, but serving data to the DMT viewer. The procedure is as follows. Firstly, a monitor is launched, and at the same time, monitor registers its name, socket information and monitored data types to the name server. The name server receives monitors' registration information and starts to keep track of the status of monitors. Then users can launch a DMT viewer. The DMT viewer will connect to the name server and requests for the list of monitors currently running with corresponding information. Then the DMT viewer will visualize these monitors' names and data objects. Users can select monitors in the DMT viewer, and the DMT viewer then requests a list of data objects' names directly from the selected monitors. Users select one or more data objects in the DMT viewer, the data objects are transferred from the selected monitor to the DMT viewer for displaying.

3.2. *DMT features*

To adapt large data volume situation and distributed environments, DMT is designed to have the following features:

- **Extensibility:** DMT is written in C++ using the object-oriented method. DMT has developed several sub-packages which consist of some basic classes. Take the MonServer class as an example. Users can easily develop their own monitors by inheriting the MonServer class without considering how their monitors keep running, but focusing on which data type they want to monitor and which data object they want the monitor to generate and send to the DMT viewer. And in DMT the frequently monitored data types have already been written in C++ classes, users can expediently invoke them. Even if there is a new data type to monitor, it only needs write a new C++ class and add it once in DMT. So DMT provides good extensibility for developing new monitors. Besides, DMT has defined multiple standard output file formats, such as html, xml. The external programs can easily access the monitors' output file in seamless way, which also improves DMT's extensibility.
- **Flexibility and scalability:** The name server's port number is set in advance and this port number is defined in basic classes related to monitors and the DMT viewer. So all the monitors and the DMT viewer can always communicate with the name server via this port number. It can be inferred that all the monitors, the name server and the DMT viewer can be at totally different locations. They can communicate with each other via the TCP/IP protocol. Users scattered all over the world can freely start or shutdown their own DMT viewers in their local hosts at any time without influence to each other. And DMT supports multiple users in different places to see a same monitor at the same time. This greatly improves the user experience in a distributed environment.
- **Light-weight and fine-grained:** Since DMT is required to support continuous data monitoring, its overhead on monitoring must be low enough to follow large data streams. The low overhead is achieved by light-weight monitor design and fine-grained division of monitoring jobs. Light-weight design means only some simple statistic calculations can be kept in the monitor. Fine-grained division of monitoring jobs is to assign as few data types as possible to each monitor. The optimized assignment is that each monitor has only one data type to monitor, and then the load of each monitor is the lowest.

4. OmegaMon for monitoring burst triggers

In this section, we introduce in details a DMT monitor, OmegaMon, which monitors statistical outputs of a gravitational-wave burst search pipeline, namely Omega Pipeline[15].

4.1. *Overview of Omega Pipeline*

Gravitational wave bursts (GWBs) are one of the most interesting classes of signals being sought by GW detectors[16]. The core collapse of supernovae, the merger of binaries, the gamma ray bursts and other relativistic systems could be the sources of GWBs[17]. Figuratively speaking, a burst is just like a pulse signal, the duration of which is typically shorter than 1 second.

In LIGO, a search pipeline called Omega Pipeline is developed for GWB detection. For each interferometer of LIGO, it performs time-frequency decomposition by filtering data against bisquare-enveloped sine waves, in what amounts to an over-sampled wavelet transform[18]. The filtering procedure will generate a value for each tile of a time-frequency plane. This value is called the normalized energy Z[19]. A tile is called a trigger if its Z energy exceeds a predefined threshold. The Omega Pipeline writes every trigger's attributes as an item into a plain text file, called trigger files. Currently each trigger is characterized with 5 attributes: the central time, the central frequency, the duration, the bandwidth and the normalized energy Z. All trigger files of different interferometers are sent to the central location of LIGO. Omega Pipeline on the central location will perform post-processing to triggers in these trigger files.

Since in a single interferometer Omega Pipeline performs a time-frequency decomposition for each 64 seconds interferometer data block, and every two adjacent 64 seconds interferometer data blocks has a 32 seconds overlap. So the relationship between each trigger file is shown in Figure 3.

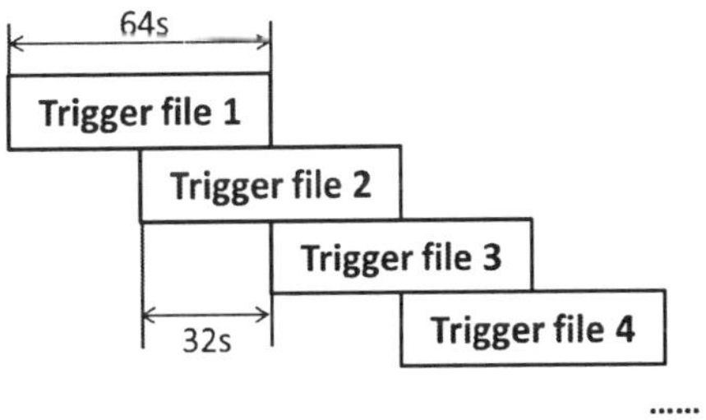

Figure 3. The relationship between trigger files.

There are many environment and equipment sources that could be related with trigger statistics, for example, an earthquake happens near an interferometer or the laser intensity fluctuates due to the unstable power of the laser generator in the interferometer. Data during the time interval in which the earthquake happens or the laser intensity fluctuates will be corrupted or contaminated, which may lead to changes of some trigger statistics. For example, if the average trigger rate in the past one hour is 500 per minute, which is two times larger than the normal trigger rate, it implies that the interferometer is highly possible not working well. This reminds operators or scientists to check the interferometer immediately and to discard the triggers in the past one hour in the post-processing of triggers. Based on the above consideration, OmegaMon is developed to monitor the triggers generated from the Omega Pipeline.

4.2. *Implementation of OmegaMon*

OmegaMon is a typical DMT monitor, which inherits the MonServer class, and has the following four typical functions[20].

- **OmegaMon Constructor:** This function initializes OmegaMon. It reads command line arguments and initialize the monitor process, such as serving OmegaMon's name and data objects' names to the name server, etc.
- **OmegaMon Destructor:** This function cleans up after OmegaMon is shut down. It typically flushes the pending output.
- **Process Data:** This is the primary function of OmegaMon for calculating trigger statistics. It is called repeatedly over a predefined period. In each cycle, firstly it reads trigger files in a certain time interval, the length of which is defined in the command line arguments. And then it calculates the statistics of triggers in the time interval. At last, the time interval is updated for next cycle to use.
- **Attention:** This function handles interrupt requests, e.g. the quit request.

4.3. *Key issues in OmegaMon*

During the development of OmegaMon, several key issues are considered and addressed:

- **Multi-resolution monitoring:** Monitoring requirement on the time dimension ranges from the long term trend to the short term fluctuation. Take the trigger rate as an example. We concern about the number of triggers per hour, and we also concern about the number of triggers per minute since it represents latest status. In the multi-resolution monitoring case, OmegaMon uses the maximum resolution value to set the time interval in the process data function, so that OmegaMon can get all data objects' values for all resolutions. In this trigger rate example, the time interval is set to 1 minute.

- **Name conflicts in the name server:** When there are multiple same monitors running simultaneously, only the first registered monitor's information is kept in the name server, the others are ignored due to name conflicts. In a distributed environment since users don't know each other at all; several users could all start a same monitor. But since the monitor name is set in OmegaMon's constructor function, we can solve name conflicts by adding suffix to monitor names. User can set their preferred suffix in the command line arguments of OmegaMon. When user starts a new OmegaMon, this OmegaMon will tell the name server its name with suffix. It is almost impossible to have the same suffix for different users, so it effectively avoids name conflicts in the name server.

4.4. *Monitoring results of OmegaMon in the DMT viewer*

Figure 4 shows four OmegaMon data objects illustrated in the DMT viewer. They are the maximum normalized energy (top left corner), the minimum normalized energy (top right corner), the mean normalized energy (lower left corner) and the number of triggers (lower right corner).

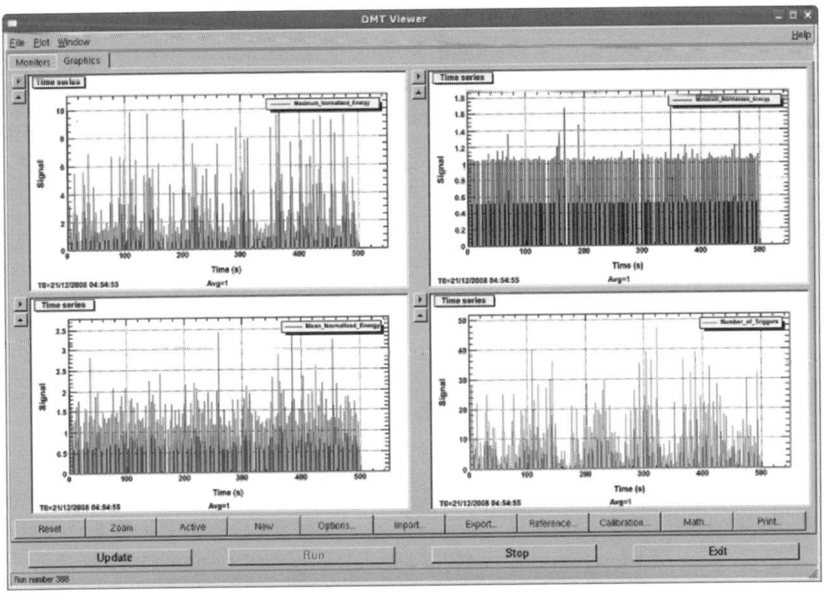

Figure 4. Monitoring results of OmegaMon in the DMT viewer.

For example, we can see from Figure 4 that the maximum normalized energy in one second interval varies acutely in the sub graph of the top left corner. Some values are obviously too high. It can be due to large transients of gravitational waves or some non-gravitational wave origins, such as loud noises. So in the post-processing, triggers in intervals of which the maximum normalized energy is too high should be discarded or tagged for further detailed analysis.

5. Conclusions and future work

We have introduced a C++ based real-time data monitoring toolkit dedicated for LIGO. It has the advantages of extensibility, flexibility and scalability. The architecture of the toolkit is introduced, and it is very suitable for distributed usage. An example DMT monitor, OmegaMon, is also introduced in details to show statistical tracking of GWB triggers generated by the Omega Pipeline.

More data objects are being developed in OmegaMon so as to provide more diagnostic information for the reference of trigger post-processing and interferometer maintenance. The OmegaMon is also being integrated into a real-time LIGO burst data computing infrastructure currently being developed at Tsinghua University.

Acknowledgments

This work is supported by National Science Foundation of China (grant No. 60803017), Ministry of Science and Technology of China under the national 863 high-tech R&D program (grants No. 2008AA01Z118 and No. 2008BAH32B03).

The authors would like to show their gratitude to Erik Katsvounidis at LIGO Laboratory of Massachusetts Institute of Technology and John Zweizig at LIGO Laboratory of California Institute of Technology for their extensive supports on this work.

References

1. P. J. Napier, D. S. Bagri et al, *Proc. IEEE.* **VOL. 82**, NO. 5 (1994).
2. A. S. Szalay et al, *ACM SIGMOD Rec.* **VOL. 29**, Issue 2 (2000).
3. J. A. Hinton, *New Astron. Rev.* **48**, 331 (2004).
4. Data Monitor Tool Project. http://www.ligo.caltech.edu/~jzweizig/DMT-Project.html
5. Overview of LIGO. http://www.ligo-la.caltech.edu/LLO/overviewsci.htm
6. B. Abbott et al, *Phys. Rev.* **D77**, 062002 (2008).

7. A. Abramovici et al, *Science.* **VOL. 256,** No. 5055 (1992).
8. B. Abbott et al, *Phys. Rev.* **D69**, 122001 (2004).
9. G. Mendell, http://www.ligo.caltech.edu/docs/G/G070127-00/G070127-00.ppt
10. B. Willke, P. Ajith, *Class. Quantum Grav.* **23**. S207-S214 (2006).
11. VIRGO project. http://www.virgo.infn.it/
12. F. Acernese et al, *IEEE Trans. Nucl. Sci.* **VOL. 55**, No. 1 (2008).
13. *K. Kotter et al, Class. Quantum Grav.* **19**, 1399-1407 (2002).
14. DMT system diagram. http://www.ldas-sw.ligo.caltech.edu/cgi-bin/cvsweb.cgi/~checkout~/gds/doc/quickstart/MonitorAPI/monserv1.jpg?rev=1.1;content-type=image%2Fjpeg;cvsroot=GDS
15. Omega Pipeline. https://geco.phys.columbia.edu/omega
16. P. J. Sutton et al, *arXiv.org.* 0908.3665 (2009).
17. C. Cutler and K. S. Thorne, *arXiv.org.* gr-qc/0204090 (2002).
18. B. P. Abbott et al, *arXiv.org.* 0905.0020 (2009).
19. B. P. Abbott et al, *arXiv.org.* 0904.4910 (2009).
20. hMon status update. http://emvogil-3.mit.edu/~shourov/fan/dmt/update_20080822.html

TESTING GRAVITATIONAL WAVES WITH TOTAL-PHASE-COUNT DOPPLER TRACKING IN CHINESE MARS MISSION

KUN SHANG*, CHUN-LI DAI and JIN-SONG PING

Shanghai Astronomical Observatory, 80 Nandan Road, Shanghai 200030, China
**E-mail: shangkun@shao.ac.cn*

Using the radio telescopes in Chinese VLBI Network (CVN) and the K5/VSSP32 VLBI system of NICT in Japan, we have developed Total-Phase-Count algorithms that extract open-loop Doppler information from the Chang'E-1 tracking data. Our latest results indicate that Doppler residuals of three-way Doppler are about 3mm/s in 1s integration time, relative to an 2.3GHz carrier. This system has also been used in ESA's Mars Express observation in Nov. 2008. In the Chinese-Russian Mars mission (Yinghuo-1) of 2011, the current software algorithms and hardware performance will be improved, and we will attempt to use Total-Phase-Count observable and high precision Doppler shift information to test gravitational waves.

Keywords: Yinghuo-1; Total-Phase-Count; Open-Loop Doppler; Chang'E-1.

1. Introduction

Lunar and the planetary exploration is to recover the origin of the Moon and the planets, to explore their evolution, and to look for the possible life in other planetary bodies. Since the beginning of the new century, Mars exploration has attracted the huge attention from space communities. As a beginner in this area, China has launched the first lunar orbiter Chang'E-1 (CE-1) successfully, and has got some new scientific results from this exploration. Beyond this, a joint Russian-Chinese Martian mission, YingHuo-1 (YH-1) sub-satellite and Phobos-Grunt Spacecraft (FGSC), has been developed and promoted solidly. The two probes will be launched together in October 2011. Then the joint spacecraft YH-1 and FGSC will be sent to a transfer orbit flying to Mars. After 10 - 11 months, the joint craft will arrive in the Martian system, and will be ejected into an equatorial orbit of 800 km periapsis altitude and 80,000 km apoapsis altitude, with a period of about 72 hours, inclination of 1°- 5°. The joint craft will fly in this orbit for about 3 circles, then they will be separated. FGSC will change its orbit in order to land on Phobos, and then take some soil (0.1 kg - 0.2 kg) back to the Earth. YH-1 will free-fly in this large elliptical orbit for 1 year.

YH-1 is a small sub-satellite focused on investigating the Martian space environment and the solar wind-Mars interaction. It will be combined with FGSC to form a two-point measurement configuration in the Martian environment. The two spacecrafts will also carry out satellite-to-satellite radio link, so as to study the Martian ionosphere by using radio occultation links.

2. Experiments and Results

2.1. *Tracking system and Open-loop Doppler experiments*

Considering that the Chinese deep space tracking system is still under construction, there will not be any uplink system in China which can meet the power requirement of uplink communication for a distance of AU. To simplify and minimize the design, an X-band receiver and X-band transmitter system have been adopted for onboard communication. There is not a common transponder used for tracking. To solve the tracking and orbit determination problem, one-way open loop link will be used in YH-1, and the ground astronomical Very Long Baseline Interferometry system (Chinese VLBI Network) will be used to receive the radio signal, so as to retrieve the Doppler information and then the differential Doppler information[1]. The open loop Doppler observable will be applied in the positioning and orbit determination of YH-1.

Chinese VLBI Network (CVN), composed of four radio telescopes, acts as the deep space tracking system in China. Four radio telescopes located in four long-distance stations with different sizes, including: Shanghai station with a 25m telescope built in 1987, Urumqi station with a 25m telescope built in 1993, Beijing station with a 50m telescope built in 2006, and Kunming station with a 40m telescope built also in 2006. Recording devices have already been installed in two CVN stations - Shanghai and Urumqi. We use K5/VSSP32 sampling board as recording device, with a local H-maser atomic clock. K5/VSSP32 is a high-performance sampler dedicated to a geodetic VLBI system developed by the Kashima VLBI group of National Institute of Information and Communications Technology in Japan.

2.2. *Total-Phase-Count algorithm and result*

A software receiver using the post-processing method is developed to retrieve the Doppler frequency from the data recorded in K5/VSSP32 sampler. The kernel of this algorithm is so-called Total-Phase-Count [2], which can extract the precise main frequency from the raw data. Comparing the calculated received main frequency with the original transmitted frequency, we obtain the frequency shift in a certain integral time. The Doppler effect presents the relationship between the frequency shift and the radial velocities:

$$f_R \approx (1 - \frac{v_1}{c})(1 - \frac{v_2}{c})M \cdot f_T \approx (1 - \frac{v_1 + v_2}{c})M \cdot f_T \quad (1)$$

Where, f_R represents the calculated received frequency, f_T represents the original transmitted frequency, v_1 represents the range rate of the transmitting station and spacecraft, v_2 represents the range rate of spacecraft and receiving station, and M represents the transmit ratio. Define the average of v_1 and v_2 as the three-way velocity, which means $v_{3w} = (v_1 + v_2)/2$.

During the nominal mission period of Chinese lunar orbiter CE-1, we used the three-way method to test the open loop tracking ability as a preview of future one-way method in YH-1 mission. Three-way method means that a tracking station sends uplink S-Band signal generated by Rb atomic clock to CE-1, and then the transponder in the orbiter locks the uplink carrier wave and sends it back to the Earth. The reason why a one-way Doppler experiment was not carried in CE-1 is that there is no stable oscillator. Similar experiments have been done together with ESOC deep space tracking stations for tracking the Mars Express. The radio telescopes used as the receiver station in three-way doppler experiment are located in Shanghai and Urumqi. Figure 1 and Figure 2 show the open loop three-way Doppler results of CE-1 mission and Mars Express mission, including the three-way velocity and the residuals after orbit determination.

3. Testing Gravitational Waves With Total-Phase-Count Observable

After we obtain Doppler time series above, we could extract a number of scientific subjects from the properties of radio wave, such as frequency, phase, amplitude and polarization, which is called radio science. One important application of radio science is to detect gravitational waves from high precise Doppler frequency time series[3]. In the forthcoming YH-1 mission, we are going to use Total-Phase-Count Observable to detect gravitational waves instead of traditional Doppler method.

A gravitational wave incident can cause a frequency shift in Doppler time series. Define this frequency shift y as $(f_R - f_T)/f_T$. If we assume

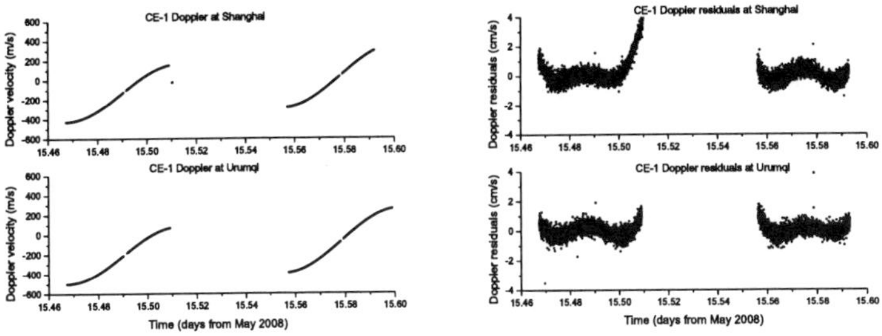

Fig. 1. Open loop three-way Doppler result of CE-1 mission on 15 May 2008. The left panel shows the three-way Doppler velocity of Shanghai and Urumqi station. The right panel shows the Orbit Determination Doppler residual. The root-mean-square (r.m.s) of 1 second integration time is about 3mm/s, relative to a 2.3 GHz carrier.

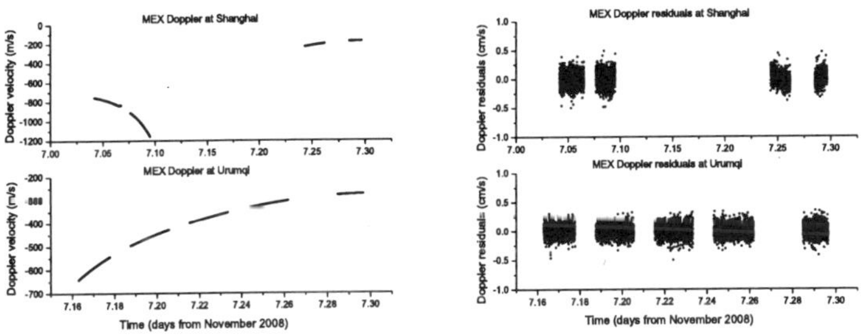

Fig. 2. Open loop three-way Doppler result of MEX mission on 7 November 2008. The left panel shows the three-way Doppler velocity of Shanghai and Urumqi station. The right panel shows the Orbit Determination Doppler residual. The r.m.s of 1 second integration time is about 1mm/s, relative to an 8.4 GHz carrier.

the signal of gravitational wave into a continuous wave: $M \cdot e^{i(kz-\omega t)}$, the frequency shift y will be[4]:

$$y = M \cdot [(1-\cos\theta) + (2\cos\theta)e^{i\cdot(klc\cos\theta+\omega l)} - (1+\cos\theta)e^{i\cdot 2\omega l}]e^{-i\cdot\omega t_R} \quad (2)$$

where ω is the frequency of gravitational wave, the large separation of Yinghuo-1 and earth makes this method compared with millihertz gravitational waves. l is the distance between earth and spacecraft, and θ is the angle from the direction of gravitational wave to the vector of earth-spacecraft. Here we can see the amplitude of traditional Doppler frequency time series due to gravitational wave is :

$$M \cdot [(1-\cos\theta) + (2\cos\theta)e^{i\cdot(klc\cos\theta+\omega l)} - (1+\cos\theta)e^{i\cdot 2\omega l}] \quad (3)$$

However, in the new algorithm, we use continuous phase information from the so-called Total-Phase-Count observable, which is :

$$\Phi = \int_{t_1}^{t_2} 2\pi f_R dt_R = 2\pi f_T(t_2 - t_1) + 2\pi f_T \int_{t_1}^{t_2} y dt_R \tag{4}$$

The real part of $\int_{t_1}^{t_2} e^{-i\cdot\omega t_R} dt_R$ equals to $\frac{\sin\omega t_2 - \sin\omega t_1}{\omega}$. Hence, the amplitude of Total-Phase Count observable series due to gravitational wave is:

$$\frac{2\pi f_T}{\omega} \cdot M \cdot [(1-\cos\theta) + (2\cos\theta)e^{i\cdot(klc\cos\theta + \omega l)} - (1+\cos\theta)e^{i\cdot 2\omega l}] \tag{5}$$

which means, if the frequency of gravitational wave is 10^{-3} Hz, the sensitivity for the amplitude of gravitational wave in Total-Phase-Count observable will be 10^3 higher than the traditional Doppler observable.

4. Prospects

At present, the Doppler accuracy of $\Delta f/f$ for Chinese tracking system is about 3mm/s in S-band and 1mm/s in X-band, both in 1 second integration time, which means the root-mean-square of frequency residuals is about $\delta(\Delta f/f) \approx 3 \times 10^{-11}$, in 1 second integration time. If we improve the integration time to 1000 seconds, the frequency residuals would be $\delta(\Delta f/f) \approx 10^{-12}$.

The limit of accuracy is restricted by the size of the radio telescopes in CVN. In 2014, a 65 meters radio telescope will be built by Shanghai Astronomical Observatory, which will improve the sensitivity and accuracy in the Chinese deep space missions, especially the Doppler tracking accuracy.

High accuracy Doppler tracking can be used for other test of gravitational theories, such as testing the PPN parameter. And considering the large elliptical orbit of YH-1 satellite, other relativistic Doppler shift such as second order Doppler shift and gravitational shift can also be test.

Acknowledgments

This work was supported by the VLBI system of Chinese Academic of Science and GEODYNII software provided by NASA GFSC.

References

1. J. Ping, W. Frank, and K. Yusuke, *Journal of Planetary Geodesy* **36**, 15 (2001).
2. T. D. Moyer, *Formulationo for Observed and Computed Values of Deep Space Network Data Types for Navigation*, JPL Publication 00-7 (2000).
3. J. W. Armstrong, *Living Rev. Relativity* **9**, (2006), 1.
4. R. W. Hellings, *Physical Review D* **17**, 3158 (1978).

Gravitation

SHEAR VISCOSITY FROM THE EFFECTIVE COUPLING OF GRAVITONS

RONG-GEN CAI, ZHANG-YUE NIE and YA-WEN SUN

Institute of Theoretical Physics, Chinese Academy of Sciences,
Beijing 100190, China
E-mail: cairg,niezy,sunyw@itp.ac.cn

We review the progress in the holographic calculation of shear viscosity for strongly coupled field theories. We focus on the calculation of shear viscosity from the effective coupling of transverse gravitons and present some explicit examples.

Keywords: Shear viscosity; AdS/CFT.

1. Introduction

The AdS/CFT correspondence[1–4] provides a new way to solve strongly coupled field theory problems. Especially, we can calculate the transport coefficients in the hydrodynamic limit of the field theory from the gravity side. Shear viscosity is one of these coefficients that have been vastly studied through the holographic method.[5–8] In the early days, the shear viscosity was calculated in a lot of theories dual to Einstein gravity with or without chemical potential.[9–12] It was found that the value of the shear viscosity over the entropy density ratio η/s is always $1/4\pi$. After considering large N effects, this ratio gets positive corrections. Then it was conjectured that there is a universal lower bound on the ratio η/s in nature(KSS bound[8]). Known experiments such as the Quark-Gluon Plasma(QGP) in RHIC also satisfy this conjecture. Then more attention has been paid to the study of the universality of η/s through holographic method, and the η/s was found to be still above $1/4\pi$ in Ref. 13–20. However, it was later found that the KSS bound can be slightly violated in theories with gravity dual of more general theories of gravity.[21]

In the study of the η/s through holographic method, it was gradually found that the shear viscosity of the boundary field theory only depends on the effective coupling of the transverse gravitons valued on the horizon in the gravity side.[22–26] This relation applies to systems dual to more general theories beyond Einstein gravity, and may shed some light on the understanding of the AdS/CFT correspondence in another point of view. We will shortly review the progress of studies on this problem in this article, and also introduce some interesting results using the language of the effective coupling.

The plan of this paper is as follows. In the next section, we review the progress in the study of shear viscosity using holographic method. In Sec. 3, we introduce recent studies on the calculation of shear viscosity from the effective coupling of transverse gravitons, and give the result in AdS Gauss-Bonnet gravity. In Sec. 4 we present some examples from the effective coupling.

2. Shear viscosity from AdS/CFT

The AdS/CFT correspondence[1–4] is a duality between the anti-de Sitter (AdS) spacetime and a conformal field theory(CFT) living on the boundary of the AdS spacetime. In details, suppose that we have a field ϕ moving in the AdS spacetime and a corresponding operator $\mathcal{O}$ in the CFT side, and the boundary value ϕ_0 of ϕ coupled to the boundary operator $\mathcal{O}$ as a source, then the partition function of ϕ with boundary value ϕ_0 in the gravity side is equal to the generating functional of $\mathcal{O}$ with a source ϕ_0 in the CFT side

$$Z(\phi_0) = \left\langle e^{i\int \phi_0 \mathcal{O}} \right\rangle_{CFT}. \tag{1}$$

In order to get some information in the CFT, we only need to study the partition function $Z(\phi_0)$ in the gravity side. In the AdS/CFT correspondence, the gravity system in the AdS side is classical at low energies, so the partition function $Z(\phi_0)$ can be just determined by the classical solution of the field ϕ

$$Z(\phi_0) = e^{iS(\phi)}, \tag{2}$$

where $S(\phi)$ is the classical action of the classical solution ϕ with boundary value ϕ_0. Then with this correspondence, we can easily get the correlation functions of the operator $\mathcal{O}$ in the strongly coupled field theory by the standard functional formulas.

In the point of view of the field theory, knowing the correlation functions of operators, one can extract transport coefficients using the Kubo formula. Applying this formula to the case of energy momentum tensor, we can calculate the shear viscosity by[27,28]

$$\eta = \lim_{\omega\to 0}\frac{1}{2\omega i}\Big(G^A_{xy,xy}(\omega,0) - G^R_{xy,xy}(\omega,0)\Big), \tag{3}$$

where η is the shear viscosity, and the retarded Green's function is defined by

$$G^R_{\mu\nu,\lambda\rho}(k) = -i\int d^4x e^{-ik\cdot x}\theta(t)\langle[T_{\mu\nu}(x), T_{\lambda\rho}(0)]\rangle. \tag{4}$$

The advanced Green's function can be related to the retarded Green's function by $G^A_{\mu\nu,\lambda\rho}(k) = G^R_{\mu\nu,\lambda\rho}(k)^*$.

Using the AdS/CFT duality and the Kubo formula, the shear viscosity η can be calculated from the gravity dual, and interesting results have been obtained using this holographic method. It was found that the ratio of shear viscosity over entropy η/s calculated from gravity dual of Einstein gravity is always $1/4\pi$. After considering some corrections from large N effect, the ratio gets positive corrections. Notice that in all the known experiments, the fluids measured all have a ratio η/s larger than $1/4\pi$, The authors of Ref. 8 conjectured that this ratio may have a universal lower bound in real fluids in nature, called the KSS bound.

3. Shear viscosity and the effective coupling of transverse gravitons

It has recently been conjectured that the shear viscosity depends on the effective coupling of transverse gravitons valued on the horizon.[23] In Ref. 22–24 the authors conjectured that this ratio is equal to a quotient of effective couplings of two different polarizations of gravitons, κ_{xy} and κ_{rt} valued on the horizon. Then the dependence of shear viscosity on the effective coupling of transverse gravitons is confirmed in Ref. 25 using a membrane paradigm and in Ref. 26 by calculating the on-shell action of the transverse gravitons. In Sec. 3.1 we show the relation of the shear viscosity and the effective coupling of transverse gravitons. In Sec. 3.2 we discuss the case of Gauss-Bonnet gravity.

3.1. *Transverse Graviton*

For simplicity, we consider a five dimensional asymptotic AdS black hole solution with a Ricci flat horizon to some kind of gravity theory with or without matter fields. We write the metric as

$$ds^2 = -g(u)(1-u)dt^2 + \frac{1}{h(u)(1-u)}du^2 + \frac{r_h^2}{ul}(dx^2+dy^2+dz^2), \tag{5}$$

where r_h is the horizon radius and $u = r_h/r$. Then the horizon is located at $u = 1$ and the boundary lives at $u = 0$. $g(u)$ and $h(u)$ are two functions regular at $u = 1$ and l is the AdS radius which is related to the cosmological constant by $\Lambda = -6/l^2$.

The boundary of this AdS spacetime is a $\mathbb{R}^4$, on which a CFT of some kind lives. To calculate the shear viscosity of this CFT, we need the Green functions of the energy momentum tensor, which we can get by making a small perturbation to the metric tensor in the gravity side. For simplicity, we choose spatial coordinates x, y, z, so that the momentum of the perturbations points along the z-axis. Then the perturbations can be written as $h_{\mu\nu} = h_{\mu\nu}(t, z, u)$. In this basis there are three groups of gravity perturbations, each of which is decoupled from others: the scalar, vector and tensor perturbations.[29] Here we use the simplest one that can be used to calculate the shear viscosity, the tensor perturbation h_{xy}. We use ϕ to denote this perturbation $\phi = h_y^x$ and write ϕ in a basis as $\phi(t, u, z) = \phi(u)e^{-i\omega t+ipz}$. Then by expanding the gravity action to the second order with the background metric (5), one can get the effective action of the transverse graviton

$$S = \frac{1}{16\pi G}\int d^5x\sqrt{-g}K_{eff}(u)(\nabla_\mu\phi\nabla^\mu\phi), \tag{6}$$

where we assume the graviton to be a minimal coupled massless scalar with an effective coupling K_{eff}. Then it can be proved that the shear viscosity depends only on the effective coupling valued on the horizon $K_{eff}(u = 1)$.[25,26]

In Ref. 25, the authors used the membrane paradigm to analysis the universality of shear viscosity over entropy ratio, and proved that the shear viscosity depends on the coupling of the transverse gravitons valued on the horizon. This shows some relations between the fluid on the near horizon slicing and the long wavelength low frequency limit of the boundary CFT. The boundary theory can be seen as flowing from the horizon physics throw some spacetime structures.

In the paper,[26] we directly calculated the on-shell action of the transverse gravitons, and obtained the shear viscosity expressed by the effective coupling of transverse gravitons. In Einstein gravity, this effective coupling is always $-1/2$, so the shear viscosity over entropy density is always $1/4\pi$ in theories with Einstein gravity dual. But in modified gravity theories, the effective coupling of transverse gravitons could be different. For example, in gravity theory with R^2 corrections, the effective coupling of transverse gravitons gets corrections, thus in the dual field theory, the universal low bound $1/4\pi$ can be violated.[21,30,31]

3.2. *KSS bound violation in Gauss-Bonnet gravity*

The R^2 corrections come from string theory or some other UV completion of Einstein gravity. After field redefinitions these theories reduce to Gauss-Bonnet gravity up to the R^2 order. The action of Gauss-Bonnet gravity is

$$S = \frac{1}{16\pi G}\int d^5x\sqrt{-g}\Big(R - 2\Lambda + \frac{\lambda l^2}{2}R_{GB}\Big), \qquad (7)$$

in which

$$R_{GB} = (R^2 - 4R_{\mu\nu}R_{\mu\nu} + R^{\mu\nu\rho\sigma}R_{\mu\nu\rho\sigma})$$

is the Gauss-Bonnet term, λ is the Gauss-Bonnet parameter with a small positive value.

With the Gauss-Bonnet term, the action of the transverse gravitons is still that of minimal coupled massless scalar, but the effective coupling differs from the value in pure Einstein gravity(it is no longer a constant and depends on the radial coordinate u). Thus the ratio of shear viscosity over entropy density η/s in the dual field theory also differs from the universal value with Einstein gravity dual. The ratio in the Gauss-Bonnet case is calculated to be

$$\frac{\eta}{s} = \frac{1}{4\pi}(1 - 4\lambda). \qquad (8)$$

With a positive Gauss-Bonnet correction $\lambda > 0$, the universal lower bound of $1/4\pi$ will obviously be violated. But with causality consideration from the CFT side, the Gauss-Bonnet parameter λ is constrained as $\lambda < 0.09$,[30] and then the ratio η/s will have a new lower bound as

$$\frac{\eta}{s} \geq \frac{4}{25\pi}. \qquad (9)$$

The calculation of η/s from the effective coupling of the transverse gravitons is convenient, and we will give more examples of studying the ratio η/s with this method. More discussions on the shear viscosity from the effective coupling of transverse gravitons could be found in Ref. 32–37.

4. More examples from effective coupling

Since the violation of the KSS bound was discovered, the universality of the bound of η/s needs more attention, it would be valuable to examine whether the ratio will get more corrections in more general gravity duals.

In our previous work,[26,39] we calculated the shear viscosity in the case of AdS Gauss-Bonnet gravity with F^4 term corrections of Maxwell field, and in AdS Gauss-Bonnet gravity with dilaton coupling. In both the two cases, the ratio of shear viscosity over entropy density gets more corrections from the AdS Gauss-Bonnet case. We also considered the shear viscosity in the extremal case in a recent work.[40] We give our results here.

4.1. *Gauss-Bonnet Gravity with F^4 term corrections of Maxwell field*

In Ref. 41, when Maxwell field is added to the Gauss-Bonnet gravity, the ratio η/s gets positive corrections to the pure AdS Gauss-Bonnet gravity case. Then in our work,[26] we further studied the effect of non-linear term of Maxwell field on the shear viscosity in the setup of Gauss-Bonnet gravity dual. We found that the F^4 term has some effects on the ratio η/s. We further found that with Maxwell field, with or without F^4 terms, the ratio η/s depends on the temperature T in the way

$$\frac{\eta}{s} = \frac{1}{4\pi}(1 - \frac{4\lambda\pi l^2}{r_h}T). \tag{10}$$

This is very interesting, for that the effect of the Maxwell field on η/s is simply relating this ratio to the temperature of the black hole.

4.2. *AdS Gauss-Bonnet gravity with dilaton coupling*

The dilaton field coupled to the Gauss-Bonnet gravity has non-trivial contributions, for that the pure Gauss-Bonnet black hole solution without the dilaton is not a solution of the case with dilaton by simply requiring the dilaton being a constant. So it would be interesting to see whether the dilaton field has contributions to the shear viscosity in the dual field theory. In Ref. 39, we found that in the case of Gauss-Bonnet gravity with dilaton coupling the effective action of the transverse gravitons is still the type of minimally coupled massless scalar, so the shear viscosity depends on the effective coupling of the transverse gravitons. We calculated the ratio of shear viscosity over entropy density, and the result is

$$\frac{\eta}{s} = \frac{1}{4\pi}\left(1 - 4\lambda e^{(\tau-\gamma)\phi_d(1)}\left(1 - 24\gamma^2\lambda e^{(\tau-\gamma)\phi_d(1)} + 3\gamma\tau\right)\right). \tag{11}$$

After considering the causality restriction, we can see that with dilaton coupling, the new lower bound in AdS Gauss-Bonnet case of the η/s can be further violated in the parameter space of the black hole solution.

4.3. *Shear viscosity at zero temperature*

It is well known that the extremal black hole is very different from the ordinary black holes with a finite temperature. In a recent paper,[42] shear viscosity and other transport coefficients of field theories with nonzero chemical potentials were calculated at zero temperature in the background of extremal AdS RN black holes using similar methods as in Ref. 43 where properties of non-Fermi fluid[44–46] are discussed from AdS/CFT, there the IR physics plays an important role in deriving the UV physics on the boundary. In the extremal case of Einstein gravity, the shear viscosity over entropy density is still $1/4\pi$.

We calculated the shear viscosity in field theories dual to Gauss-Bonnet gravity at zero temperature with non-zero chemical potential. We confirmed that in the extremal case, the shear viscosity also depends on the effective coupling of transverse gravitons valued on horizon. Our results show that, the shear viscosity over entropy density in field theory dual to Gauss-Bonnet gravity at zero temperature is still $1/4\pi$, the same as the in Einstein gravity.

5. Conclusion

In this paper we gave a brief review on the progress of calculating the shear viscosity in strongly coupled field theory through holographic method, with special attention to the connection between the shear viscosity and the effective coupling of transverse gravitons in the dual gravity side. This connection is applicable for a wide class of gravity theories. We presented some examples of this connection in Gauss-Bonnet gravity with Maxwell field, and the case with dilaton coupling. We also discussed results with extremal black holes.

Acknowledgments

We would like to thank N. Ohta for useful discussions and collaborations in some relevant studies. RGC thanks the organizers ICGA9 for a warm hospitality during the conference. This work was supported partially by grants from NSFC, China (No. 10535060, No. 10821504 and No. 10975168) and a grant from MSTC, China (No. 2010CB833004).

References

1. J. M. Maldacena, "The large N limit of superconformal field theories and supergravity," Adv. Theor. Math. Phys. **2**, 231 (1998) [Int. J. Theor. Phys. **38**, 1113 (1999)] [arXiv:hep-th/9711200].

2. S. S. Gubser, I. R. Klebanov and A. M. Polyakov, "Gauge theory correlators from non-critical string theory," Phys. Lett. B **428**, 105 (1998) [arXiv:hep-th/9802109].
3. E. Witten, "Anti-de Sitter space and holography," Adv. Theor. Math. Phys. **2**, 253 (1998) [arXiv:hep-th/9802150].
4. O. Aharony, S. S. Gubser, J. M. Maldacena, H. Ooguri and Y. Oz, "Large N field theories, string theory and gravity," Phys. Rept. **323**, 183 (2000) [arXiv:hep-th/9905111].
5. G. Policastro, D. T. Son and A. O. Starinets, "The shear viscosity of strongly coupled N = 4 supersymmetric Yang-Mills plasma," Phys. Rev. Lett. **87**, 081601 (2001) [arXiv:hep-th/0104066].
6. A. Buchel and J. T. Liu, "Universality of the shear viscosity in supergravity," Phys. Rev. Lett. **93**, 090602 (2004) [arXiv:hep-th/0311175].
7. P. Kovtun, D. T. Son and A. O. Starinets, "Viscosity in strongly interacting quantum field theories from black hole physics," Phys. Rev. Lett. **94**, 111601 (2005) [arXiv:hep-th/0405231].
8. P. Kovtun, D. T. Son and A. O. Starinets, "Holography and hydrodynamics: Diffusion on stretched horizons," JHEP **0310**, 064 (2003) [arXiv:hep-th/0309213].
9. K. Maeda, M. Natsuume and T. Okamura, "Viscosity of gauge theory plasma with a chemical potential from AdS/CFT," Phys. Rev. D **73**, 066013 (2006) [arXiv:hep-th/0602010].
10. O. Saremi, "The viscosity bound conjecture and hydrodynamics of M2-brane theory at finite chemical potential," JHEP **0610**, 083 (2006) [arXiv:hep-th/0601159].
11. D. T. Son and A. O. Starinets, "Hydrodynamics of R-charged black holes," JHEP **0603**, 052 (2006) [arXiv:hep-th/0601157].
12. J. Mas, "Shear viscosity from R-charged AdS black holes," JHEP **0603**, 016 (2006) [arXiv:hep-th/0601144].
13. A. Buchel, R. C. Myers, M. F. Paulos and A. Sinha, "Universal holographic hydrodynamics at finite coupling," Phys. Lett. B **669**, 364 (2008) [arXiv:0808.1837 [hep-th]].
14. A. Buchel, "Resolving disagreement for eta/s in a CFT plasma at finite coupling," Nucl. Phys. B **803**, 166 (2008) [arXiv:0805.2683 [hep-th]].
15. R. C. Myers, M. F. Paulos and A. Sinha, "Quantum corrections to eta/s," Phys. Rev. D **79**, 041901 (2009) [arXiv:0806.2156 [hep-th]].
16. A. Buchel, "Shear viscosity of CFT plasma at finite coupling," Phys. Lett. B **665**, 298 (2008) [arXiv:0804.3161 [hep-th]].
17. A. Buchel, "Shear viscosity of boost invariant plasma at finite coupling," Nucl. Phys. B **802**, 281 (2008) [arXiv:0801.4421 [hep-th]].
18. A. Buchel, J. T. Liu and A. O. Starinets, "Coupling constant dependence of the shear viscosity in N=4 supersymmetric Yang-Mills theory," Nucl. Phys. B **707**, 56 (2005) [arXiv:hep-th/0406264].
19. P. Benincasa and A. Buchel, "Transport properties of N = 4 supersymmetric Yang-Mills theory at finite coupling," JHEP **0601**, 103 (2006) [arXiv:hep-th/0510041].

20. R. G. Cai and Y. W. Sun, "Shear Viscosity from AdS Born-Infeld Black Holes," JHEP **0809**, 115 (2008) [arXiv:0807.2377 [hep-th]].
21. M. Brigante, H. Liu, R. C. Myers, S. Shenker and S. Yaida, "Viscosity Bound Violation in Higher Derivative Gravity," Phys. Rev. D **77**, 126006 (2008) [arXiv:0712.0805 [hep-th]].
22. R. Brustein, D. Gorbonos and M. Hadad, "Wald's entropy is equal to a quarter of the horizon area in units of the effective gravitational coupling," Phys. Rev. D **79**, 044025 (2009) [arXiv:0712.3206 [hep-th]].
23. R. Brustein and A. J. M. Medved, "The ratio of shear viscosity to entropy density in generalized theories of gravity," Phys. Rev. D **79**, 021901 (2009) [arXiv:0808.3498 [hep-th]].
24. R. Brustein and A. J. M. Medved, "The shear diffusion coefficient for generalized theories of gravity," Phys. Lett. B **671**, 119 (2009) [arXiv:0810.2193 [hep-th]].
25. N. Iqbal and H. Liu, "Universality of the hydrodynamic limit in AdS/CFT and the membrane paradigm," Phys. Rev. D **79**, 025023 (2009) [arXiv:0809.3808 [hep-th]].
26. R. G. Cai, Z. Y. Nie and Y. W. Sun, "Shear Viscosity from Effective Couplings of Gravitons," Phys. Rev. D **78**, 126007 (2008) [arXiv:0811.1665 [hep-th]].
27. G. Policastro, D. T. Son and A. O. Starinets, "From AdS/CFT correspondence to hydrodynamics," JHEP **0209**, 043 (2002) [arXiv:hep-th/0205052].
28. D. T. Son and A. O. Starinets, "Viscosity, Black Holes, and Quantum Field Theory," Ann. Rev. Nucl. Part. Sci. **57**, 95 (2007) [arXiv:0704.0240 [hep-th]].
29. P. K. Kovtun and A. O. Starinets, "Quasinormal modes and holography," Phys. Rev. D **72**, 086009 (2005) [arXiv:hep-th/0506184].
30. M. Brigante, H. Liu, R. C. Myers, S. Shenker and S. Yaida, "The Viscosity Bound and Causality Violation," Phys. Rev. Lett. **100**, 191601 (2008) [arXiv:0802.3318 [hep-th]].
31. Y. Kats and P. Petrov, "Effect of curvature squared corrections in AdS on the viscosity of the dual gauge theory," JHEP **0901**, 044 (2009) [arXiv:0712.0743 [hep-th]].
32. S. S. Pal, "η/s at finite coupling," arXiv:0910.0101 [hep-th].
33. N. Banerjee and S. Dutta, "Near-Horizon Analysis of η/s," arXiv:0911.0557 [hep-th].
34. D. W. Pang, "On Charged Lifshitz Black Holes," arXiv:0911.2777 [hep-th].
35. N. Banerjee and S. Dutta, "Higher Derivative Corrections to Shear Viscosity from Graviton's Effective Coupling," JHEP **0903**, 116 (2009) [arXiv:0901.3848 [hep-th]].
36. N. Banerjee and S. Dutta, "Shear Viscosity to Entropy Density Ratio in Six Derivative Gravity," JHEP **0907**, 024 (2009) [arXiv:0903.3925 [hep-th]].
37. D. W. Pang, "R^2 Corrections to Asymptotically Lifshitz Spacetimes," JHEP **0910**, 031 (2009) [arXiv:0908.1272 [hep-th]].
38. M. F. Paulos, "Transport coefficients, membrane couplings and universality at extremality," arXiv:0910.4602 [hep-th].
39. R. G. Cai, Z. Y. Nie, N. Ohta and Y. W. Sun, "Shear Viscosity from Gauss-

Bonnet Gravity with a Dilaton Coupling," Phys. Rev. D **79**, 066004 (2009) [arXiv:0901.1421 [hep-th]].

40. R. G. Cai, Y. Liu and Y. W. Sun, "Transport Coefficients from Extremal Gauss-Bonnet Black Holes," arXiv:0910.4705 [hep-th].
41. X. H. Ge, Y. Matsuo, F. W. Shu, S. J. Sin and T. Tsukioka, "Viscosity Bound, Causality Violation and Instability with Stringy Correction and Charge," JHEP **0810**, 009 (2008) [arXiv:0808.2354 [hep-th]].
42. M. Edalati, J. I. Jottar and R. G. Leigh, "Transport Coefficients at Zero Temperature from Extremal Black Holes," arXiv:0910.0645 [hep-th].
43. T. Faulkner, H. Liu, J. McGreevy and D. Vegh, "Emergent quantum criticality, Fermi surfaces, and AdS2," arXiv:0907.2694 [hep-th].
44. S. J. Rey, "String Theory On Thin Semiconductors: Holographic Realization Of Fermi Points And Surfaces," Prog. Theor. Phys. Suppl. **177**, 128 (2009) [arXiv:0911.5295 [hep-th]].
45. S. S. Lee, "A Non-Fermi Liquid from a Charged Black Hole: A Critical Fermi Ball," Phys. Rev. D **79**, 086006 (2009) [arXiv:0809.3402 [hep-th]].
46. H. Liu, J. McGreevy and D. Vegh, "Non-Fermi liquids from holography," arXiv:0903.2477 [hep-th].

PRINCIPLE OF RELATIVITY, 24 POSSIBLE KINEMATICAL ALGEBRAS AND NEW GEOMETRIES WITH POINCARÉ SYMMETRY

C.-G. HUANG*

Institute of High Energy Physics, Chinese Academy of Sciences, Beijing 100049, and Theoretical Physics Center for Science Facilities, Chinese Academy of Sciences, Beijing 100049
**E-mail: hangcg@ihep.ac.cn*

From the principle of relativity with two universal invariant parameters c and l, 24 possible kinematical (including geometrical and static) algebras can be obtained. Each algebra is of 10 dimensional, generating the symmetry of a 4 dimensional homogeneous space-time or a pure space. In addition to the ordinary Poincaré algebra, there is another Poincaré algebra among the 24 algebras. New 4d geometries with the new Poincaré symmetry are presented. The motion of free particles on one of the new space-times is discussed.

Keywords: Kinematical algebras, Poincaré symmetry; 4d degenerate geometry.

1. Introduction

The principle of relativity, laws of non-gravitational physics having the same form in all inertial frames, is valid not only in the Minkowski space-time but also in the de Sitter (dS) space-time.[1,2] Based on the principle of relativity and the postulate of the two universal invariant parameters (c, l), dS invariant special relativity can be established in a dS space-time,[1,2] where c is the vacuum speed of light at the origin and l is the dS radius. For brevity, the principle of relativity and the postulate of the two universal invariant parameters are known as the principle of relativity with two universal invariant parameters, denoted by $PoR_{c,l}$.[3,4]

Very recently, in the study of the principle of relativity with two invariant parameters, we construct 24 kinematical algebras, including purely geometrical ones and static one.[4] Each algebra is of 10 dimensional. The 11 of them are the algebras in Bacry-Lévy-Leblond theorem.[5] They are (Anti-)dS ($\mathfrak{d}_{\pm}$), Poincaré ($\mathfrak{p}$), (Anti-)Newton-Hooke ($\mathfrak{n}_{\pm}$), Inhomogeneous SO(4) and para-Poincaré ($\mathfrak{p'}_{\pm}$), Galilei ($\mathfrak{g}$), Carroll ($\mathfrak{c}$), para-Galilei ($\mathfrak{g'}$) and static ($\mathfrak{s}$) algebras. The 3 of them correspond to the Euclid geometry ($\mathfrak{e}$), Riemann geometry ($\mathfrak{r}$), and Lobachevski geometry ($\mathfrak{l}$), which can be obtained by relaxing the third assumption in Bacry-Lévy-Leblond theorem. It is remarkable that among the 10 new kinematical or purely geometrical algebras, there is another Poincaré algebra[a].

It is well known that the Poincaré symmetry is the foundation of Einstein's special relativity, relativistic field theories in Minkowski space-time, particle physics, as well as the Poincaré gauge theories of gravity, etc. Conventionally, only the Minkowski space-time is invariant under global Poincaré transformations. It is natural to ask: what is the role played by the new Poincaré symmetry.

The aim of the present talk is twofold. One is to exhibit 24 kinematical algebras, including purely geometrical ones and static one. The other is to first present the nontrivial 4d geometries which are invariant under the new Poincaré transformations. The structure of the new nontrivial 4d geometries will be explored briefly. The motion of free particles on the one of the new space-times is also discussed.

The talk is divided into 6 parts. After the introduction, I shall review the inertial motion and Umov-Weyl-Fock-Hua (UWFH) transformations, show all possible kinematical and geometrical algebras, present the new nontrivial geometries, and study the motion of free particles, successively. Finally, I shall end my talk with the summary.

2. Inertial Motions and UWFH transformations

In a Cartesian coordinate system in a flat space-time (no matter whether it is relativistic one or not), the motions satisfying

$$\begin{cases} x^i = x_0^i + v^i(t - t_0), \\ v^i = \dfrac{dx^i}{dt} = \text{consts.} \qquad i = 1, 2, 3 \end{cases} \tag{1}$$

or

$$\frac{d^2 x^i}{dt^2} = 0 \tag{2}$$

[a]An algebra is said to be Poincaré one if (1) it is isomorphic to $\mathfrak{iso}(1,3)$ algebra, (2) the unique Abelean ideal of the $\mathfrak{iso}(1,3)$ algebra is regarded as a translation sub-algebra and is divided into the time translation and space translations as a 1d and a 3d representation, respectively, of $\mathfrak{so}(3)$ sub-algebra of the $\mathfrak{so}(1,3)$ sub-algebra and (3) the algebra is invariant under the suitably defined parity and time-reversal operation.[5]

are called the uniform rectilinear motions or inertial motions. The set of observers in the space-time, moving according to Eq.(1) or Eq. (2), make up an inertial frame, denoted by $\mathcal{F}$. In a given flat space-time, the forms of Eq.(1) and Eq.(2) are unchanged under the linear coordinate transformation with 10 parameters.

It has been shown that the forms of Eq.(1) and Eq.(2) are unchanged under the linear fractional transformations with a common denominator in the Beltrami coordinate system in a(n) (A)dS space-time.[1,2] In other words, the above concept of inertial motions and inertial frame can be generalized to the (A)dS space-times.[1,2] Then, the principle of relativity can be generalized to the two space-times. Obviously, in the (A)dS space-time, the (A)dS radius l is an invariant parameter in addition to the invariant speed of light c. Therefore, the postulate of invariant speed of light c in Einstein's special relativity should be replaced by the postulate of two invariant parameters. The principle of relativity and the postulate of an invariant parameter should be replaced by the principle of relativity with two universal invariant parameters, $PoR_{c,l}$.

Furthermore, the forms of Eq.(1) and Eq.(2) are also unchanged under the linear fractional transformations with a common denominator in (Anti-)Newton-Hooke ((A)NH) space-times.[6] Then, a simple question appears: what is the most general transformation $\mathcal{T}$ s.t.

$$\mathcal{T}: \quad x'^{\mu} = f^{\mu}(x), x^0 = ct, \mu = 0, \cdots, 3, \tag{3}$$

preserving the form of Eq.(1) and Eq.(2)?

The following theorem answers the question.

Theorem 2.1. *The most general transformations are the linear fractional transformations:*

$$\mathcal{T}: \quad l^{-1}x'^{\mu} = \frac{A^{\mu}{}_{\nu} l^{-1} x^{\nu} + a^{\mu}}{b^t l^{-1} x + d} \tag{4}$$

and

$$det\ \mathcal{T} = \begin{vmatrix} A & a \\ b^t & d \end{vmatrix} = 1, \tag{5}$$

where $A = \{A^{\mu}{}_{\nu}\}$ *a* 4×4 *matrix,* a, b 4×1 *matrixes,* $d \in R$ *and* $b^t = \eta b$ *with* $\eta_{\mu\nu} = \mathrm{diag}(1, -1, -1, -1)$.

The proof of the theorem can be found in Ref. 7–9. The question was first raised and answered by Umov and Weyl.[10] Fock and Hua also studied the question in details in their books.[7,8] Therefore, we name the transformations by Umov-Weyl-Fock-Hua (UWFH) transformations.

Clearly, all UWFH transformations form a group. The number of generators of the group is 24. It implies that more possible space-times admit Eq.(1) and Eq.(2), as expected.

3. Possible Kinematical Groups

In order to clarify how many space-times admit Eq.(1) and Eq.(2), we begin with the Beltrami model of (A)dS spacetime

$$ds_{\pm}^2 = \left(\frac{\eta_{\mu\nu}}{\sigma_{\pm}(x)} \pm \frac{x_\mu x_\nu}{l^2\sigma_{\pm}^2(x)}\right) dx^\mu dx^\nu, \tag{6}$$

where $x_\mu = \eta_{\mu\lambda}x^\lambda$ and

$$\sigma_{\pm}(x) = \sigma_{\pm}(x, x) = 1 \mp l^{-2}x^\mu x_\mu > 0. \tag{7}$$

In the above equations, the upper sign corresponds to dS space-time and the lower sign to AdS space-time. The (A)dS space-time is invariant under the (A)dS transformations, respectively. The generators of (A)dS group are

$$\begin{aligned}\mathbf{P}_\mu^{\pm} &= (\delta_\mu^\nu \mp l^{-2}x_\mu x^\nu)\partial_\nu,\\ \mathbf{L}_{\mu\nu} &= x_\mu \mathbf{P}_\nu - x_\nu \mathbf{P}_\mu = x_\mu\partial_\nu - x_\nu\partial_\mu \in \mathfrak{so}(1,3),\end{aligned} \tag{8}$$

or

$$\begin{aligned}&H^{\pm} = \partial_t \mp \nu^2 t x^\sigma \partial_\sigma,\ \mathbf{P}_i^{\pm} = \partial_i \mp l^{-2}x_i x^\sigma\partial_\sigma, \quad \nu = c/l,\\ &\mathbf{K}_i = t\partial_i - c^{-2}x_i\partial_t,\ \mathbf{J}_i = \tfrac{1}{2}\epsilon_i^{\ jk}L_{jk} = \tfrac{1}{2}\epsilon_i^{\ jk}(x_j\partial_k - x_k\partial_j),\end{aligned} \tag{9}$$

where $H^{\pm}$ are called the Beltrami-time-translation generators, $\mathbf{P}_i^{\pm}$ are known as the Beltrami-space-translation generators, $\mathbf{K}_i$ and $\mathbf{J}_i$ are the Lorentz boost generators and the space-rotation generators, as usual. $H^{\pm}$ are the scalar representation of the $\mathfrak{so}(3)$ spanned by $\mathbf{J}_i$. $\mathbf{P}_i^{\pm}$ and $\mathbf{K}_i$ are vector representations of the $\mathfrak{so}(3)$.

Now, we can write down the 24 generators for the group which keeps Eq.(1) and Eq.(2). They are:[4] 4 kinds of generators for time translation

$$H^{\pm} = \partial_t \mp \nu^{-2}tx^\mu\partial_\mu, \qquad H := \partial_t, \qquad H' = -\nu^{-2}tx^\mu\partial_\mu; \tag{10}$$

4 kinds of generators for space translation

$$\mathbf{P}_i^{\pm} = \partial_i \mp l^{-2}x_i x^\mu\partial_\mu, \qquad \mathbf{P}_i := \partial_i, \qquad \mathbf{P}_i' = -l^{-2}x_i x^\mu\partial_\mu; \tag{11}$$

4 kinds of generators for boost

$$\begin{aligned}&\mathbf{K}_i := t\partial_i - c^{-2}x_i\partial_t, &&\mathbf{K}_i^{\mathfrak{g}} = t\partial_i,\\ &\mathbf{K}_i^{\mathfrak{c}} = -c^{-2}x_i\partial_t, &&\mathbf{N}_i = t\partial_i + c^{-2}x_i\partial_t;\end{aligned} \tag{12}$$

3 generators of rotation $\mathbf{J}_i$ as shown in Eq.(9) and

$$\begin{array}{l} R_{ij} = R_{ji} = x_i\partial_j + x_j\partial_i, \qquad (i<j) \\ M_0 = t\partial_t, \quad M_1 = x^1\partial_1, \quad M_2 = x^2\partial_2, \quad M_3 = x^3\partial_3. \end{array} \tag{13}$$

Among them, two time-translation generators, two sets of space-translation generators and two sets of boost generators are independent, respectively. $H = \frac{1}{2}(H^+ + H^-)$ and $\mathbf{P}_i = \frac{1}{2}(\mathbf{P}_i^+ + \mathbf{P}_i^-)$ are ordinary time- and space-translation generators, respectively. $H' = \frac{1}{2}(H^+ - H^-)$ and $\mathbf{P}'_i = \frac{1}{2}(\mathbf{P}_i^+ - \mathbf{P}_i^-)$ are known as pseudo-time- and pseudo-space-translation generators, respectively.[3,4] $\mathbf{K}_i^\mathfrak{g}$ are the Galilei-boost generators. $\mathbf{K}_i^\mathfrak{c} = \mathbf{K}_i - \mathbf{K}_i^\mathfrak{g}$ are the Carroll-boost generators, $\mathbf{N}_i = 2\mathbf{K}_i^\mathfrak{g} - \mathbf{K}_i$ are the geometrical-boost generators.[3,4] The set $\{T\} := (H^\pm, \mathbf{P}_i^\pm, \mathbf{J}_i, \mathbf{K}_i, \mathbf{N}_i, M_0, M_i, R_{ij})$ spans a closed algebra,

$$\begin{array}{ll} [\mathbf{P}_i^+, \mathbf{P}_j^-] = (1-\delta_{(i)(j)})l^{-2}R_{(i)(j)} - 2l^{-2}\delta_{i(j)}(M_{(j)} + \Sigma_\kappa M_\kappa), & \\ [\mathbf{P}_i^\pm, M_j] = \delta_{i(j)}\mathbf{P}_{(j)}^\mp, & [\mathbf{P}_i^\pm, R_{jk}] = -\delta_{ij}\mathbf{P}_k^\mp - \delta_{ik}\mathbf{P}_j^\mp, \\ [H^+, H^-] = 2\nu^2\left(M_0 + \Sigma_\kappa M_\kappa\right), & [H^\pm, M_0] = H^\mp, \\ [\mathbf{K}_i, M_0] = -\mathbf{N}_i, & [\mathbf{K}_i, M_j] = \delta_{i(j)}\mathbf{N}_{(j)}, \\ [\mathbf{K}_i, R_{jk}] = -\delta_{ij}\mathbf{N}_k \quad \delta_{ik}\mathbf{N}_j, & [\mathbf{N}_i, M_0] = -\mathbf{K}_i, \\ [\mathbf{N}_i, M_j] = \delta_{i(j)}\mathbf{K}_{(j)}, & [\mathbf{N}_i, R_{jk}] = -\delta_{ij}\mathbf{K}_k - \delta_{ik}\mathbf{K}_j, \\ [\mathbf{K}_i, \mathbf{N}_j] = (\delta_{(i)(j)} - 1)c^{-2}R_{(i)(j)} - 2\delta_{(i)j}c^{-2}\left(M_0 - M_{(i)}\right), & \\ [L_{ij}, M_k] = \delta_{j(k)}R_{i(k)} - \delta_{i(k)}R_{j(k)}, & [R_{ij}, M_k] = \delta_{i(k)}L_{j(k)} + \delta_{j(k)}L_{i(k)}, \\ [L_{ij}, R_{kl}] = 2(\delta_{ik}\delta_{jl} + \delta_{il}\delta_{jk})(M_i - M_j) + \delta_{ik}R_{jl} + \delta_{il}R_{jk} & \\ \qquad -\delta_{jk}R_{il} - \delta_{jl}R_{ik}, & \\ [R_{ij}, R_{kl}] = -\delta_{ik}L_{jl} - \delta_{il}L_{jk} - \delta_{jk}L_{il} - \delta_{jl}L_{ik}, & \\ \text{dS, AdS, Riemann, and Lobachevski algebraic relations,} & \end{array} \tag{14}$$

where no summation is taken for the repeated indexes in brackets.

It has been shown that there are 24 possible kinematical (including geometrical) algebras,[4] in which $\mathbf{J}_i$ serve as the space-rotation generators. They include 4 pure geometrical algebras and 1 static algebra. The algebras, the sets of generators, and the commutators are listed in Table 1. In Table 1, $\mathcal{H}$, $\mathbf{P}$, $\mathbf{K}$ are the shorthands for the time-translation, space-translation and boost generators, respectively. $[\mathbf{P}, \mathbf{P}] = l^{-2}\mathbf{J}$ implies $[\mathbf{P}_i^+, \mathbf{P}_j^+] = -l^{-2}\epsilon_{ij}{}^k\mathbf{J}_k$, etc. ($\epsilon_{123} = -\epsilon_{12}{}^3 = 1$, $\eta_{ij} = -\delta_{ij}$.) The commutators between $\mathbf{J}$s and the commutators between $\mathbf{J}$ and $\mathcal{H}$, $\mathbf{P}$, $\mathbf{K}$ are not included in Table 1 because they have the same form for different algebras. Apart from the three classical geometrical algebras, there are 10 more possible algebras than those in BLL paper.

In particular, both the set of generators $(H, \mathbf{P}_i, \mathbf{K}_i, \mathbf{J}_i)$ and the set of

Table 1. All possible relativistic, geometrical and non-relativistic kinematical algebras

Algebra	Symbol	Generator set[b]	$[\mathcal{H},\mathbf{P}]$	$[\mathcal{H},\mathbf{K}]$	$[\mathbf{P},\mathbf{P}]$	$[\mathbf{K},\mathbf{K}]$	$[\mathbf{P},\mathbf{K}]$
dS	$\mathfrak{d}_+$	$(H^+,\mathbf{P}_i^+,\mathbf{K}_i,\mathbf{J}_i)$	$\nu^2\mathbf{K}$	$\mathbf{P}$	$l^{-2}\mathbf{J}$	$-c^{-2}\mathbf{J}$	$c^{-2}\mathcal{H}$
AdS	$\mathfrak{d}_-$	$(H^-,\mathbf{P}_i^-,\mathbf{K}_i,\mathbf{J}_i)$	$-\nu^2\mathbf{K}$	$\mathbf{P}$	$-l^{-2}\mathbf{J}$	$-c^{-2}\mathbf{J}$	$c^{-2}\mathcal{H}$
Poincaré	$\mathfrak{p}$ $\mathfrak{p}_2$	$(H,\mathbf{P}_i,\mathbf{K}_i,\mathbf{J}_i)$ $\{H',\mathbf{P}'_i,\mathbf{K}_i,\mathbf{J}_i\}$	0	$\mathbf{P}$	0	$-c^{-2}\mathbf{J}$	$c^{-2}\mathcal{H}$
Riemann	$\mathfrak{r}$	$(H^-,\mathbf{P}_i^+,\mathbf{N}_i,\mathbf{J}_i)$	$-\nu^2\mathbf{K}$	$\mathbf{P}$	$l^{-2}\mathbf{J}$	$c^{-2}\mathbf{J}$	$-c^{-2}\mathcal{H}$
Lobachevsky	$\mathfrak{l}$	$(H^+,\mathbf{P}_i^-,\mathbf{N}_i,\mathbf{J}_i)$	$\nu^2\mathbf{K}$	$\mathbf{P}$	$-l^{-2}\mathbf{J}$	$c^{-2}\mathbf{J}$	$-c^{-2}\mathcal{H}$
Euclid	$\mathfrak{e}$ $\mathfrak{e}_2$	$(H,\mathbf{P}_i,\mathbf{N}_i,\mathbf{J}_i)$ $(-H',\mathbf{P}'_i,\mathbf{N}_i,\mathbf{J}_i)$	0	$\mathbf{P}$	0	$c^{-2}\mathbf{J}$	$-c^{-2}\mathcal{H}$
Galilei	$\mathfrak{g}$ $\mathfrak{g}_2$	$(H,\mathbf{P}_i,\mathbf{K}_i^{\mathfrak{g}},\mathbf{J}_i)$ $(H',\mathbf{P}'_i,\mathbf{K}_i^{\mathfrak{c}},\mathbf{J}_i)$	0	$\mathbf{P}$	0	0	0
Carroll	$\mathfrak{c}$ $\mathfrak{c}_2$	$(H,\mathbf{P}_i,\mathbf{K}_i^{\mathfrak{c}},\mathbf{J}_i)$ $(H',\mathbf{P}'_i,\mathbf{K}_i^{\mathfrak{g}},\mathbf{J}_i)$	0	0	0	0	$c^{-2}\mathcal{H}$
NH_+	$\mathfrak{n}_+$ $\mathfrak{n}_{+2}$	$(H^+,\mathbf{P}_i,\mathbf{K}_i^{\mathfrak{g}},\mathbf{J}_i)$ $(H^+,\mathbf{P}'_i,\mathbf{K}_i^{\mathfrak{c}},\mathbf{J}_i)$	$\nu^2\mathbf{K}$	$\mathbf{P}$	0	0	0
NH_-	$\mathfrak{n}_-$ $\mathfrak{n}_{-2}$	$(H^-,\mathbf{P}_i,\mathbf{K}_i^{\mathfrak{g}},\mathbf{J}_i)$ $(-H^-,\mathbf{P}'_i,\mathbf{K}_i^{\mathfrak{c}},\mathbf{J}_i)$	$-\nu^2\mathbf{K}$	$\mathbf{P}$	0	0	0
para-Galilei	$\mathfrak{g}'$ $\mathfrak{g}'_2$	$(H',\mathbf{P},\mathbf{K}^{\mathfrak{g}},\mathbf{J}_i)$ $(H,\mathbf{P}'_i,\mathbf{K}_i^{\mathfrak{c}},\mathbf{J}_i)$	$\nu^2\mathbf{K}$	0	0	0	0
HN_+[c]	$\mathfrak{h}_+$ $\mathfrak{h}_{+2}$	$(H,\mathbf{P}_i^+,\mathbf{K}_i^{\mathfrak{c}},\mathbf{J}_i)$ $(H',\mathbf{P}_i^+,\mathbf{K}_i^{\mathfrak{g}},\mathbf{J}_i)$	$\nu^2\mathbf{K}$	0	$l^{-2}\mathbf{J}$	0	$c^{-2}\mathcal{H}$
HN_-	$\mathfrak{h}_-$ $\mathfrak{h}_{-2}$	$(H,\mathbf{P}_i^-,\mathbf{K}_i^{\mathfrak{c}},\mathbf{J}_i)$ $(-H',\mathbf{P}_i^-,\mathbf{K}_i^{\mathfrak{g}},\mathbf{J}_i)$	$-\nu^2\mathbf{K}$	0	$-l^{-2}\mathbf{J}$	0	$c^{-2}\mathcal{H}$
Static	$\mathfrak{s}$ $\mathfrak{s}_2$	$(H^{\mathfrak{s}},\mathbf{P}'_i,\mathbf{K}_i^{\mathfrak{c}},\mathbf{J}_i)$[d] $(H^{\mathfrak{s}},\mathbf{P}_i,\mathbf{K}_i^{\mathfrak{g}},\mathbf{J}_i)$	0	0	0	0	0

Note: [b]All commutators of generators have right dimensions expressed by the universal constants c, l or ν. [c]We named isomorphic to ISO(1,3) and para-Poincaré by the Hooke-Newton and anti-Hooke-Newton algebras because of the relation between the generators of the two algebras and the generators of Newton-Hooke and anti-Newton-Hooke algebras. [d]The generator $H^{\mathfrak{s}}$ is meaningful only when the central extension is considered.

generators $(H', \mathbf{P}'_i, \mathbf{K}_i, \mathbf{J}_i)$ satisfy the Poincaré algebra

$$I: \begin{array}{lll} [\mathbf{P}_i, \mathbf{P}_j] = 0, & [\mathbf{K}_i, \mathbf{K}_j] = -c^{-2}\epsilon_{ijk}\mathbf{J}_k, & [\mathbf{P}_i, \mathbf{K}_j] = c^{-2}H, \\ [H, \mathbf{P}_i] = 0, & [H, \mathbf{K}_i] = \mathbf{P}, & [\mathbf{J}_i, H] = 0, \\ [\mathbf{J}_i, \mathbf{P}_j] = \epsilon_{ijk}\mathbf{P}_k, & [\mathbf{J}_i, \mathbf{K}_j] = \epsilon_{ijk}\mathbf{K}_k, & [\mathbf{J}_i, \mathbf{J}_j] = \epsilon_{ijk}\mathbf{J}_k, \end{array} \tag{15}$$

$$II: \begin{array}{lll} [\mathbf{P}'_i, \mathbf{P}'_j] = 0, & [\mathbf{K}_i, \mathbf{K}_j] = -c^{-2}\epsilon_{ijk}\mathbf{J}_k, & [\mathbf{P}'_i, \mathbf{K}_j] = c^{-2}H', \\ [H', \mathbf{P}'_i] = 0, & [H', \mathbf{K}_i] = \mathbf{P}', & [\mathbf{J}_i, H'] = 0, \\ [\mathbf{J}_i, \mathbf{P}'_j] = \epsilon_{ijk}\mathbf{P}'_k, & [\mathbf{J}_i, \mathbf{K}_j] = \epsilon_{ijk}\mathbf{K}_k, & [\mathbf{J}_i, \mathbf{J}_j] = \epsilon_{ijk}\mathbf{J}_k, \end{array} \tag{16}$$

respectively. The former set of generators are the generators of the ordinary Poincaré transformations:

$$x'^{\mu} = L^{\mu}{}_{\nu}x^{\nu} + la^{\mu}, \qquad L \in SO(1,3) \tag{17}$$

which can be realized by 5×5 matrix

$$\begin{pmatrix} L & a \\ 0 & 1 \end{pmatrix}. \tag{18}$$

The ordinary Poincaré transformation preserves the metric of the Minkowski space-time. The latter set of generators are the generators of transformations:

$$x'^{\mu} = \frac{L^{\mu}{}_{\nu} x^{\nu}}{1 + b_{\mu} x^{\mu}}, \tag{19}$$

which can be expressed in terms of matrixes,

$$\begin{pmatrix} L & 0 \\ b^t & 1 \end{pmatrix}. \tag{20}$$

The set of all matrixes of type (20) are the transpose of the set of all matrixes of type (18) in the Minkowski space-time. Therefore, the set of all matrix (20) also form a Poincaré group. It should be noted that the new Poincaré group does not preserves the metric of the Minkowski space-time. Instead, it preserves the light cone at origin in the Minkowski space-time.[3,4]

4. New Geometries with Poincaré Symmetry

Now that the new Poincaré group does not preserve the metric of the Minkowski space-time, what geometry does the new Poincaré group preserve?

It can be checked that $(M^{\pm}, \boldsymbol{g}^{\pm}, \boldsymbol{h}_{\pm}, \nabla^{\pm})$ are invariant under the new Poincaré transformations, where $\boldsymbol{g}^{\pm}$ is a 4d type-(0,2) degenerate symmetric tensor field

$$\boldsymbol{g}^{\pm} = g^{\pm}_{\mu\nu} dx^{\mu} \otimes dx^{\nu} = \pm \frac{l^2}{(x \cdot x)^2} (\eta_{\mu\rho}\eta_{\nu\tau} - \eta_{\mu\nu}\eta_{\rho\tau}) x^{\rho} x^{\tau} dx^{\mu} dx^{\nu}, \tag{21}$$

$\boldsymbol{h}_{\pm}$ is a 4d type-(2,0) degenerate symmetric tensor field

$$\boldsymbol{h}_{\pm} = h^{\mu\nu}_{\pm} \partial_{\mu} \otimes \partial_{\nu} = l^{-4} (x \cdot x) x^{\mu} x^{\nu} \partial_{\mu} \partial_{\nu}, \tag{22}$$

and $\nabla^{\pm}$ is a connection compatible to $\boldsymbol{g}^{\pm}$ and $\boldsymbol{h}_{\pm}$, i.e.

$$\nabla^{\pm}_{\lambda} g^{\pm}_{\mu\nu} = \partial_{\lambda} g^{\pm}_{\mu\nu} - \Gamma^{\kappa}_{\pm\lambda\nu} g^{\pm}_{\mu\kappa} - \Gamma^{\kappa}_{\pm\mu\lambda} g^{\pm}_{\kappa\nu} = 0 \tag{23}$$

and

$$\nabla^{\pm}_{\lambda} h^{\mu\nu}_{\pm} = \partial_{\lambda} h^{\mu\nu}_{\pm} + \Gamma^{\nu}_{\pm\lambda\kappa} h^{\mu\kappa}_{\pm} + \Gamma^{\mu}_{\pm\lambda\kappa} h^{\kappa\nu}_{\pm} = 0, \tag{24}$$

respectively, with connection coefficients,

$$\Gamma^{\mu}_{\pm\nu\lambda} = -\frac{(x_{\nu}\delta^{\mu}_{\lambda} + \delta^{\mu}_{\nu} x_{\lambda})}{x \cdot x}. \tag{25}$$

In the above equations,

$$x \cdot x = \eta_{\mu\nu} x^\mu x^\nu \begin{cases} < 0 & \text{for upper sign,} \\ > 0 & \text{for lower sign.} \end{cases} \tag{26}$$

Clearly, $|g| = |h| = 0$. The ranks of $\boldsymbol{g}$ and $\boldsymbol{h}$ are 3 and 1, respectively. It can be shown that when and only when $\forall \boldsymbol{\xi} \in \mathfrak{p}_2 \subset TM^\pm$,

$$\begin{cases} \mathcal{L}_{\boldsymbol{\xi}} g^\pm_{\mu\nu} = g^\pm_{\mu\nu,\lambda} \boldsymbol{\xi}^\lambda + g^\pm_{\mu\lambda} \partial_\nu \boldsymbol{\xi}^\lambda + g^\pm_{\lambda\nu} \partial_\mu \boldsymbol{\xi}^\lambda = 0, \\ \mathcal{L}_{\boldsymbol{\xi}} h^{\pm\mu\nu} = h^{\pm\mu\nu}{}_{,\lambda} \boldsymbol{\xi}^\lambda - h^{\pm\mu\lambda} \partial_\lambda \boldsymbol{\xi}^\nu - h^{\pm\lambda\nu} \partial_\lambda \boldsymbol{\xi}^\mu = 0, \\ [\mathcal{L}_{\boldsymbol{\xi}}, \nabla^\pm] = 0 \end{cases} \tag{27}$$

are valid simultaneously. By definition, the curvature is

$$R^\mu_{\pm\nu\lambda\sigma} = l^{-2}(g^\pm_{\nu\lambda} \delta^\mu_\sigma - g^\pm_{\nu\sigma} \delta^\mu_\lambda) \tag{28}$$

and

$$R^\pm_{\mu\nu} = R^\lambda_{\pm\mu\nu\lambda} = 3l^{-2} g^\pm_{\mu\nu}. \tag{29}$$

In order to see the structures of the manifolds more transparently, we consider the coordinate transformations,

$$\left.\begin{aligned} x^0 &= l^2 \rho^{-1} \sinh(\psi/l) \\ x^1 &= l^2 \rho^{-1} \cosh(\psi/l) \sin\theta \cos\phi \\ x^2 &= l^2 \rho^{-1} \cosh(\psi/l) \sin\theta \sin\phi \\ x^3 &= l^2 \rho^{-1} \cosh(\psi/l) \cos\theta \end{aligned}\right. \quad \text{for } x \cdot x < 0, \tag{30}$$

$$\left.\begin{aligned} x^0 &= l^2 \eta^{-1} \cosh(r/l) \\ x^1 &= l^2 \eta^{-1} \sinh(r/l) \sin\theta \cos\phi \\ x^2 &= l^2 \eta^{-1} \sinh(r/l) \sin\theta \sin\phi \\ x^3 &= l^2 \eta^{-1} \sinh(r/l) \cos\theta \end{aligned}\right. \quad \text{for } x \cdot x > 0, \tag{31}$$

respectively. Under the coordinate transformations, Eqs.(21), (22), and (25) become, respectively

$$\boldsymbol{g}^\pm = \begin{cases} d\psi^2 - l^2 \cosh^2(\psi/l) d\Omega_2^2 & \text{for } x \cdot x < 0 \\ -dr^2 - l^2 \sinh^2(r/l) d\Omega_2^2 & \text{for } x \cdot x > 0, \end{cases} \tag{32}$$

$$\boldsymbol{h}_\pm = \begin{cases} -\left(\dfrac{\partial}{\partial\rho}\right)^2 & \text{for } x \cdot x < 0 \\ \left(\dfrac{\partial}{\partial\rho}\right)^2 & \text{for } x \cdot x < 0, \end{cases} \tag{33}$$

$$\begin{aligned}
\bar{\Gamma}^{\psi}_{+\theta\theta} &= l\sinh(\psi/l)\cosh(\psi/l), \quad \bar{\Gamma}^{\psi}_{+\phi\phi} = \bar{\Gamma}^{\psi}_{+\theta\theta}\sin^2\theta \\
\bar{\Gamma}^{\theta}_{+\theta\psi} &= \bar{\Gamma}^{\theta}_{+\psi\theta} = \bar{\Gamma}^{\phi}_{+\phi\psi} = \bar{\Gamma}^{\phi}_{+\psi\phi} = l^{-1}\tanh(\psi/l) \\
\bar{\Gamma}^{\theta}_{+\phi\phi} &= -\sin\theta\cos\theta, \quad \bar{\Gamma}^{\phi}_{+\theta\phi} = \bar{\Gamma}^{\phi}_{+\phi\theta} = \cot\theta \\
\bar{\Gamma}^{\rho}_{+\alpha\beta} &= -l^{-2}\rho g_{\alpha\beta}, \qquad \text{others vanish,}
\end{aligned} \qquad \text{for } x\cdot x < 0, \quad (34)$$

where $(\bar{x}^{\alpha};\bar{x}^{3}) = (\psi,\theta,\phi;\rho)$, and

$$\begin{aligned}
\bar{\Gamma}^{\eta}_{-ij} &= -l^{-2}\eta g_{ij} \\
\bar{\Gamma}^{r}_{-\theta\theta} &= -l\sinh(r/l)\cosh(r/l), \quad \bar{\Gamma}^{r}_{-\phi\phi} = \bar{\Gamma}^{r}_{-\theta\theta}\sin^2\theta \\
\bar{\Gamma}^{\theta}_{-r\theta} &= \bar{\Gamma}^{\theta}_{-\theta r} = \bar{\Gamma}^{\phi}_{-r\phi} = \bar{\Gamma}^{\phi}_{-\phi r} = l^{-1}\mathrm{ctanh}(r/l) \\
\bar{\Gamma}^{\theta}_{-\phi\phi} &= -\sin\theta\cos\theta, \quad \bar{\Gamma}^{\phi}_{-\theta\phi} = \bar{\Gamma}^{\phi}_{-\phi\theta} = \cot\theta \\
&\text{others vanish,}
\end{aligned} \qquad \text{for } x\cdot x > 0, \quad (35)$$

where $(\bar{x}^{0};\bar{x}^{i}) = (\eta; r,\theta,\phi)$. The Ricci curvature (29) read

$$R^{\pm}_{\mu\nu} = \begin{cases} 3l^{-2}\,\mathrm{diag}(1, -\cosh^2(\psi/l), -\cosh^2(\psi/l)\sin^2\theta, 0) & x\cdot x < 0, \\ -3l^{2}\,\mathrm{diag}(0, -1, -\sinh^2(r/l), -\sinh^2(r/l)\sin^2\theta) & x\cdot x > 0. \end{cases} \tag{36}$$

They show that the space-times are the homogeneous spaces and that

$$M^{+} = ISO(1,3)/ISO(1,2) = \mathbb{R}\times dS_3 \qquad \text{for} \qquad x\cdot x < 0, \tag{37}$$

and

$$M^{-} - ISO(1,3)/ISO(3) = \mathbb{R}\times H_3 \qquad \text{for} \qquad x\cdot x > 0. \tag{38}$$

It is obvious that on M^{+} there is no spatial SO(3) isotropy at each point though there exists the algebraic SO(3) isotropy, but on M^{-} there exists the spatial SO(3) isotropy at each point.

Further studies show that $(M^{-}, \boldsymbol{g}^{-}, \boldsymbol{h}_{-}, \nabla^{-})$ satisfies the three basic assumptions in the Theorem in Ref. 5. The kinematics on it will have better behaviors than on $(M^{+}, \boldsymbol{g}^{+}, \boldsymbol{h}_{+}, \nabla^{+})$. Therefore, we shall study the kinematics briefly.

5. Motion of a Free Particle on $(M^{-}, \boldsymbol{g}^{-}, \boldsymbol{h}_{-}, \nabla\ \ )$

The motion for a free particle is still supposed to be determined by the geodesic equation

$$\frac{d^2x^{\mu}}{d\lambda^2} + \Gamma^{\mu}_{-\nu\lambda}\frac{dx^{\nu}}{d\lambda}\frac{dx^{\lambda}}{d\lambda} = 0, \tag{39}$$

as usual. It gives rise to the 'uniform rectilinear' motion

$$x^{i} = a^{i}x^{0} + lb^{i} \tag{40}$$

if x^0 and x^i are regarded as the 'temporal' and 'spatial' coordinates, respectively, where a^i and b^i are two dimensionless constants. The result is consistent with Eq.(1) and Eq.(2), which are the start points of our work.

The 'uniform rectilinear' motion (40) can also be obtained from the Lagrangian

$$L = \frac{mlc}{x \cdot x}\sqrt{(\eta_{\mu\nu}\eta_{\eta\tau} - \eta_{\mu\eta}\eta_{\nu\tau})x^\eta x^\tau \dot{x}^\mu \dot{x}^\nu}. \tag{41}$$

The Euler-Lagrangian equation is equivalent to

$$\begin{aligned}&[(x\cdot x)(\dot{x}\cdot\dot{x}) - (x\cdot\dot{x})^2]\ddot{x}_\kappa + (\dot{x}\cdot\ddot{x})[(x\cdot\dot{x})x_\kappa - (x\cdot x)\dot{x}_\kappa]\\ &\quad +(x\cdot\ddot{x})[(x\cdot\dot{x})\dot{x}_\kappa - (\dot{x}\cdot\dot{x})x_\kappa] = 0.\end{aligned}$$

The nonzero determinant of its coefficients for $\ddot{x}$ requires $\ddot{x}_\kappa = 0$.

It should be noted that the coordinate system x^0/c and x^i are not the intrinsic coordinates of the time and space, respectively. Therefore, Eq.(40) cannot be interpreted as the uniform rectilinear motion or inertial motion in the space-time $(M^-, \boldsymbol{g}^-, \boldsymbol{h}_-, \nabla^-)$ in the usual sense.

6. Summary

There are 24 different possible kinematical groups, including geometrical ones and static one. Each has 10 parameters. Among the 24 possible kinematical groups, there exists a new Poincaré symmetry in addition to the ordinary Poincaré symmetry.

The new Poincaré symmetry does not preserve the metric of the Minkowski space-time. Instead, it preserves the degenerate geometries $(M^\pm, \boldsymbol{g}^\pm, \boldsymbol{h}_\pm, \nabla^\pm)$ presented in the talk. The degenerate geometries and their topology are dramatically different from those of the Minkowski space-time. But, they are still homogeneous spaces. The physical applications of the degenerate space-times need to be explored.

From the study on the degenerate geometries with Poincaré symmetry, we can see that algebraic SO(3) isotropy does not always imply the geometrical space isotropy. Whether a kinematics possesses the space isotropy or space-time isotropy should be determined by the underlying geometry.

The Lagrangian for a free particle can be defined on the new geometry. In the coordinate system x^μ, the motion takes the form of the uniform rectilinear motion. Unfortunately, the coordinate system x^i and x^0/c do not the intrinsic coordinates of the space and time. Therefore, 'the uniform rectilinear motion' is not in the usual sense.

Acknowledgments

I am grateful to H.-Y. Guo, Y. Tian, H.-t. Wu, X.-N. Wu, Z. Xu, and B. Zhou for the cooperation in the works related to the talk. I would like to thank Z.-N. Hu, W. T. Ni, J. Xu, and H.-X. Yang for helpful discussion. The work is supported in part by NSFC under Grant Nos. 10775140, 10975141, and KIFCAS (KJCX3-SYW-S03).

References

1. K. H. Look (Qi-Keng Lu), *Why the Minkowski metric must be used?*, 1970, unpublished; K. H. Look, C. L. Tsou (Z. L. Zou), and H. Y. Kuo (H. Y. Guo), *Acta Phys. Sinica* 23 225 (1974) (in Chinese).
2. H.-Y. Guo, C.-G. Huang, Z. Xu, and B. Zhou, *Mod. Phys. Lett. A* **19** 1701 (2004); *Phys. Lett. A* **331** 1 (2004); *Chin. Phys. Lett.* **22** 2477 (2055); H.-Y. Guo, C.-G. Huang, Y. Tian, Z. Xu, and B. Zhou, *Acta Phys. Sin.* **54** 2494 (2005) (in Chinese).
3. H.-Y. Guo, H.-T. Wu, and B. Zhou, *Phys. Lett. B* **670** 437 (2009).
4. H.-Y. Guo, C.-G. Huang, H.-T. Wu, and B. Zhou, *The principle of relativity, kinematics and algebraic relations*, arXiv: 0812.0871, to appear in *Sci. China G*.
5. H. Bacry and J.-M. Lévy-Leblond, *J. Math. Phys.* **9** 1605 (1968).
6. C.-G. Huang, H.-Y. Guo, Y. TIAN, Z. Xu, and B. Zhou, *Int. Journ. Mod. Phys. A* **22** 2535 (2007); Y. Tian, H.-Y. Guo, C.-G. Huang, Z. Xu, and B. Zhou, *Phys. Rev. D* **71** 044030 (2005).
7. V. Fock, *The Theory of Space Time and Gravitation*, (Pergamon, Oxford, 1964).
8. L. K. Hua (L. G. Hua), *Starting with Unit Circle*, (Lecture notes, 1962, in Chinese), (Science Press, Beijing, 1977, in Chinese), (Springer-Verlag, New York, 1981).
9. H.-Y. Guo, C.-G. Huang, Y. Tian, H.-T. Wu, Z. Xu, and B. Zhou, *Class. Quantum Grav.* **24** 4009 (2007).
10. N. A. Umov, *Phys. Z.* **11** 908 (1910); H. Weyl, *Mathematische Analyse des Raumproblems*, (Springer, Berlin, 1923).

PHYSICAL DECOMPOSITION OF THE GAUGE AND GRAVITATIONAL FIELDS

XIANG-SONG CHEN* and BEN-CHAO ZHU

Department of Physics, Huazhong University of Science and Technology
Wuhan 430074, China
**E-mail: cxs@hust.edu.cn*

Physical decomposition of the non-Abelian gauge field has recently solved the two-decade-lasting problem of a gauge-invariant gluon spin. Here we extend this approach to gravitation and attack the century-lasting problem of a covariant gravitational energy density. Counterpart of the gauge field in gravity, the affine connection, is unambiguously separated into a pure geometric term which does not contribute to the curvature tensor, and a gauge-independent physical term which always vanishes in a flat space-time. By this decomposition the conventional pseudo-tensors of the gravitational energy are easily converted to be true tensors, and a physical meaning does exist for the energy density.

Keywords: Gauge Invariance; Field Decomposition; Gravitational Energy.

1. Introduction

Gauge invariance is the most elegant and efficient principle for constructing interactions in the present field theories of physics. By requiring field equations to be gauge invariant, the manner of the couplings (and self-couplings) of various fields are strongly constrained. This applies both to the standard model of the strong and electro-weak interactions, and to Einstein's gravitational theory. For the latter case gauge invariance refers to general covariance under arbitrary coordinate transformation. It is rather annoying, however, that a theory built uniquely out of the gauge-invariance requirement does not seem to guarantee gauge invariance for all physical quantities. In hadron physics, e.g., in the two-decade efforts to understand how the nucleon spin originates from the spin and orbital motion of its quark and gluon constituents, one encounters severe difficulty in finding a gauge-invariant description of the gluon spin and orbital angular momentum. It was not until 2008 that a solution was obtained in Ref.,[1] and further developed in Ref.[2]

A more celebrated and still unsolved gauge-dependence problem is the energy density of the gravitational field. After countless attempts of nearly a century, a convincing solution is still lacking. A reflection of this desperation is the often heard argument that gravitational energy is intrinsically nonlocalizable and can at best be quasi-local to a closed two-surface,[3,4] since by transiting to a free-fall reference frame the effect of gravity at any point can be eliminated. While it might be reasonable to expect that the energy property of a physical system changes when viewed in another reference frame with different acceleration, it is definitely unacceptable that the energy density at a point may be eliminated or created when one simply shifts between the Cartesian and spherical coordinates in the same reference frame.

The key obstacle to constructing all physical quantities gauge invariantly is the inevitable involvement of the gauge field or the affine connection together with their ordinary derivatives, which are all intrinsically gauge dependent. The idea in Refs.[1,2] is to decompose the gauge field: $A_\mu \equiv \hat{A}_\mu + \bar{A}_\mu$. The aim that is that $\hat{A}_\mu$ will be a physical term which is gauge-covariant and always vanishes in the vacuum, and $\bar{A}_\mu$ is a pure-gauge term which solely carries the gauge freedom and has no essential physical effects (particularly, it does not contribute to the electric or magnetic field strength). Equipped with the separate $\hat{A}_\mu$ and $\bar{A}_\mu$, a naively gauge-dependent quantity (such as the gluon spin $\vec{S} = \vec{E} \times \vec{A}$) can easily be rescued to be gauge-covariant, simply by replacing A_μ with $\hat{A}_\mu$, and by replacing the ordinary derivative with the pure-gauge covariant derivative constructed with $\bar{A}_\mu$ instead of A_μ.

Mathematically, a well-defined separation $A_\mu = \hat{A}_\mu + \bar{A}_\mu$ means an unambiguous prescription for constructing $\hat{A}_\mu$ and $\bar{A}_\mu$ out of a given A_μ. The properties (especially, gauge transformations) of $\hat{A}_\mu$ and $\bar{A}_\mu$ are then inherently determined via their mathematical expressions in terms of A_μ. In Refs.,[1,2] it was found that $\hat{A}_\mu$ and $\bar{A}_\mu$ can indeed be solved in terms of A_μ by setting up proper differential equations and boundary conditions, which lead to unique solutions for $\hat{A}_\mu$ and $\bar{A}_\mu$ with desired physical properties. In this paper, we show that this method can be generalized to gravitational theory. The affine connection $\Gamma^\rho_{\sigma\mu}$, which is a counterpart of A_μ, is unambiguously decomposed into the sum of a physical term $\hat{\Gamma}^\rho_{\sigma\mu}$, which represents the true gravitational effect, and a pure geometric term $\bar{\Gamma}^\rho_{\sigma\mu}$, which represents the spurious gravitational effect associated with coordinate choice. Gauge-dependence of the gravitational energy density originates exactly from the fact that the affine connection may contain a spurious gravitational effect. While in a flat space-time the Cartesian coordinate in which $\Gamma^\rho_{\sigma\mu} = 0$ seems a natural choice, in an intrinsically curved space-time no coordinate is obviously more natural than others, hence it is no longer a trivial task to get rid of the spurious gravitational effect. In our formalism, the task is to seek a prescription to identify $\bar{\Gamma}^\rho_{\sigma\mu}$ for a given gravitational system in any given reference frame.

As for gauge theories, we find that the prescription is again a set of defining differential equations for $\hat{\Gamma}^\rho_{\sigma\mu}$ and $\bar{\Gamma}^\rho_{\sigma\mu}$. Let us first display these equations: (Greek indices run from 0 to 3, Latin indices run from 1 to 3, and repeated indices are summed over)

$$\bar{R}^\rho_{\sigma\mu\nu} \equiv \partial_\mu \bar{\Gamma}^\rho_{\sigma\nu} - \partial_\nu \bar{\Gamma}^\rho_{\sigma\mu} + \bar{\Gamma}^\rho_{\alpha\mu}\bar{\Gamma}^\alpha_{\sigma\nu} - \bar{\Gamma}^\rho_{\alpha\nu}\bar{\Gamma}^\alpha_{\sigma\mu} = 0 \tag{1a}$$

$$\bar{D}_i \hat{\Gamma}^\rho_{\sigma i} \equiv \partial_i \hat{\Gamma}^\rho_{\sigma i} + \bar{\Gamma}^\rho_{i\alpha}\hat{\Gamma}^\alpha_{\sigma i} - \bar{\Gamma}^\alpha_{i\sigma}\hat{\Gamma}^\rho_{\alpha i} - \bar{\Gamma}^\alpha_{ii}\hat{\Gamma}^\rho_{\sigma\alpha} = 0 \tag{1b}$$

2. Decomposition of the Gauge Fields

To comprehend how Eqs. (1) are chosen, how they give solutions for $\hat{\Gamma}^\rho_{\sigma\mu}$ and $\bar{\Gamma}^\rho_{\sigma\mu}$ with desired properties, and how the solutions in turn are employed to solve the gauge-dependence problem of the gravitational energy density, it is most helpful to recall the parallel constructions for gauge theories in Refs.[1,2] In Abelian case, the gauge field A_μ transforms as $A_\mu \to A'_\mu = A_\mu - \partial_\mu \omega$, which leaves the field strength invariant: $F_{\mu\nu} = \partial_\mu A_\nu - \partial_\nu A_\mu \to F'_{\mu\nu} = F_{\mu\nu}$. The defining equations for the separation $A_\mu = \hat{A}_\mu + \bar{A}_\mu$ are

$$\bar{F}_{\mu\nu} \equiv \partial_\mu \bar{A}_\nu - \partial_\nu \bar{A}_\mu = 0 \tag{2a}$$

$$\partial_i \hat{A}_i = 0 \tag{2b}$$

Eq. (2a) has very clear physical meaning: the pure-gauge term $\bar{A}_\mu$ gives null field strength. Eq. (2b) can be regarded as the transversity condition for a physical photon with zero mass. But to avoid confusion with the Coulomb gauge condition $\partial_i A_i = 0$ for the full A_i, it is more helpful to think in a mathematical way that Eqs. (2) are the needed differential equations to solve $\hat{A}_\mu$ and $\bar{A}_\mu$. Since $\hat{A}_\mu + \bar{A}_\mu = A_\mu$, it suffices to examine $\hat{A}_\mu$. We arrange Eqs. (2) into the more transparent form:

$$\partial_i \hat{A}_j - \partial_j \hat{A}_i = F_{ij} \tag{3a}$$

$$\partial_i \hat{A}_i = 0 \tag{3b}$$

$$\partial_i \hat{A}_0 = \partial_0 \hat{A}_i + F_{i0} \tag{3c}$$

To pin down the solutions, we still need the boundary conditions. For a finite physical system, it is natural to require that at infinity the physical part $\hat{A}_\mu$ approaches zero (as does the field strength $F_{\mu\nu}$), and thus

$\bar{A}_\mu = A_\mu - \hat{A}_\mu$ approaches A_μ. For an infinite system, such as the magnetic field produced by an infinitely long solenoid, it is not obvious at all to specify a natural choice of boundary conditions. While an infinite system is hardly relevant for particle physics, in gravitational theory the universe does present a real example. In this paper we restrict out attention to finite systems, and leave the discussion for infinite systems to future studies.

With $\hat{A}_\mu$ vanishing at spatial infinity, the solutions to Eqs. (3) are familiar:

$$\hat{A}_i = \partial_j (\partial_k \partial_k)^{-1} F_{ji} \tag{4a}$$

$$\hat{A}_0 = \int_\infty^x dx^i (\partial_0 \hat{A}_i + F_{i0}) \tag{4b}$$

In (4b) the index i takes any value of 1 to 3, and is not summed; and $\hat{A}_i$ is to be substituted by the expression in (4a). By subtracting the solved $\hat{A}_\mu$ from A_μ, one then obtains the expression of $\bar{A}_\mu$ in terms of A_μ. The explicit expressions in Eqs. (4) indicate clearly that the physical field $\hat{A}_\mu$ is gauge invariant, and hence the pure-gauge field $\bar{A}_\mu = A_\mu - \hat{A}_\mu$ carries all the gauge freedom and transforms in the same manner as does the full A_μ. Moreover, the solution in Eqs. (4) tell us that the physical term $\hat{A}_\mu$ vanishes as the field strength $F_{\mu\nu} = 0$.

In non-Abelian case, the gauge transformation is more complicated: $A'_\mu = U A_\mu U^\dagger - \frac{i}{g} U \partial_\mu U^\dagger$. The field strength now contains a self-interaction term, and transforms covariantly instead of invariantly: $F_{\mu\nu} = \partial_\mu A_\nu - \partial_\nu A_\mu + ig[A_\mu, A_\nu] \to F'_{\mu\nu} = U F_{\mu\nu} U^\dagger$. It is fairly non-trivial to choose proper defining equations for the non-Abelian $\hat{A}_\mu$ and $\bar{A}_\mu$. They were originally proposed in Ref.,[1] and further developed in Ref.[2] to be:

$$\bar{F}_{\mu\nu} \equiv \partial_\mu \bar{A}_\nu - \partial_\nu \bar{A}_\mu + ig[\bar{A}_\mu, \bar{A}_\nu] = 0 \tag{5a}$$

$$\bar{\mathcal{D}}_i \hat{A}_i \equiv \partial_i \hat{A}_i + ig[\bar{A}_i, \hat{A}_i] = 0 \tag{5b}$$

We will shortly show that Eqs. (5) give solutions for $\hat{A}_\mu$ and $\bar{A}_\mu$ with desired gauge-transformation properties:

$$\hat{A}'_\mu = U \hat{A}_\mu U^\dagger, \quad \bar{A}'_\mu = U \bar{A}_\mu U^\dagger - \frac{i}{g} U \partial_\mu U^\dagger. \tag{6}$$

Therefore, $\bar{\mathcal{D}}_\mu = \partial_\mu + ig[\bar{A}_\mu, \ ]$ is a pure-gauge covariant derivative for a field in the adjoint representation, and Eqs. (5) are invariant under non-Abelian gauge transformations. Analogous to the Abelian theory, Eqs. (5) say that $\bar{A}_\mu$ is a pure-gauge field which gives null field strength, and the physical field $\hat{A}_\mu$ satisfies a "covariant transversity condition". However, as we remarked in the Abelian case, the real justification for Eqs. (5) is that they are the right mathematic equations to solve $\hat{A}_\mu$ and $\bar{A}_\mu$ in terms of A_μ, with desired gauge transformations in (6). Again, we examine $\hat{A}_\mu$ with trivial boundary condition, and arrange Eqs. (5) into decoupled forms for $\hat{A}_i$ and $\hat{A}_0$:

$$\partial_i \hat{A}_j - \partial_j \hat{A}_i = F_{ij} + ig([\hat{A}_i, \hat{A}_j] - [A_i, \hat{A}_j] - [\hat{A}_i, A_j]) \tag{7a}$$

$$\partial_i \hat{A}_i = ig[\hat{A}_i, A_i] \tag{7b}$$

$$\partial_i \hat{A}_0 = \partial_0 \hat{A}_i + F_{i0} + ig([\hat{A}_i, \hat{A}_0] - [A_i, \hat{A}_0] - [\hat{A}_i, A_0]) \tag{7c}$$

Due to their non-linearity, Eqs. (7) are not really easy to solve. To proceed, we employ the usual technique of perturbative expansion, which applies when either the coupling constant g or the field amplitudes are small. For a small g, e.g., we write $\hat{A}_\mu = \hat{A}_\mu^{(0)} + g\hat{A}_\mu^{(1)} + g^2\hat{A}_\mu^{(2)} + \cdots$. Eqs. (7) can then be solved order by order. The zeroth-order term $\hat{A}_\mu^{(0)}$ satisfy the same equations as Eqs. (3) for the Abelian case. Its solution is given by Eqs. (4), and can in turn be used to solve the equations for the leading non-trivial term $\hat{A}_\mu^{(1)}$:

$$\partial_i \hat{A}_j^{(1)} - \partial_j \hat{A}_i^{(1)} = i[\hat{A}_i^{(0)}, \hat{A}_j^{(0)}] - i[A_i, \hat{A}_j^{(0)}] - i[\hat{A}_i^{(0)}, A_j] \tag{8a}$$

$$\partial_i \hat{A}_i^{(1)} = i[\hat{A}_i^{(0)}, A_i] \tag{8b}$$

$$\partial_i \hat{A}_0^{(1)} = \partial_0 \hat{A}_i^{(1)} + i[\hat{A}_i^{(0)}, \hat{A}_0^{(0)}] - i[A_i, \hat{A}_0^{(0)}] - i[\hat{A}_i^{(0)}, A_0] \tag{8c}$$

Solution of these equations is then straightforward, and can be further employed to solve the next-order term $\hat{A}_\mu^{(2)}$, and so on. Given validity of this perturbative expansion, the solution to Eqs. (5) or (7) is unique. This uniqueness has important implications: a) $F_{\mu\nu} = 0$ is necessary and sufficient for $\hat{A}_\mu = 0$, and b) $\hat{A}_\mu$ and $\bar{A}_\mu$ have the gauge transformations as in (6). The proof of b) is as follows: Eqs. (6) are solution of Eqs. (5) with $\hat{A}_\mu$ and $\bar{A}_\mu$ replaced by $\hat{A}'_\mu$ and $\bar{A}'_\mu$, and since the solution to Eqs. (5) is unique, Eqs. (6) are the right gauge transformations.

3. Decomposition of the Gravitational Field

We now turn to Einstein's gravitational theory which has a deep analogy with the non-Abelian gauge theory. The covariant Riemann curvature tensor, $R^\rho_{\sigma\mu\nu}$, is built out of the non-covariant affine connection $\Gamma^\rho_{\sigma\mu}$ in very much the same manner as the covariant field strength $F_{\mu\nu}$ is built out of the non-covariant gauge potential: $R^\rho_{\sigma\mu\nu} = \partial_\mu \Gamma^\rho_{\sigma\nu} - \partial_\nu \Gamma^\rho_{\sigma\mu} + \Gamma^\rho_{\alpha\mu}\Gamma^\alpha_{\sigma\nu} - \Gamma^\rho_{\alpha\nu}\Gamma^\alpha_{\sigma\mu}$. It is a true tensor which is always zero in an intrinsically flat space-time.

In contrast, the affine connection carries a gauge term in coordinate transformation $x^\mu \to x'^\mu$:

$$\Gamma'^\rho_{\sigma\mu} = \frac{\partial x'^\rho}{\partial x^\tau}\frac{\partial x^\alpha}{\partial x'^\sigma}\frac{\partial x^\beta}{\partial x'^\mu}\Gamma^\tau_{\alpha\beta} + \frac{\partial x'^\rho}{\partial x^\tau}\frac{\partial^2 x^\tau}{\partial x'^\sigma \partial x'^\mu}. \tag{9}$$

Unlike the curvature tensor $R^\rho_{\sigma\mu\nu}$, the affine connection $\Gamma^\rho_{\sigma\mu}$ can be made non-zero in an intrinsically flat space-time simply by using curvilinear coordinates, or eliminated away at any given point in an intrinsically curved space-time by choosing a suitable reference frame. This property makes the notion of gravitational energy density in the traditional pseudo-tensor constructions meaningless.

Analogous to the physical decomposition of the gauge field, we write $\Gamma^\rho_{\sigma\mu} \equiv \bar{\Gamma}^\rho_{\sigma\mu} + \hat{\Gamma}^\rho_{\sigma\mu}$, with the aim that in an intrinsically flat space time the physical term $\hat{\Gamma}^\rho_{\sigma\mu}$ always vanishes as does the curvature tensor $R^\rho_{\sigma\mu\nu}$, while the pure-geometric term $\bar{\Gamma}^\rho_{\sigma\mu}$ gives zero curvature tensor and solely carries the gauge freedom in coordinate transformation. Namely, $\hat{\Gamma}^\rho_{\sigma\mu}$ transforms covariantly and $\bar{\Gamma}^\rho_{\sigma\mu}$ transforms in the same non-covariant manner as does the full affine connection $\Gamma^\rho_{\sigma\mu}$: (see further discussion at the end of the paper)

$$\bar{\Gamma}'^\rho_{\sigma\mu} = \frac{\partial x'^\rho}{\partial x^\tau}\frac{\partial x^\alpha}{\partial x'^\sigma}\frac{\partial x^\beta}{\partial x'^\mu}\bar{\Gamma}^\tau_{\alpha\beta} + \frac{\partial x'^\rho}{\partial x^\tau}\frac{\partial^2 x^\tau}{\partial x'^\sigma \partial x'^\mu}. \tag{10}$$

This property enables us to use $\bar{\Gamma}^\rho_{\sigma\mu}$ instead of $\Gamma^\rho_{\sigma\mu}$ to construct pure-gauge covariant derivatives on a tensor V^ρ_μ:

$$\bar{D}_\alpha V^\rho_\mu = \partial_\alpha V^\rho_\mu + \bar{\Gamma}^\rho_{\alpha\beta} V^\beta_\mu - \bar{\Gamma}^\beta_{\alpha\mu} V^\rho_\beta. \tag{11}$$

As we remarked for the gauge-field separation, a well-defined separation $\Gamma^\rho_{\sigma\mu} = \Gamma^\rho_{\sigma\mu} + \hat{\Gamma}^\rho_{\sigma\mu}$ is a prescription to unambiguously express $\hat{\Gamma}^\rho_{\sigma\mu}$ and $\bar{\Gamma}^\rho_{\sigma\mu}$ in terms of the full $\Gamma^\rho_{\sigma\mu}$. We now demonstrate that the defining equations (1), together with the boundary condition that $\hat{\Gamma}^\rho_{\sigma\mu}$ approaches zero for a finite system, provide the desired separation. To this end we arrange Eqs. (1) into differential equations for $\hat{\Gamma}^\rho_{\sigma\mu}$:

$$\begin{aligned}
\partial_i \hat{\Gamma}^\rho_{\sigma j} - \partial_j \hat{\Gamma}^\rho_{\sigma i} &= R^\rho_{\sigma ij} + \hat{\Gamma}^\rho_{\alpha i}\hat{\Gamma}^\alpha_{\sigma j} - \Gamma^\rho_{\alpha i}\hat{\Gamma}^\alpha_{\sigma j} - \hat{\Gamma}^\rho_{\alpha i}\Gamma^\alpha_{\sigma j} \\
&\quad - \hat{\Gamma}^\rho_{\alpha j}\hat{\Gamma}^\alpha_{\sigma i} + \Gamma^\rho_{\alpha j}\hat{\Gamma}^\alpha_{\sigma i} + \hat{\Gamma}^\rho_{\alpha j}\Gamma^\alpha_{\sigma i} && \text{(12a)} \\
\partial_i \hat{\Gamma}^\rho_{\sigma i} &= \hat{\Gamma}^\rho_{i\alpha}\hat{\Gamma}^\alpha_{\sigma i} - \hat{\Gamma}^\alpha_{i\sigma}\hat{\Gamma}^\rho_{\alpha i} - \hat{\Gamma}^\alpha_{ii}\hat{\Gamma}^\rho_{\sigma\alpha} - \Gamma^\rho_{i\alpha}\hat{\Gamma}^\alpha_{\sigma i} + \Gamma^\alpha_{i\sigma}\hat{\Gamma}^\rho_{\alpha i} + \Gamma^\alpha_{ii}\hat{\Gamma}^\rho_{\sigma\alpha} && \text{(12b)} \\
\partial_i \hat{\Gamma}^\rho_{\sigma 0} &= \partial_0 \hat{\Gamma}^\rho_{\sigma i} + R^\rho_{\sigma i0} + \hat{\Gamma}^\rho_{\alpha i}\hat{\Gamma}^\alpha_{\sigma 0} - \Gamma^\rho_{\alpha i}\hat{\Gamma}^\alpha_{\sigma 0} - \hat{\Gamma}^\rho_{\alpha i}\Gamma^\alpha_{\sigma 0} \\
&\quad - \hat{\Gamma}^\rho_{\alpha 0}\hat{\Gamma}^\alpha_{\sigma i} + \Gamma^\rho_{\alpha 0}\hat{\Gamma}^\alpha_{\sigma i} + \hat{\Gamma}^\rho_{\alpha 0}\Gamma^\alpha_{\sigma i} && \text{(12c)}
\end{aligned}$$

Like Eqs. (7), there equations are non-linear, and we have to rely on perturbative expansion. But unlike Eqs. (7), there is no coupling constant here. So we have to assume smallness for all field amplitudes, to which we associate a small parameter χ. Namely, we write $\Gamma^{\rho}_{\sigma\mu} = \chi\tilde{\Gamma}^{\rho}_{\sigma\mu}$, $R^{\rho}_{\sigma\mu\nu} = \chi\tilde{R}^{\rho}_{\sigma\mu\nu}$, and expand the unknown $\hat{\Gamma}^{\rho}_{\sigma\mu}$ as: $\hat{\Gamma}^{\rho}_{\sigma\mu} = \chi\hat{\Gamma}^{(1)\rho}_{\sigma\mu} + \chi^2\hat{\Gamma}^{(2)\rho}_{\sigma\mu} + \cdots$. (There is no zeroth-order term since $\hat{\Gamma}^{\rho}_{\sigma\mu}$ is assumed to be small.) The equations for the leading-order term $\hat{\Gamma}^{(1)\rho}_{\sigma\mu}$ are:

$$\partial_i\hat{\Gamma}^{(1)\rho}_{\sigma j} - \partial_j\hat{\Gamma}^{(1)\rho}_{\sigma i} = \tilde{R}^{\rho}_{\sigma ij} \tag{13a}$$

$$\partial_i\hat{\Gamma}^{(1)\rho}_{\sigma i} = 0 \tag{13b}$$

$$\partial_i\hat{\Gamma}^{(1)\rho}_{\sigma 0} = \partial_0\hat{\Gamma}^{(1)\rho}_{\sigma i} + \tilde{R}^{\rho}_{\sigma i0} \tag{13c}$$

The solutions are similar to that in Eqs. (4):

$$\hat{\Gamma}^{(1)\rho}_{\sigma i} = \partial_j(\partial_k\partial_k)^{-1}\tilde{R}^{\rho}_{\sigma ji} \tag{14a}$$

$$\hat{\Gamma}^{(1)\rho}_{\sigma 0} = \int_{\infty}^{x} dx^i(\partial_0\hat{\Gamma}^{(1)\rho}_{\sigma i} + \tilde{R}^{\rho}_{\sigma i0}) \tag{14b}$$

As in (4b), the index i in (14b) takes any value of 1 to 3, and is not summed; and $\hat{\Gamma}^{(1)\rho}_{\sigma i}$ is to be substituted by the expression in (14a). The solution for $\hat{\Gamma}^{(1)\rho}_{\sigma\mu}$ is then used to solve $\hat{\Gamma}^{(2)\rho}_{\sigma\mu}$, and so on.

Eqs. (12) indicate that if $R^{\rho}_{\sigma\mu\nu} = 0$ (namely, the space-time is intrinsically flat), $\hat{\Gamma}^{\rho}_{\sigma\mu} = 0$ is always a solution. And given the validity of the above perturvative expansion, it is also the unique solution.

To examine the arbitrariness in the energy density of gravitational field associated with coordinate choices, it is elucidating to arrange Eqs. (1) to the form for the pure-geometric term $\bar{\Gamma}^{\rho}_{\sigma\mu}$:

$$\partial_i\bar{\Gamma}^{\rho}_{\sigma j} - \partial_j\bar{\Gamma}^{\rho}_{\sigma i} = \bar{\Gamma}^{\rho}_{\alpha j}\bar{\Gamma}^{\alpha}_{\sigma i} - \bar{\Gamma}^{\rho}_{\alpha i}\bar{\Gamma}^{\alpha}_{\sigma j} \tag{15a}$$

$$\begin{aligned}\partial_i\bar{\Gamma}^{\rho}_{\sigma i} &= \partial_i\Gamma^{\rho}_{\sigma i} + \bar{\Gamma}^{\rho}_{i\alpha}(\Gamma^{\alpha}_{\sigma i} - \bar{\Gamma}^{\alpha}_{\sigma i}) - \bar{\Gamma}^{\alpha}_{i\sigma}(\Gamma^{\rho}_{\alpha i} - \bar{\Gamma}^{\rho}_{\alpha i}) \\ &\quad - \bar{\Gamma}^{\alpha}_{ii}(\Gamma^{\rho}_{\sigma\alpha} - \hat{\Gamma}^{\rho}_{\sigma\alpha})\end{aligned} \tag{15b}$$

$$\partial_i\bar{\Gamma}^{\rho}_{\sigma 0} = \partial_0\bar{\Gamma}^{\rho}_{\sigma i} + \bar{\Gamma}^{\rho}_{\alpha 0}\bar{\Gamma}^{\alpha}_{\sigma i} - \bar{\Gamma}^{\rho}_{\alpha i}\bar{\Gamma}^{\alpha}_{\sigma 0} \tag{15c}$$

Eqs. (15a) and (15b) indicate that in order to have the pure-geometric term $\bar{\Gamma}^{\rho}_{\sigma\mu} = 0$ (so that the spurious gravitational effect is absent), it is a necessary condition that $\partial_i\Gamma^{\rho}_{\sigma i} = 0$. On the other hand, if all field amplitudes are small and the perturvative expansion as we did above for $\hat{\Gamma}^{\rho}_{\sigma\mu}$ is valid, $\partial_i\Gamma^{\rho}_{\sigma i} = 0$ will lead to the unique solution of $\bar{\Gamma}^{\rho}_{\sigma\mu} = 0$. We therefore name a coordinate in which $\partial_i\Gamma^{\rho}_{\sigma i} = 0$ the "pertinent coordinate". (Similarly, in gauge theories, the gauge $\partial_i A_i = 0$ gives the solution for the pure-gauge field $\bar{A}_\mu = 0$, and can be termed the "pertinent gauge".[5]) It is straightforward to verify that the spherical coordinate in a flat space-time does not satisfy $\partial_i \Gamma^{\rho}_{\sigma i} = 0$, thus is not "pertinent". This explains why the traditional formulae for gravitational energy give unreasonable result in the spherical coordinate.

It is very interesting and important to note that $\partial_i \Gamma^{\rho}_{\sigma i} = 0$ is a fairly non-trivial condition. E.g., while the Cartesian coordinate in flat space-time gives $\Gamma^{\rho}_{\sigma i} = 0$ and is clearly pertinent, the quasi-Cartesian coordinate in a curved space-time is not necessary pertinent. This can be verified by the simplest Schwarzschild solution:

$$ds^2 = \left(\frac{1 - MG/2r}{1 + MG/2r}\right)^2 dt^2 - \left(1 + \frac{MG}{2r}\right)^4 d\vec{r}^2. \tag{16}$$

Furthermore, it is not trivial to convert this coordinate to be a pertinent one.

Once the pure-geometric connection $\bar{\Gamma}^{\rho}_{\mu\nu}$ has been obtained, we can use it to define a flat background metric $\bar{g}_{\mu\nu}$:

$$\bar{\Gamma}^{\rho}_{\mu\nu} = \frac{1}{2}\bar{g}^{\rho\sigma}(\partial_\mu \bar{g}_{\sigma\nu} + \partial_\nu \bar{g}_{\sigma\mu} - \partial_\sigma \bar{g}_{\mu\nu}, \tag{17}$$

where $\bar{g}^{\mu\nu}$ is the inverse of $\bar{\Gamma}^{\rho}_{\mu\nu}$: $\bar{g}^{\mu\nu}\bar{g}_{\nu\upsilon} = \delta^{\mu}_{\upsilon}$); and $\bar{\Gamma}^{\rho}_{\mu\nu}$ is set to approach the Mincowski metric at infinity.

We are now in the position to explain how to calculate a physically meaningful energy density of the gravitational field, for a finite and weak gravitating system. First, we choose a coordinate system in which the affine connection $\Gamma^{\rho}_{\sigma\mu}$ is small and vanishing at infinity. Eq. (16) is such an example. This allows us to choose vanishing boundary values, and apply perturbative expansions for both $\hat{\Gamma}^{\rho}_{\upsilon\mu}$ and $\bar{\Gamma}^{\rho}_{\sigma\mu}$. The pure geometric term $\bar{\Gamma}^{\rho}_{\sigma\mu}$ is especially easy to solve:

$$\bar{\Gamma}^{\rho}_{\sigma i} = \partial_i (\partial_k \partial_k)^{-1} \partial_j \Gamma^{\rho}_{\sigma j} + \text{higher order terms} \tag{18a}$$

$$\bar{\Gamma}^{\rho}_{00} = \int_{\infty}^{x} dx^i \partial_0 \bar{\Gamma}^{\rho}_{0i} + \text{higher order terms} \tag{18b}$$

Again, the index i in (18b) takes any value from 1 to 3, and is not summed. If we find $\bar{\Gamma}^{\rho}_{\sigma\mu} = 0$, it means that the spurious gravitational effect in this coordinate is absent, thus to compute the energy density of the gravitational field we can use directly the traditional formulae, e.g., that of Landau and Lifshitz:[6]

$$t^{\mu\nu} = \frac{1}{16\pi G}\{(2\Gamma^{\alpha}_{\rho\sigma}\Gamma^{\beta}_{\alpha\beta} - \Gamma^{\alpha}_{\rho\beta}\Gamma^{\beta}_{\sigma\alpha} - \Gamma^{\alpha}_{\rho\alpha}\Gamma^{\beta}_{\sigma\beta}) \times (g^{\mu\rho}g^{\nu\sigma} - g^{\mu\nu}g^{\rho\sigma}) + \cdots\}, \tag{19}$$

where the omitted terms are also products of the metric and the affine connection.

If a coordinate gives non-zero $\bar{\Gamma}^{\rho}_{\sigma\mu}$, it means that it contains spurious gravitational effect, and one should not use pseudo-tensor formulae like Eq. (19) directly. However, these pseudo-tensor formulae can easily be ungraded to be physical ones. Concerning Eq. (19), the procedure is simply replacing the full connection $\Gamma^{\rho}_{\sigma\mu}$ with the physical connection $\hat{\Gamma}^{\rho}_{\sigma\mu} = \Gamma^{\rho}_{\sigma\mu} - \bar{\Gamma}^{\rho}_{\sigma\mu}$.

4. Discussion

(i) The traditional energy-momentum pseudo-tensors of various types show a high degeneracy concerning the total energy of a gravitating body. It would be very interesting to examine whether such degeneracy persists to the level of a meaningful density.

(ii) In gauge theories, gauge transformation and Lorentz transformation are two different manipulations. Therefore, in Eq. (2b)/(5b), $\hat{A}_i$ is gauge invariant/covariant so as to make the equation gauge invariant/covariant. However, to make the equation hold in any Lorentz frame, the physical field $\hat{A}_i$ must not transform as a four-vector. This is an inevitable physical feature of a massless particle with spin-1 or higher.[7] In general relativity, however, gauge transformation and coordinate transformation mean the same thing. Therefore, to make Eq. (1b) hold in any coordinate, the physical term $\hat{\Gamma}^{\rho}_{\sigma\mu}$ is a true tensor only under spatial transformations, and must not transform covariantly under four-dimensional transformations, even linear (Lorentz) ones. This is not a signal of partial failure in our construction, but rather a successful representation of the masslessness of the gravitational field.

(iii) Similar to the gauge-invariant gluon spin,[1,2] the graviton spin can very hopefully be given a gauge-independent description by our approach.

Acknowledgement

This work is supported by China NSF under Grant No. 10875082, and by China Program for NCET.

References

1. X.S. Chen, X.F. Lü, W.M. Sun, F. Wang, and T. Goldman, Phys. Rev. Lett. **100**, 232002 (2008).
2. X.S. Chen, W.M. Sun, X.F. Lü, F. Wang, and T. Goldman, Phys. Rev. Lett. **103**, 062001 (2009).

3. M.T. Wang and S.T. Yau, Phys. Rev. Lett. **102**, 021101 (2009).
4. L.B. Szabados, Living Rev. Relativ. **12**, 4 (2009).
5. This may easily lead to a confusion that $\hat{A}_\mu$ is just the field $A_\mu^{\rm Coul}$ defined in Coulomb gauge $\partial_i A_i^{\rm Coul} = 0$. It should be clarified that $A_\mu^{\rm Coul}$ so defined is a *fixed* (in this sense gauge-*invariant*) quantity, while in non-Abelian theories $\hat{A}_\mu$ is gauge-*covariant*. Only in Coulomb gauge we have $\hat{A}_\mu = A_\mu^{\rm Coul}$. In other gauges $\hat{A}_\mu \neq A_\mu^{\rm Coul}$.
6. L.D. Laudau and E.M. Lifshitz, *The Classical Theory of Fields* (Butterworth-Heinermann, Oxford, 1975) 4th ed..
7. S. Weinberg, *The Quantum Theory of Fields* (Cambridge, New York, 1995), section 5.9.

PHYSICAL DECOMPOSITION OF GAUGE FIELDS IN QED AND IN YANG-MILLS GRAVITY WITH TRANSLATION GAUGE SYMMETRY

DANIEL C. KATZ[a], XIANG-SONG CHEN[b,c,d] and JONG-PING HSU[a]

[a] *Department of Physics, University of Massachusetts Dartmouth North Dartmouth, MA 02747-2300, USA*

[b] *Department of Physics, Huazhong University of Science and Technology, Wuhan 430074, China*

[c] *Department of Physics, Schuan University, Chengdu 610064, China*

[d] *Kavli Institute for Theoretical Physics China, Chinese Academy of Science, Beijing 100190, China*

E-mail: jhsu@umassd.edu

We discuss a covariant decomposition of gauge fields into gauge-invariant physical components and gauge-dependent unphysical components. In particular, we consider QED with nonlinear gauge conditions and Yang-Mills gravity with 4-dimensional translational gauge symmetry (T(4) group) in flat space-time. The consistency and intimate relations among decomposition of gauge fields for all times, and gauge conditions and Faddeev-Popov ghost fields are investigated. The gauge invariance of QED and Yang-Mills gravity are revealed by the invariance of physical components under gauge transformations.

1. Introduction

A generalized Yang-Mills theory for gravity with external space-time translation group T(4) in flat space-time involves a massless tensor field $\phi^{\mu\nu}$ and the associated gauge curvature $C^{\mu\nu\alpha}$.[1] It differs from the usual gauge theory with internal groups U(1) or SU(n) which leads to a vector potential field and an anti-symmetric gauge curvature (i.e., the field strength). The main reason is that the generators of the external translation gauge group T(4) are differential operators, which do not have constant matrix representations like internal gauge groups. Consequently, the coupling constant of the tensor field $\phi^{\mu\nu}$ must have the dimension of length (in natural units) and the associated force cannot be both attractive and repulsive, in contrast to QED and Yang-Mills theories with internal gauge groups.

We compare the covariant decomposition of the electromagnetic gauge field A^μ and the massless tensor field $\phi^{\mu\nu}$ into two observable transverse components and other unphysical components. The decomposition into these physical and unphysical components of a gauge field is important for the understanding of hadron spin in terms of the spin and angular motion of quarks and gluons.[2] The method of physical decomposition is based on explicit solutions of differential equations. We discuss the gauge transformations and contributions to gauge curvatures $F^{\mu\nu}$ and $C^{\mu\nu\alpha}$ by the physical components of massless gauge fields in electromagnetic theory and Yang-Mills gravity.

2. Covariant decomposition of the electromagnetic field

Let us first consider covariant decomposition of the familiar electromagnetic gauge field A^μ for massless photons in any inertial frame (with the metric tensor $\eta_{\mu\nu} = (+,-,-,-)$). The massless physical photon is a spin-1 vector particle that consists of two physical transverse components (denoted by $\widehat{A}^\mu$) in any inertial frame and shows up in the two observable polarizations of electromagnetic waves. However, for an explicit covariant formulation of electrodynamics, we have to introduce a four-component vector field A^μ to satisfy the requirement of Poincaré-Lorentz invariance (or 4-dimensional symmetry). The extra two components correspond to the longitudinal and time-like photons, which are not observable and, hence, unphysical. They are denoted by $\overline{A}^\mu = A^\mu - \widehat{A}^\mu$.

Since the U(1) gauge curvature (i.e., the electromagnetic field strength) $F^{\mu\nu}$ is observable, it must contain only the physical components $\widehat{A}^\mu$,[2]

$$F^{\mu\nu} = \partial^\mu A^\nu - \partial^\nu A^\mu = \partial^\mu \widehat{A}^\nu - \partial^\nu \widehat{A}^\mu \equiv \widehat{F}^{\mu\nu}, \qquad A^\mu = \widehat{A}^\mu + \overline{A}^\mu. \tag{1}$$

It follows from (1) that

$$\overline{F}^{\mu\nu} = \partial^\mu \overline{A}^\nu - \partial^\nu \overline{A}^\mu = 0, \tag{2}$$

which assures that the unphysical components do not contribute to the U(1) gauge curvature $F^{\mu\nu}$.

Following Ogievetski and Polubarinov,[3] a simple covariant solution of the unphysical components $\overline{A}^\mu$ satisfying (2) can be expressed in terms of the usual 4-vector potential A^μ

$$\overline{A}^\mu(x) = \partial^\mu \partial_\alpha \partial^{-2} A^\alpha(x) \tag{3}$$

$$= \partial^\mu \partial_\alpha \int \frac{-1}{k_\lambda k^\lambda} exp[ik_\sigma(x^\sigma - x'^\sigma)] A^\alpha(x') \frac{d^4k}{(2\pi)^4} d^4x'.$$

From equation (3), we obtain a constraint for $\widehat{A}^\mu$,

$$\partial_\mu \widehat{A}^\mu(x) = \partial_\mu A^\mu(x) - \partial_\mu(\partial^\mu \partial_\alpha \partial^{-2} A^\alpha(x)) = 0. \tag{4}$$

This indicates that the four components of $\widehat{A}^\mu$ are not all independent, and that it corresponds to a spin-1 vector field.[3,4]

The U(1) gauge transformation of the electromagnetic field A^μ is given by

$$A'^\mu(x) = A^\mu(x) + \partial^\mu \Lambda(x), \tag{5}$$

where $\Lambda(x)$ is an arbitrary infinitesimal function. From (3) and (5) we obtain the following gauge transformations for the physical components $\widehat{A}^\mu(x)$ and unphysical components $\overline{A}^\mu(x)$:

$$\widehat{A'}^\mu(x) = A^\mu(x) + \partial^\mu \Lambda(x) - \partial^\mu \partial_\alpha \partial^{-2}(A^\alpha(x) + \partial^\alpha \Lambda(x)) = \widehat{A}^\mu(x),$$

$$\overline{A'}^\mu(x) = \overline{A}^\mu(x) - \partial^\mu \Lambda(x). \tag{6}$$

The gauge invariance of $\widehat{A}^\mu$ can also be seen from the following relation,

$$\widehat{A}^\mu = \partial^{-2}(\partial^2 A^\mu - \partial^\mu \partial_\alpha A^\alpha) = \partial^{-2}\partial_\alpha F^{\alpha\mu}, \tag{7}$$

where $F^{\alpha\mu}$ is gauge invariant and we have used $\partial^2 \partial^{-2} A^\mu = A^\mu$. The result (6) reveals the essence of gauge invariance of electrodynamics and quantum electrodynamics: Namely, the physical and covariant components of A^μ, i.e., $\widehat{A}^\mu$, remain the same under the gauge transformations in any inertial frame. Therefore, the theory is gauge invariant in all inertial frames because only $\widehat{A}^\mu$ contributes to observable results in electrodynamics.

3. Preservation of physical decompositions and gauge conditions

The above discussion of the physical decomposition of electromagnetic gauge fields in (1) and (3) are satisfactory for QED as long as one imposes a linear gauge condition for quantization. In this case, once the linear gauge condition is imposed at a certain time, it holds for all times. However, if one imposes a nonlinear gauge condition for QED at a certain time, the non-linear gauge condition cannot hold for all times because there will be interaction between the longitudinal and time-like photons and the 2 transverse photons (introduced through this non-linear gauge condition).[4] Such an additional interaction of unphysical photons can upset the gauge invariance (and unitarity) of the theory.

Such a new photon self-interaction due to the nonlinear gauge condition suggests that the stability and preservation of the physical decomposition (1) and (3) could be interlocked to the property whether a gauge condition can be imposed for all times. To see the details, let us consider the Lagrangian of QED with a non-linear gauge condition,

$$\partial_\mu A^\mu - \beta' A_\mu A^\mu = a(x), \tag{8}$$

where $a(x)$ is a suitable function independent of A_μ and $\Lambda(x)$. Such a Lagrangian is given by[4]

$$L = -\frac{1}{4} F_{\mu\nu} F^{\mu\nu} - \frac{1}{2\alpha}(\partial_\mu A^\mu - \beta' A_\mu A^\mu)^2; \tag{9}$$

where α and β' are two arbitrary parameters. In particular, the choices (i) $\alpha \to 0$ and $\beta' \neq 0$ and (ii) $\alpha = 1$ and $\beta' \neq 0$ correspond to a "nonlinear Landau gauge" and a "nonlinear Feynman gauge" respectively. The field equation derived from (9) is

$$\partial_\mu(\partial^\mu A^\lambda - \partial^\lambda A^\mu) - \frac{1}{\alpha}\partial^\lambda(\partial_\mu A^\mu - \beta' A_\mu A^\mu) - \frac{2\beta'}{\alpha} A^\lambda(\partial_\mu A^\mu - \beta' A_\mu A^\mu) = 0. \tag{10}$$

First, let consider the linear gauge condition, i.e., $\beta' = 0$. The field equation (10) implies that the gauge condition, $\partial_\mu A^\mu = a_1(x)$, satisfies a free field equation,

$$\partial_\lambda \partial^\lambda (\partial_\mu A^\mu) = 0. \tag{11}$$

Similarly, field equation (10) implies that the physical photon $\widehat{A}^\mu$ also satisfies a free equation (i.e., without the sources or interaction) terms

$$\partial^2 \widehat{A}^\mu + \frac{1}{\alpha}\partial^\mu \partial_\nu A^\nu = 0. \tag{12}$$

Thus, based on (11) and (12), one concludes that both the linear gauge condition and the physical decomposition (1) can be preserve for all times.

Second, when $\beta' \neq 0$, the field equation (10) implies that the gauge condition satisfies the following equations,

$$\partial^2 \chi(x) = -2\beta' \partial_\mu (A^\mu \chi(x)), \qquad \chi(x) \equiv (\partial_\mu A^\mu - \beta' A_\mu A^\mu). \tag{13}$$

The physical photon $\widehat{A}^\mu$ now satisfies a complicated equation involving the unphysical photons $\overline{A}_\mu$

$$\partial^2 \widehat{A}^\mu + \frac{1}{\alpha}\partial^\mu \partial_\nu \widehat{A}^\nu = -\frac{1}{\alpha}\partial^\mu[\partial_\nu \overline{A}^\nu - \beta' \widehat{A}_\nu \widehat{A}^\nu - 2\beta' \widehat{A}_\nu \overline{A}^\nu - \beta' \overline{A}_\nu \overline{A}^\nu]$$

$$-\frac{2\beta'}{\alpha}[\widehat{A}^\mu + \overline{A}^\mu][\partial_\nu \widehat{A}^\nu + \partial_\nu \overline{A}^\nu - \beta' \widehat{A}_\nu \widehat{A}^\nu - 2\beta' \widehat{A}_\nu \overline{A}^\nu - \beta' \overline{A}_\nu \overline{A}^\nu]. \tag{14}$$

The source terms in (13) and (14) indicate that both (i) the physical component $\widehat{A}$ in the decomposition (1) with the property (6) and (ii) the nonlinear gauge condition (8) cannot be preserved for all times, in contrast to that of linear gauge.

Nevertheless, according to Faddeev and Popov's formulation of gauge theory,[5,6] one can still insist and enforce that the nonlinear gauge condition holds for all time by, say, including a delta function involving the gauge condition in the path integral. To wit, the vacuum-to-vacuum amplitude of QED can be expressed by the generating functional $W(a(x))$ of Green's function in the nonlinear gauge (8),[4]

$$W(a) = \int d[A_\lambda] exp\left[i \int \frac{-1}{4} F_{\mu\nu}F^{\mu\nu} d^4x\right] \delta(\partial_\mu A^\mu - \beta' A_\mu A^\mu - a(x))\Delta_\chi, \tag{15}$$

$$\Delta_\chi = det[(\partial_\mu - 2\beta' A_\mu)\partial^\mu], \tag{16}$$

where the external source terms $iA_\mu J^\mu$ in (15) are neglected. The functional determinant in (16) is equivalent to a Lagrangian involving Faddeev-Popov ghost fields that obey Fermi statistics. The ghost fields do not have external sources, by definition of the physical subspace of the S-matrix. The vacuum-to-vacuum amplitude (15) (in the presence of an external source) determines completely the physics of QED. This generating functional $W(a)$ implies that the S-matrix of QED with a nonlinear gauge condition is gauge invariant and unitary.[4] In this sense, the QED with nonlinear gauge condition can also effectively preserve the gauge condition because the gauge non-invariant amplitudes generated by the unphysical photons are all cancelled by the extra amplitudes produced by the ghost particles.[4] In other words, the unphysical (longitudinal and time-like) photons are 'effectively free particles' in QED based on the vacuum-to-vacuum amplitude (15) because they do not produce amplitudes to upset the gauge invariance of QED.

With the help of the physical decomposition (1), the gauge-invariant total angular momentum density, $\propto \mathbf{r} \times (\mathbf{E} \times \mathbf{B})$, in vacuum can be split into two gauge invariant parts:

$$\int d^3x\, [\mathbf{r} \times (\mathbf{E} \times \mathbf{B})] = \int d^3x \left[\mathbf{E} \times \widehat{\mathbf{A}} + \sum_{k=1}^{3} E_k (\mathbf{r} \times \nabla)\widehat{A}_k\right]. \tag{17}$$

The first term involving $\mathbf{E} \times \widehat{\mathbf{A}}$ is sometimes identified with the spin angular momentum, and the second term involving $\mathbf{r} \times \nabla$ is identified with the orbital angular momentum of the electromagnetic field.[7]

4. Physical decomposition of tensor gauge fields in Yang-Mills gravity

Let us consider the covariant decomposition of the tensor fields $\phi^{\mu\nu}$ in Yang-Mills gravity within the framework of flat space-time. The Lagrangian of the T(4) tensor field is given by

$$L_\phi = \frac{1}{4g^2}\left(C_{\mu\nu\alpha}C^{\mu\nu\alpha} - 2C_{\mu\alpha}{}^{\alpha}C^{\mu\beta}{}_{\beta}\right), \tag{18}$$

$$C^{\mu\nu\alpha} = J^{\mu\sigma}\partial_\sigma J^{\nu\alpha} - J^{\nu\sigma}\partial_\sigma J^{\mu\alpha} = -C^{\nu\mu\alpha}, \tag{19}$$

$$J_{\mu\nu} = \eta_{\mu\nu} + g\phi_{\mu\nu} = J_{\nu\mu}, \tag{20}$$

where $C^{\mu\alpha\beta}$ is the T(4) gauge curvature in inertial frames. The Lagrangian L_ϕ changes only by a divergence under the translation gauge transformation, and the action functional $S_\phi = \int L_\phi d^4x$ is invariant under the space-time translation gauge transformation. This invariance property can be shown in a general non-inertial frames[8,9,10] with arbitrary coordinates in flat space-time.

Suppose the physical and unphysical components of $\phi^{\mu\nu}$ are denoted by $\widehat{\phi}^{\mu\nu}$ and $\overline{\phi}^{\mu\nu}$ respectively. Then we have the relation,

$$\phi^{\mu\nu} = \widehat{\phi}^{\mu\nu} + \overline{\phi}^{\mu\nu}. \tag{21}$$

We can express the T(4) gauge curvature in terms of the tensor field $\phi^{\mu\nu}$ as follows

$$C^{\mu\nu\alpha} = g[\partial^\mu\phi^{\nu\alpha} - \partial^\nu\phi^{\mu\alpha} + g\epsilon^{\mu\nu}_{\lambda\rho}\phi^{\lambda\sigma}\partial_\sigma\phi^{\rho\alpha}], \tag{22}$$

$$\epsilon^{\mu\nu}_{\lambda\rho} = \delta^{\mu}_{\lambda}\delta^{\nu}_{\rho} - \delta^{\nu}_{\lambda}\delta^{\mu}_{\rho}.$$

Similar to eq. (1), it is natural to assume that the T(4) gauge curvature contain only the physical components $\widehat{\phi}^{\mu\nu}$:

$$C^{\mu\nu\alpha} = g[\partial^\mu\widehat{\phi}^{\nu\alpha} - \partial^\nu\widehat{\phi}^{\mu\alpha} + g\epsilon^{\mu\nu}_{\lambda\rho}\widehat{\phi}^{\lambda\sigma}\partial_\sigma\widehat{\phi}^{\rho\alpha}] = \widehat{C}^{\mu\nu\alpha}. \tag{23}$$

From equations (21) and (22), we obtain the relation between $C^{\mu\nu\alpha}$ and $\widehat{C}^{\mu\nu\alpha}$

$$C^{\mu\nu\alpha} = \widehat{C}^{\mu\nu\alpha} + \overline{C}^{\mu\nu\alpha}, \tag{24}$$

$$\overline{C}^{\mu\nu\alpha} = g[\partial^\mu\overline{\phi}^{\nu\alpha} - \partial_\nu\overline{\phi}^{\mu\alpha} + g\epsilon^{\mu\nu}_{\lambda\rho}(\overline{\phi}^{\lambda\sigma}\partial_\sigma\overline{\phi}^{\rho\alpha} + \widehat{\phi}^{\lambda\sigma}\partial_\sigma\overline{\phi}^{\rho\alpha} + \overline{\phi}^{\lambda\sigma}\partial_\sigma\widehat{\phi}^{\rho\alpha})]. \tag{25}$$

It follows from (23) and (24) that

$$\overline{C}^{\mu\nu\alpha} = 0. \tag{26}$$

This equation corresponds to (2) in electromagnetic gauge field. Its physical meaning is clear, namely, the T(4) gauge curvature $C^{\mu\nu\alpha}$ in the Lagrangian (18) must not contain any term which involves the unphysical tensor component. But the solution for the unphysical components $\overline{\phi}^{\nu\alpha}$ in (25) and (26) are complicated in general because of the self-coupling of the tensor field, as indicated in (18) and (19). The situation resembles to those of QED with a nonlinear gauge condition and Yang-Mills theory.[11,12]

Let us consider the gauge transformation of $\overline{\phi}^{\mu\nu}$ and $\widehat{\phi}^{\mu\nu}$. Similar to the result (5) for gauge invariant electrodynamics, the T(4) gauge invariant Yang-Mills gravity must have the property that the physical components $\widehat{\phi}^{\mu\nu}$ are unchanged under the T(4) gauge transformation,

$$\widehat{\phi'}^{\mu\nu} = \widehat{\phi}^{\mu\nu}. \tag{27}$$

The infinitesimal gauge transformation of the tensor field $\phi^{\mu\nu}$ is[1]

$$\phi'^{\mu\nu} = \phi^{\mu\nu} - \Lambda^\lambda \partial_\lambda \phi^{\mu\nu} + \phi^{\lambda\nu}\partial_\lambda \Lambda^\mu + \phi^{\mu\lambda}\partial_\lambda \Lambda^\nu, \tag{28}$$

where $\Lambda(x)$ is an arbitrary infinitesimal gauge function. It follows from (21) and (27) that the gauge transformation for the unphysical components $\overline{\phi'}^{\mu\nu}$ must be

$$\overline{\phi'}^{\mu\nu} = \overline{\phi}^{\mu\nu} - \Lambda^\lambda \partial_\lambda \phi^{\mu\nu} + \phi^{\lambda\nu}\partial_\lambda \Lambda^\mu + \phi^{\mu\lambda}\partial_\lambda \Lambda^\nu. \tag{29}$$

In analogy with the photon self-coupling in QED with nonlinear gauge condition discussed previously, the self-coupling of the tensor field indicates that the physical component $\widehat{\phi}^{\mu\nu}$ satisfying (27) cannot hold for all times. In order to see this property in details, let us consider Yang-Mills gravity with a specific gauge function, which is necessary for quantization of the tensor field. Suppose we choose the following gauge condition,

$$\partial_\mu \phi^{\mu\nu} - \frac{1}{2}\partial^\nu \phi = a^\nu(x), \qquad \phi = \phi^\lambda_\lambda, \tag{30}$$

where $a^\mu(x)$ is a suitable function independent of $\phi^{\mu\nu}$ and the gauge function Λ^μ. In this case the Lagrangian of Yang-Mills gravity is

$$L = L_\phi + \frac{\xi}{2g^2} Y^\nu Y_\nu, \qquad Y^\nu \equiv \partial_\mu \phi^{\mu\nu} - \frac{1}{2}\partial^\nu \phi. \tag{31}$$

One can derive the field equation[13] and show that $Y^\nu(x)$ does not satisfy a free field equation, $\partial^2 Y^\nu \neq 0$. This implies that if one imposes the gauge condition (30) initially, it cannot hold for all times.[11]

Nevertheless, we can apply the Faddeev-Popov method to enforce the gauge condition (30). The generate functional $W_\xi(Y^\nu)$ of Green's function in the gauge condition (30) can be written as[11,13]

$$W_\xi(Y^\nu) = \int d[\phi_{\alpha\beta}]exp\left[i\int d^4x(L_\phi + \phi^{\mu\nu}j_{\mu\nu})\right]$$

$$\times \prod_\alpha \delta(\partial_\lambda J'^{\lambda\alpha} - \frac{1}{2}\partial^\alpha J' - Y^\alpha)\ detQ, \tag{32}$$

$$detQ = \int d[\chi^\alpha, \chi'^\beta]exp\left[i\int d^4x L_{\chi'\chi}\right], \tag{33}$$

$$L_{\chi'\chi} = \chi'_\alpha[\partial_\lambda\partial^\lambda\chi^\alpha + \partial^\alpha\partial_\mu\chi^\mu - g\partial_\mu((\partial_\lambda\phi^{\mu\alpha})\chi^\lambda)$$

$$+g\partial_\mu(\phi^{\lambda\alpha}\partial_\lambda\chi^\mu) + g\partial_\mu(\phi^{\mu\lambda}\partial_\lambda\chi^\alpha) + g\frac{1}{2}\partial^\alpha((\partial_\lambda\phi)\chi^\lambda)]. \tag{34}$$

The matrix Q in (14) can be obtained by expressing the ghost Lagrangian (34) in the form, $L_{\chi'\chi} = \chi'^\mu Q_{\mu\nu}\chi^\nu$.[13] In equation (32), $j_{\mu\nu}$ are external source and arbitrary functions, while $J'^{\mu\nu}$ are the T(4) gauge transformation of $J^{\mu\nu}$. The δ-function in (32) enforces the gauge condition (28) for all times. The vacuum-to -vacuum amplitude (30) completely determines the dynamics of Yang-Mills gravity, which is unitary and gauge invariant.[7] Thus, even though there are complicated interactions between physical $\widehat{\phi}^{\mu\nu}$ and unphysical $\overline{\phi}^{\mu\nu}$, the unwanted amplitudes produced by such interactions are completely cancelled by the extra amplitudes generated by the interaction of the Faddeev-Popov ghost fields χ and χ' in (33). In this sense, it is gratifying that although we cannot solve for the unphysical components $\overline{\phi}^{\mu\nu}$ and explicitly extract them from (21) and (24), Yang Mills gravity effectively contains only two physical transverse components of gravitons, so that the constraint (26) is guaranteed by the vacuum-to-vacuum amplitude (30).

5. Discussion

There is a hidden subtlety in the covariant decomposition of $A^\mu(x)$: The physical component is not completely gauge invariant. This can be seen in Eq.(4) that $\partial_\mu\widehat{A}'^\mu(x) = \partial_\mu(\widehat{A}^\mu + \partial^\mu\Lambda(x)) = 0$ can be maintained with a non-zero $\Lambda(x)$ satisfying $\partial^2\Lambda(x) = 0$.[14] Thus, the covariant decomposition works fine only if this residual gauge freedom does not make serious trouble. Otherwise, we may choose instead a non-covariant decomposition, by replacing Eq.(4) with the three-dimensional constraint $\nabla \cdot \widehat{\mathbf{A}} = 0$. In this scheme, $\widehat{A}^\mu$ contains no more gauge freedom because $\nabla^2\Lambda(x)=0$ has only a trivial solution for a vanishing boundary value.

The non-covariant scheme can also lead to an explicit solution for the decomposition of the tensor field $\phi^{\mu\nu}$. To this end, we impose a constraint on the physical component $\widehat{\phi}^{\mu\nu}$ that $\partial^i\widehat{\phi}^{i\nu}=0$, which resembles $\nabla\cdot\widehat{\mathbf{A}}=0$. Together with Eq.(23), this can lead to a solution of $\widehat{\phi}^{\mu\nu}$ by power expansion in g: $\widehat{\phi}^{\mu\nu}\equiv\widehat{\phi}^{\mu\nu}_{(0)}+g\widehat{\phi}^{\mu\nu}_{(1)}+g^2\widehat{\phi}^{\mu\nu}_{(2)}+\cdots$. The equations for the leading term $\widehat{\phi}^{\mu\nu}_{(0)}$ can be written in the form:

$$\partial^i\widehat{\phi}^{j\alpha}_{(0)}-\partial^j\widehat{\phi}^{i\alpha}_{(0)}=C^{ij\alpha}/g,\quad i,j=1,2,3; \tag{35}$$

$$\partial^i\widehat{\phi}^{i\alpha}_{(0)}=0, \tag{36}$$

$$\partial^i\widehat{\phi}^{0\alpha}_{(0)}=\partial^0\widehat{\phi}^{i\alpha}_{(0)}+C^{i0\alpha}/g. \tag{37}$$

Equations (35) and (36) lead to a familiar solution for $\widehat{\phi}^{i\alpha}_{(0)}$, which can then be used in (36) to solve for $\widehat{\phi}^{0\alpha}_{(0)}$. The solution for $\widehat{\phi}^{\mu\nu}_{(0)}$ can further be employed to solve the equations for the next-order term $\widehat{\phi}^{\mu\nu}_{(1)}$:

$$\partial^i\widehat{\phi}^{j\alpha}_{(1)}-\partial^j\widehat{\phi}^{i\alpha}_{(1)}=\epsilon^{ij}_{\lambda\rho}\widehat{\phi}^{\lambda\sigma}_{(0)}\partial_\sigma\widehat{\phi}^{\rho\alpha}_{(0)}, \tag{38}$$

$$\partial^i\widehat{\phi}^{i\alpha}_{(1)}=0, \tag{39}$$

$$\partial^i\widehat{\phi}^{0\alpha}_{(1)}=\partial^0\widehat{\phi}^{i\alpha}_{(1)}+\epsilon^{i0}_{\lambda\rho}\widehat{\phi}^{\lambda\sigma}_{(0)}\partial_\sigma\widehat{\phi}^{\rho\alpha}_{(0)}, \tag{40}$$

and so on.

We have discussed decompositions of gauge fields into gauge-invariant physical components and gauge-dependent unphysical components. We demonstrated that QED with nonlinear gauge conditions and Yang-Mills gravity with 4-dimensional translational gauge symmetry are more complicated in the decomposition due to self-coupling of the fields. We argued that the consistency of decomposition of such gauge fields for all times can be effectively achieved by choosing a gauge condition and using the corresponding Faddeev-Popov ghost field.

Acknowledgements

The work was supported in part by the Jing Shin Research Fund and Prof. Leung Memorial Fund of the UMass Dartmouth Foundation, and in part by the National Science Foundation of China under Grant No. 10875082.

References

1. J. P. Hsu, Int. J. Mod. Phys. A, **21**, 5119 (2006); ibid, **24** 5217 (2009).
2. X.S. Chen,X. F. Lü, W.M. Sun, F. Wang and T. Goldman, Phys. Rev. Lett. **100** 232002 (2008); X.S. Chen, W.M. Sun, X. F. Lü, F. Wang and T. Goldman, Phys. Rev. Lett. **103** 062001 (2009); X. S. Chen and B. C Zhu, 'Physical decomposition of the gauge and gravitational fields', preprint, HUST and Kavli Inst. for Theore. Phys. China, CAS, Beijing, (2009).
3. V. I. Ogievetski and I. V. Polubarinov, Nuovo Cimento, **XXIII**, 173 (1962).
4. J. P. Hsu, Phys. Rev. D**8**, 2609 (1973). For quantization with other gauge conditions, see M. Kaku. *Quantum Field Theory* (Oxford Univ. Press, 1993) pp. 106-114.
5. L. D. Faddeev and V. N. Popov, in *100 Years of Gravity and Accelerated Frames, The deepest Insights of Einstein and Yang-Mills* (Ed. J. P. Hsu and D. Fine, World Scientific, 2005) p. 325; E. S. Fradkin and I. V. Tyutin, Phys. Rev. D**2**, 2841 (1970).
6. V. N. Popov, *Functional Integrals in Quantum Field Theory and Statistical Physics* (trans. from Russian by J. Niederle and L. Hlavaty, D. Reidel Publishing Comp., Boston, 1983) chapters 2 and 3.
7. J. D. Jackson, *Classical Electrodynamics* (3rd. ed. John Wiley and Sons, 1999) p. 350.
8. L. Hsu and J. P. Hsu, Chin. J. Phys. **35**, 407 (1997).
9. D. Schmidt and J. P. Hsu, Int. J. Mod. Phys. A **20**, 5989 (2005).
10. J. P. Hsu and D. Fine, Int. J. Mod. Phys. A **20** 7485 (2005).
11. J. P. Hsu and J.A. Underwood, Phys. Rev. D**15**, 1668 (1977); J. P. Hsu and E. C. G. Sudarshan, Phys. Letters **51B**, 349 (1974).
12. J. P. Hsu and E. C. G. Sudarshan, Nucl. Phys. B **91**, 477 (1975).
13. J. P. Hsu,"The S-Matrix and Ghost Fields in Quantum Yang-Mills Gravity, III," UMassD preprint (2010); J. P. Hsu, 'Quantum Yang-Mills Gravity: The Ghost Particle and Its Interaction" in the Proceedings of ICGA9, (World Scientific, 2010).
14. The situation resembles to the Lorentz gauge for the usual vector potential. To deal with the ambiguity in the potentials, we can impose an auxiliary condition. See L. Landau and E. Lifshitz, *The Classical Theory of Fields* (Addison-Wesley, 1951) pp. 114-115; see also K. Huang, *Quarks Leptons & Gauge Fields* (World Scientific, 1982) pp. 157-159.

ON UNIQUENESS OF KERR SPACE-TIME NEAR NULL INFINITY

XIAONING WU*

Institute of Mathematics,
Academy of Mathematics and Systems Science,
Chinese Academy of Sciences,
P.O.Box 2734, Beijing, 100080, China.

We re-express the Kerr metric in standard Bondi-Saches' coordinate near null infinity $\mathcal{I}^+$. Using the uniqueness result of characteristic initial value problem, we prove the Kerr metric is the only asymptotic flat, stationary, axial symmetric, algebraic special solution of vacuum Einstein equation.

keywords: Kerr solution, Uniqueness Theorem

1. Introduction

After the work by Bondi et.al.[1], it is well-known the Bondi coordinates is a very natural choice when we want to describe the asymptotic behavior of gravitational field near null infinity $\mathcal{I}^+$. Based on works by Penrose, Newman and Unti[2,3], there is an elegant way to re-express Bondi's work in N-P formulism. This also gives us a general formulism to describe the asymptotic structure of general asymptotic flat space-time which is smooth enough near $\mathcal{I}^+$. Using characteristic initial value (CIV) problem method, many authors[8,9] have shown the existence of null infinity in general case and pointed out the degree of freedom of gravitational field near null infinity. The CIV method has many advantages in dealing with gravitational radiation problem. Recently, this method has been used in numerical relativity[10]. On the other hand, Kerr solution is a very important exact solution of Einstein equation both in theoretical area and in application. It is believed that Kerr metric describes the space-time outside a stationary rotating star. For long time, the Bondi coordinates of Kerr space-time is not very clear. For example, how to describe the Kerr space-time in Unit-Newman's general formulism[3]? The uniqueness theorem[11,12] tells us that Kerr solution is the

*e-mail : wuxn@amss.ac.cn

only asymptotic flat, stationary, axial symmetric solution of vacuum Einstein equation with regular event horizon. From the application point of view, it is very difficult to get the detail information about the event horizon of a space-time because of the infinite red-shift near horizon. A more interesting question is how to identify the Kerr solution based on information near null infinity. This is more practical in future gravitational experiments. Obviously, stationary and axial symmetric condition is not enough because there are many asymptotic flat exact solutions of Ernst equation. In next section, it is found such uncertainty comes from the homogeneous part of a control equation which comes from the Killing equation. The general solutions of that equation constrain some free constants. These unknown constants are closed related with Geroch-Hanson multi-pole moments[15]. In order to identify the Kerr solution, we use Petrov classification[4] and show that condition will help us to pick out Kerr solution finally.

2. Main theorem

Let (M, g) be an asymptotic flat space-time, (u, r, θ, φ) be the standard Bondi-Sachs coordinates.

Theorem 2.1. *Suppose (M, g) be an asymptotic flat, stationary, axial symmetric, algebraic special, vacuum space-time with in a neighborhood of null infinity, then it is isometric to Kerr space-time in the Bondi coordinates neighborhood.*

In order to make the proof clearly enough, we divide it into two subsections. In first section, we calculate the Taylor series of general stationary spacetime in Bondi coordinates. After that, with the help of algebraic special condition, we will focus on how to pick out the Kerr metric.

2.1. *Taylor series of general axial symmetric vacuum stationary spacetime*

Suppose (M, g) be a vacuum stationary axial symmetric space-time. t^a and ϕ^a are two commutative Killing vectors. Near $\mathcal{I}^+$, we use the standard B-S coordinates to do the standard asymptotic expansion. The detailed construction of these coordinates is well-known and can be found in Re.[2,3]. With this choice of coordinates, we also can choose a set of null tetrad $\{l^a, n^a, m^a, \bar{m}^a\}$, such that $l^a = \left(\frac{\partial}{\partial r}\right)^a$ and these tetrad are transported in a parallel direction along i^a. Under such choice of coordinates, $\phi^a = \left(\frac{\partial}{\partial \varphi}\right)^a$.

The time-like Killing vector t^a can be expressed in terms of null tetrad as $t^a = Tl^a + Rn^a + A\bar{m}^a + \bar{A}m^a$. $[t^a, \phi^a] = 0$ means that T, R, A are independent of φ. Using the Bondi Gauge $\kappa = \varepsilon = \pi = 0$, $\rho = \bar{\rho}$, $\tau = \bar{\alpha} + \beta$, we have

$$-DR = 0, \tag{1}$$
$$-DT - D'R + (\gamma + \bar{\gamma})R + \bar{\tau}A + \tau\bar{A} = 0, \tag{2}$$
$$DA - \delta R + \tau R + \bar{\rho}A + \sigma\bar{A} = 0, \tag{3}$$
$$-D'T - (\gamma + \bar{\gamma})T - \nu A - \bar{\nu}\bar{A} = 0, \tag{4}$$
$$-\tau T + \bar{\nu}R + D'A + (\bar{\gamma} - \gamma)A - \delta T - \tau T - \mu A - \bar{\lambda}\bar{A} = 0, \tag{5}$$
$$-\sigma T + \bar{\lambda}R + \delta A + (\bar{\alpha} - \beta)A = 0, \tag{6}$$
$$-\rho T + \mu R + \delta\bar{A} - (\bar{\alpha} - \beta)\bar{A} - \bar{\rho}T + \bar{\mu}R + \bar{\delta}A - (\alpha - \bar{\beta})A = 0. \tag{7}$$

Here we use the standard notation of [2,3,4]. Differential operators in above equations are defined as

$$D := \frac{\partial}{\partial r},$$
$$D' := \frac{\partial}{\partial u} + U\frac{\partial}{\partial r} + X\frac{\partial}{\partial \zeta} + \bar{X}\frac{\partial}{\partial \bar{\zeta}},$$
$$\delta := \omega\frac{\partial}{\partial r} + \xi^3\frac{\partial}{\partial \zeta} + \xi^4\frac{\partial}{\partial \bar{\zeta}}, \quad \zeta = e^{i\varphi}\cot\frac{\theta}{2}. \tag{8}$$

It is well known that stationary solutions to Einstein's vacuum field equations are analytic[5]. Moreover, it is also known that asymptotically flat stationary vacuum solutions are not only analytic, but even admit an analytic conformal extension through null infinity[6,7]. Keeping this result in mind, all geometric quantities (the coordinate components of null tetrad, N-P coefficients, components of time-like Killing vector and components of Weyl curvature) can be expressed in terms of power series of $\frac{1}{r}$, for example

$$T = T^0 + \frac{T^1}{r} + \cdots, \quad A = A^0 + \frac{A^1}{r} + \cdots, \tag{9}$$

some lower order Taylor coefficients of components of tetrad, N-P coefficients and components of Weyl tensor (up to 3th order) can be found in section 9.8 of [2].

First of all, let's consider the function R. Eq.(1) and axial symmetric condition tell us that $R = R(u, \theta)$. With the formal expansion of null tetrad and N-P coefficients[2,3], zero order of Eq.(2) gives

$$\frac{\partial R}{\partial u} = 0, \tag{10}$$

so $R = R(\theta)$. In order to get more information about R, higher order of Killing equation are needed. The first order of Eq.(3), Eq.(5) and Eq.(7) are

$$\delta_0 R + A^0 = 0,, \tag{11}$$

$$-\bar{\Psi}_3^0 R + \dot{A}^1 - \delta_0 T^0 + \frac{1}{2}A^0 - \dot{\sigma}^0 \bar{A}^0 = 0, \tag{12}$$

$$2T^0 - R = 0, \tag{13}$$

where $\delta_0 = \frac{(1+\zeta\bar{\zeta})}{\sqrt{2}} \frac{\partial}{\partial\zeta}$ and " . " means ∂_u . Because the space-time is stationary, there is no Bondi flux. This implies $\dot{\sigma}^0 = 0$ [2], then N-P equations tell us this leads $\Psi_3^0 = 0$. Combining this condition with (11),(12) and (13), we get

$$\dot{A}^1 + A^0 = 0. \tag{14}$$

The second order of Eq.(3) is

$$2A^1 = \frac{\zeta(1+\zeta\bar{\zeta})}{\sqrt{2}\,\bar{\zeta}} \sigma^0 \frac{\partial R}{\partial\zeta} + \sigma^0 \bar{A}^0. \tag{15}$$

It has been shown that the right hand side of above equation is independent on u, so $\dot{A}^1 = 0$. This implies $A^0 = 0$ and $R = C$. Then it is reasonable to normalize $R = 1$.

With $R = 1$, Killing equations become

$$-DT + (\gamma + \bar{\gamma}) + \bar{\tau}A + \tau\bar{A} = 0, \tag{16}$$

$$DA + \tau + \bar{\rho}A + \sigma\bar{A} = 0, \tag{17}$$

$$-D'T - (\gamma + \bar{\gamma})T - \nu A - \bar{\nu}\bar{A} = 0, \tag{18}$$

$$-\tau T + \bar{\nu} + D'A + (\bar{\gamma} - \gamma)A - \delta T - \tau T - \mu A - \bar{\lambda}\bar{A} = 0, \tag{19}$$

$$-\sigma T + \bar{\lambda} + \delta A + (\bar{\alpha} - \beta)A = 0, \tag{20}$$

$$-\rho T + \mu + \delta\bar{A} - (\bar{\alpha} - \beta)\bar{A} - \bar{\rho}T + \bar{\mu} + \bar{\delta}A - (\alpha - \bar{\beta})A = 0. \tag{21}$$

With Eq.(9), the non-trivial zero order Killing equation is

$$\dot{T}^0 = 0. \tag{22}$$

Non-trivial first order Killing equations are

$$-A^0 = 0, \tag{23}$$

$$-\dot{T}^1 + \Psi_3^0 A^0 + \bar{\Psi}_3^0 \bar{A}^0 = 0, \tag{24}$$

$$-\bar{\Psi}_3^0 + \dot{A}^1 + \frac{1}{2}A^0 - \dot{\bar{\sigma}}^0 A^0 = 0, \tag{25}$$

$$\dot{\bar{\sigma}}^0 + \delta_0 A^0 + 2\bar{\alpha}^0 A^0 = 0, \tag{26}$$

$$2T^0 - 1 = 0, \tag{27}$$

which implies $\dot{T}^1 = 0$ and $\dot{A}^1 = 0$. Other N-P equations also give $\Psi_3^0 = 0$ and $\Psi_4^0 = 0$.

Second order Killing equations are

$$T^1 - \frac{1}{2}(\Psi_2^0 + \bar{\Psi}_2^0) = 0, \tag{28}$$

$$-2A^1 = 0, \tag{29}$$

$$\dot{T}^2 = 0, \tag{30}$$

$$\frac{1}{2}\eth\Psi_2^0 + \dot{A}^2 - \delta_0 T^1 = 0, \tag{31}$$

$$\frac{1}{2}\sigma^0 = 0, \tag{32}$$

$$2T^1 - \sigma^0\dot{\bar{\sigma}}^0 - \bar{\sigma}^0\dot{\sigma}^0 - \Psi_2^0 - \bar{\Psi}_2^0 = 0, \tag{33}$$

where $\eth$ is the spin-weight operator[2] and is defined as $\eth f := (\delta_0 + 2s\bar{\alpha}^0)f$, $\alpha^0 = -\frac{\cot\theta}{2\sqrt{2}}$. From these equations, we know $\sigma^0 = 0$, $A^1 = 0$, $\dot{T}^2 = 0$, $T^1 = \frac{1}{2}(\Psi_2^0 + \bar{\Psi}_2^0)$, $\Psi_2^0 = \bar{\Psi}_2^0$, $\dot{\Psi}_2^0 = 0$, $\dot{A}^2 = -\frac{1}{2}\eth\Psi_2^0 + \frac{1}{2}\delta_0(\Psi_2^0 + \bar{\Psi}_2^0)$. It is worth to point out that the result $\dot{\sigma}^0 = 0$ tells us that the Bondi coordinates which is chosen as [3] are associated with the "good cut" of stationary spacetime. The freedom of super-translation has been rule out.

The third order killing equations are

$$2T^2 + \frac{1}{3}(\bar{\eth}\Psi_1^0 + \eth\bar{\Psi}_1^0) = 0, \tag{34}$$

$$-3A^2 - \frac{1}{2}\Psi_1^0 = 0, \tag{35}$$

$$\dot{T}^3 = 0, \tag{36}$$

$$\frac{1}{2}\Psi_1^0 + \bar{\nu}^3 + \dot{A}^3 + \frac{3}{2}A^2 - \delta_0 T^2 = 0, \tag{37}$$

$$\delta_0 A^2 + 2\bar{\alpha}^0 A^2 = 0, \tag{38}$$

$$2T^2 + \frac{1}{2}\bar{\eth}\Psi_1^0 + \frac{1}{2}\eth\bar{\Psi}_1^0 + \delta_0\bar{A}^2 - 2\bar{\alpha}^0\bar{A}^2 + \bar{\delta}_0 A^2 - 2\alpha^0 A^2 = 0. \tag{39}$$

Eq.(35),(38) imply

$$\eth\Psi_1^0 = 0. \tag{40}$$

The spin-weight of Ψ_1^0 is 1, so it is a linear combination of spin-weight harmonics $\{{}_1Y_{l,m}\}$. The axial symmetric condition implies $m = 0$. The behavior of spin-weight harmonic under the action of operators $\eth$ and $\bar{\eth}$ are

$$\eth\,{}_sY_{lm} = -\sqrt{\frac{(l+s+1)(l-s)}{2}}\;{}_{s+1}Y_{lm},$$
$$\bar{\eth}\,{}_sY_{lm} = \sqrt{\frac{(l-s+1)(l+s)}{2}}\;{}_{s-1}Y_{lm}, \qquad {}_0Y_{lm} = Y_{lm}. \tag{41}$$

So we get $\Psi_1^0 = c(u)_1Y_{1,0} = c(u)\sin\theta$. Detailed calculation on same order N-P equations also give

$$\nu^3 = -\frac{1}{12}\bar{\Psi}_1^0 - \frac{1}{6}\bar{\eth}^2\Psi_1^0,\ \Psi_3^1 = 0,\ \Psi_3^2 = \frac{1}{2}\bar{\eth}^2\Psi_1^0,$$
$$\Psi_4^1 = \Psi_4^2 = \Psi_4^3 = 0,\ \Psi_4^4 = -\frac{1}{4}\bar{\eth}\Psi_3^3. \tag{42}$$

Eq.(34) and $\dot{T}^2 = 0$ gives

$$\dot{c}\bar{\eth}\sin\theta + \dot{\bar{c}}\eth\sin\theta = 2(\dot{c}+\dot{\bar{c}})\cos\theta = 0, \tag{43}$$

which implies $\dot{c}+\dot{\bar{c}} = 0$. The first order Bianchi identities tell us $\dot{\Psi}_1^0 - \eth\Psi_2^0 = 0$, which implies $\Psi_2^0 = -\dot{c}\cos\theta + C_2$ but it is well-known that $\bar{\Psi}_2^0 = \Psi_2^0$ for stationary case[2,3], so $\dot{c} = 0$. This means $\Psi_1^0 = C_1\sin\theta$ and $\Psi_2^0 = C_2$. The Komar integral shows $-C_2$ is just the Bondi mass of the space-time. Additionally, Eq.(37) tells us that $\dot{A}^3 = 0$.

Forth order Killing equations are

$$3T^3 + (\gamma^4 + \gamma^4) = 0, \tag{44}$$
$$4A^3 = \frac{1}{3}\eth\Psi_0^0, \tag{45}$$
$$\dot{T}^4 + \frac{1}{3}(\bar{\eth}\Psi_1^0 + \eth\bar{\Psi}_1^0)(\Psi_2^0 + \bar{\Psi}_2^0) = 0, \tag{46}$$
$$\frac{1}{2}\Psi_1^0T^1 - \frac{1}{3}\bar{\eth}\Psi_0^0 + \bar{\nu}^4 + \dot{A}^4 + (\Psi_2^0 + \bar{\Psi}_2^0)A^2 + \frac{3}{2}A^3 -$$
$$\delta_0T^3 + \Psi_2^0A^2 + \frac{1}{2}A^3 = 0, \tag{47}$$
$$\frac{1}{4}\Psi_0^0 + \bar{\lambda}^4 + \eth A^3 = 0, \tag{48}$$
$$2T^3 + \mu^4 + \bar{\mu}^4 + \eth\bar{A}^3 + \bar{\eth}A^3 = 0. \tag{49}$$

where $\gamma^4 = -\frac{1}{12}(\alpha^0\bar{\eth}\Psi_0^0 - \bar{\alpha}^0\eth\bar{\Psi}_0^0) - \frac{1}{8}\bar{\eth}^2\Psi_0^0$, $\lambda^4 = -\frac{1}{12}\bar{\Psi}_0^0$, $\mu^4 = -\frac{1}{3}\Psi_2^2 = -\frac{1}{6}\bar{\eth}^2\Psi_0^0$, $\nu^4 = \frac{1}{24}(\eth\bar{\Psi}_0^0 + \bar{\eth}^3\Psi_0^0)$, $\Psi_3^3 = -\frac{1}{2}\bar{\Psi}_1^0\Psi_2^0 - \frac{1}{6}\bar{\eth}^3\Psi_0^0$ (These results are got from same order N-P equations).

The spin-weight of Ψ_0 is 2, so Eq.(45),(48) imply

$$\Psi_0^0 = c(u)\sin^2\theta, \tag{50}$$

Eq.(37),(45) eliminate time dependence of $c(u)$, i.e. $\Psi_0^0 = C_0 \sin^2\theta$. Eq.(47) is

$$\dot{A}^4 = 0 \tag{51}$$

Fifth order Killing equations are

$$4T^4 + (\gamma^5 + \bar{\gamma}^5) - \frac{1}{2}\bar{\Psi}_1^0 A^2 - \frac{1}{2}\Psi_1^0 \bar{A}^2 = 0, \tag{52}$$

$$A^4 = \frac{1}{5}\tau^5, \tag{53}$$

$$-\frac{1}{2}\sigma^5 + \bar{\lambda}^5 + \frac{1}{2}\Psi_0^0 T^1 + \frac{3}{2}\Psi_1^0 A^2 + \eth A^4 = 0, \tag{54}$$

$$-2\rho^5 + 2T^4 + (\mu^5 + \bar{\mu}^5) + \frac{3}{2}\Psi_1^0 \bar{A}^2 + \frac{3}{2}\bar{\Psi}_1^0 A^2 + \eth\bar{A}^4 + \bar{\eth} A^4 = 0, \tag{55}$$

where

$$\rho^5 = 0,\ \mu^5 = -\frac{1}{4}\Psi_2^3,\ \sigma^5 = -\frac{1}{3}\Psi_0^1,\ \lambda^5 = -\frac{1}{8}\bar{\Psi}_0^0\Psi_2^0 - \frac{1}{24}\bar{\Psi}_0^1,$$
$$\gamma^5 = -\frac{1}{40}(\alpha^0\bar{\eth}\Psi_0^1 - \bar{\alpha}^0\eth\bar{\Psi}_0^1) + \frac{1}{12}|\Psi_1^0|^2 - \frac{1}{30}\bar{\eth}^2\Psi_0^1,\quad \tau^5 = \frac{1}{8}\bar{\eth}\Psi_0^1,$$
$$\Psi_2^3 = -\frac{2}{3}|\Psi_1^0|^2 + \frac{1}{6}\bar{\eth}^2\Psi_0^1,\quad \Psi_1^2 = -\frac{1}{2}\bar{\eth}\Psi_0^1.$$

Eq.(52), (46) and Bianchi identities imply $\Psi_1^0 = iC_1 \sin\theta,\ C_1 \in \mathbf{R}$, where C_1 is the Komar angular momentum. Eq.(53),(54) give

$$\eth\bar{\eth}\Psi_0^1 + 5\Psi_0^1 = 10(\Psi_1^0)^2 - 15\Psi_0^0\Psi_2^0. \tag{56}$$

The homogeneous part of above equation is

$$\eth\bar{\eth}\Psi_0^1 + 5\Psi_0^1 = 0. \tag{57}$$

Because spin-weight of Ψ_0^1 is 2, it is a linear combination of $\{{}_2Y_{l,0}\}$. Using Eq.(41), the homogeneous equation is

$$(-l^2 - l + 12)\ {}_2Y_{l,0} = 0, \tag{58}$$

which gives $l = 3$. The general solution of Eq.(54) is

$$\Psi_0^1 = \left(\frac{10}{3}(C_1)^2 - 5C_2C_0\right)\sin^2\theta + D^1\ {}_2Y_{3,0}. \tag{59}$$

(Bianchi identities insures $\dot{\Psi}_0^1 = 0$.) Until now, we have got series expression of tetrad components up to 4th order, N-P coefficients up to 5th order and Weyl components up to 6th order. To prove this theorem, all Taylor coefficients of all geometric quantities are needed. We use inductive method to solve this problem order by order.

Suppose we have known Taylor coefficients of tetrad components up to $(k-3)^{th}$ order, Taylor coefficients of connections up to $(k-2)^{th}$ order and Taylor coefficients of Weyl curvature components up to $(k-1)^{th}$ order. The $(k-1)^{th}$ order of Killing equation (17) and (20) are

$$-(k-1)A^{k-2} + \tau^{k-1} = \cdots, \tag{60}$$

$$\cdots + \bar{\lambda}^{k-1} + \eth A^{k-2} = 0. \tag{61}$$

where "$\cdots$" means terms which only contain lower order coefficients. Based on the induction hypothesis, those terms are known. In order to solve these equations, we need coefficients λ^{k-1} and τ^{k-1}. N-P equations will help us.

$$\begin{aligned} D\Psi_1 - \bar{\delta}\Psi_0 = -4\alpha\Psi_0 + 4\rho\Psi_1 &\Rightarrow -(k-4)\Psi_1^k = \bar{\eth}\Psi_0^k + \cdots . \\ D\sigma = 2\rho\sigma + \Psi_0 &\Rightarrow -(k-3)\sigma^{k-1} = \Psi_0^k + \cdots , \\ D\lambda = \rho\lambda + \bar{\sigma}\mu &\Rightarrow -(k-2)\lambda^{k-1} = -\frac{1}{2}\,\bar{\sigma}^{k-1} + \cdots , \\ D\tau = \tau\rho + \bar{\tau}\sigma + \Psi_1 &\Rightarrow -(k-2)\tau^{k-1} = \Psi_1^k + \cdots , \end{aligned} \tag{62}$$

Combining Eq.(60), (61) and (62), we get

$$\eth\bar{\eth}\Psi_0^k + \frac{(k+4)(k+1)}{2}\Psi_0^k = \cdots . \tag{63}$$

The homogeneous part of above equation is

$$\eth\bar{\eth}\hat{\Psi}_0^k + \frac{(k+4)(k+1)}{2}\hat{\Psi}_0^k = 0. \tag{64}$$

Because of Eq.(41) and axial symmetric condition, the general solution should be

$$\Psi_0^k = \tilde{\Psi}_0^k + D^k \;{}_2Y_{k+2,0}\ , \tag{65}$$

where $\tilde{\Psi}^k$ is a special solution of eq.(63) and D^k is a constant. Obviously, Kerr solution satisfies all conditions of our theorem, so $\tilde{\Psi}_0^k$ must exist. The concrete form of $\tilde{\Psi}_0^k$ also can be got by direct calculation. One can express the "$\cdots$" terms in Eq.(63) as a linear combination of spin-weight harmonics $\{{}_2Y_{l,0}\}$. The inductive method insures the maximal value of l in that expression will be finite for any given order, then we can get $\tilde{\Psi}_0^k$ by comparing coefficients between bother sides of this equation. With the general solution of Ψ_0^k, Eq.(62) will give τ^{k-1}, σ^{k-1}, λ^{k-1} and Ψ_1^k. Further more, Cartan structure equations and Bianchi equations will help us to get other coefficients,

$$
\begin{aligned}
D\rho = \rho^2 + |\sigma|^2 &\Rightarrow -(k-3)\rho^{k-1} = \cdots, \\
D\alpha = \alpha\rho + \beta\bar{\sigma} &\Rightarrow -(k-2)\alpha^{k-1} = \cdots, \\
D\beta = \beta\rho + \alpha\sigma + \Psi_1 &\Rightarrow -(k-2)\beta^{k-1} = \Psi_1^k + \cdots, \\
D\Psi_2 - \bar{\delta}\Psi_1 = 3\rho\Psi_2 - 2\alpha\Psi_1 - \lambda\Psi_0 &\Rightarrow -(k-3)\Psi_2^k = \bar{\eth}\Psi_1^k + \cdots, \\
D\Psi_3 - \bar{\delta}\Psi_2 = 2\rho\Psi_3 - 2\lambda\Psi_1 &\Rightarrow -(k-2)\Psi_3^k = \bar{\eth}\Psi_2^k + \cdots, \\
D\Psi_4 - \bar{\delta}\Psi_3 = \rho\Psi_4 + 2\alpha\Psi_3 - 3\lambda\Psi_2 &\Rightarrow -(k-1)\Psi_4^k = \bar{\eth}\Psi_3^k + \cdots, \\
D\gamma = \tau\alpha + \bar{\tau}\beta + \Psi_2 &\Rightarrow -(k-1)\gamma^{k-1} = \Psi_2^k + \alpha_0\tau^{k-1} - \bar{\alpha}_0\bar{\tau}^{k-1} + \cdots, \\
D\mu = \mu\rho + \lambda\sigma + \Psi_2 &\Rightarrow -(k-2)\mu^{k-1} = \frac{1}{2}\rho^{k-1} + \Psi_2^k + \cdots, \\
D\nu = \tau\lambda + \bar{\tau}\mu + \Psi_3 &\Rightarrow -(k-1)\nu^{k-1} = \frac{1}{2}\bar{\tau}^{k-1} + \Psi_3^k + \cdots, \\
D\xi^3 = \rho\xi^3 + \sigma\bar{\xi}^4 &\Rightarrow -(k-3)\xi^3_{k-2} = \cdots, \\
D\xi^4 = \rho\xi^4 + \sigma\bar{\xi}^3 &\Rightarrow -(k-3)\xi^4_{k-2} = \cdots, \\
D\omega = \rho\omega + \sigma\omega - (\bar{\alpha} + \beta) &\Rightarrow -(k-3)\omega^{k-2} = -\bar{\alpha}^{k-1} - \beta^{k-1} + \cdots, \\
DX = (\bar{\alpha} + \beta)\xi^3 + (\alpha + \bar{\beta})\bar{\xi}^4 &\Rightarrow -(k-2)X^{k-2} \\
&= \frac{1 + |\zeta|^2}{\sqrt{2}}(\alpha^{k-1} + \bar{\beta}^{k-1}) + \ldots, \\
DU = (\bar{\alpha} + \beta)\bar{\omega} + (\alpha + \bar{\beta})\omega - \gamma - \bar{\gamma} &\Rightarrow \\
-(k-2)U^{k-2} &= -\gamma^{k-1} - \bar{\gamma}^{k-1} + \cdots
\end{aligned} \tag{66}
$$

From above results, we find we can express all $(k-2)^{th}$ order coefficients of tetrad components, $(k-1)^{th}$ order coefficients of connection components and k^{th} order coefficients of Weyl curvature in terms of Ψ_0^k, $\eth$ derivatives of Ψ_0^k and lower order coefficients which we have known. The form of Ψ_0^k is given in Eq.(65).

From Eq.(65), we can see that the freedom in each order coefficients are just the constant D^k. These arbitrary constants should be closely related to the famous Geroch-Hansen multi-pole moments[13,14,15]. What we want to do is to pick out the Kerr solution from these solutions, i.e. we need to fix value of $\{D^k\}$. In order to do that, we consider the Petrov classification[4]. It is well known that the Kerr solution belongs to algebraic special class, i.e.

$$I^3 = 27J^2, \tag{67}$$
$$I = \Psi_0\Psi_4 - 4\Psi_1\Psi_3 + 3(\Psi_2)^2,$$
$$J = \Psi_4\Psi_2\Psi_0 + 2\Psi_3\Psi_2\Psi_1 - (\Psi_2)^3 - (\Psi_3)^2\Psi_0 - (\Psi_1)^2\Psi_4.$$

The leading terms of I and J are

$$I = \frac{3(\Psi_2^0)^2}{r^6} + O(r^{-7}),\ J = -\frac{(\Psi_2^0)^3}{r^9} + O(r^{-10}). \tag{68}$$

Because Ψ_2^0 is the Bondi mass of the space-time, the positive mass theorem insures it is non-zero. This implies $I \neq 0$ and $J \neq 0$, i.e. an asymptotic flat stationary vacuum solution can not belong to type III and type N.

The Taylor series of I and J are

$$I = \frac{3(\Psi_2^0)^2}{r^6} + \frac{6\Psi_2^0\Psi_2^1}{r^7} + \frac{-4\Psi_1^0\Psi_3^2 + 3(\Psi_2^1)^2 + 6\Psi_2^0\Psi_2^2}{r^8} + O(r^{-9}),$$
$$J = \frac{-(\Psi_2^0)^3}{r^9} + \frac{-3(\Psi_2^0)^2\Psi_2^1}{r^{10}} + \frac{-3(\Psi_2^0)^2\Psi_2^2 - 3\Psi_2^0(\Psi_2^1)^2 + 2\Psi_3^2\Psi_1^0\Psi_2^0}{r^{11}}$$
$$+ O(r^{-12})$$

Write down Eq.(67) order by order, the first non-zero coefficient is

$$36(\Psi_1^0)^2(\Psi_2^0)^2(\Psi_3^2)^2 - 54(\Psi_2^0)^3(\Psi_3^2)^2\Psi_0^0 - 54(\Psi_1^0)^2(\Psi_2^0)^3\Psi_4^4$$
$$+81(\Psi_2^0)^4\Psi_0^0\Psi_4^4 = 0. \tag{69}$$

The constant C_0 in Ψ_0^0 can be fixed in following way : from Komar integral, we know $\Psi_2^0 = -M$, $\Psi_1^0 = \frac{3iMa}{\sqrt{2}}\sin\theta$. Submit these into Eq.(69) then get $\Psi_0^0 = 3Ma^2\sin^2\theta$ (Note : the concrete expression of Ψ_3^2, Ψ_4^4 have been got in Eq.(42)) . For general order Ψ_0^k, we can write down the $(k+17)^{th}$ order equation of Eq.(67). It takes the form

$$[81(\Psi_2^0)^4\Psi_4^4 - 54(\Psi_2^0)^3(\Psi_3^2)^2]\Psi_0^k + [81(\Psi_2^0)^4\Psi_0^0 - 54(\Psi_2^0)^3(\Psi_1^0)^2]\Psi_4^k + .. = 0 \tag{70}$$

Submit eq.(65) and (66) into above equation and suppose $\tilde{\Psi}_0^k$ to be just the special solution for Kerr space-time. It is well-known that Kerr space-time is algebraic special, so

$$[81(\Psi_2^0)^4\Psi_4^4 - 54(\Psi_2^0)^3(\Psi_3^2)^2]\tilde{\Psi}_0^k + [81(\Psi_2^0)^4\Psi_0^0 - 54(\Psi_2^0)^3(\Psi_1^0)^2]\tilde{\Psi}_4^k + . = 0. \tag{71}$$

Then (70)-(71) gives

$$[81(\Psi_2^0)^4\Psi_4^4 - 54(\Psi_2^0)^3(\Psi_3^2)^2]D^k\ {}_2Y_{k+2,0} = 0, \tag{72}$$

which means $D^k = 0$, i.e. Kerr solution is the only space-time which satisfies all requirements of this theorem.

Let's summery our proof. Although our proof is calculationally intensive, the main idea is quite simple : based on assumptions of the theorem, the Killing equation and N-P equations help us to express all Taylor coefficients of geometric quantities concretely in terms of associated Taylor coefficient Ψ_0^k of Weyl curvature. This function is controlled by Eq.(63) which comes from the Killing equation. The general solution of that equation contains a free constant D^k. The algebraic special condition fix that constant, so we know there is only one space-time fully satisfies the requirement of the theorem. On the other hand, it is well-known that Kerr space-time satisfies the theorem's requirement, so we finish the proof of the theorem.

Above method can also gives the concrete NU extension of Kerr solution in Bondi-Sachs coordinates up to any order, at least in principle. Of cause the calculation will be more and more complex for higher order. Similar topic has been considered by other works[17,18]. The concrete expression for lower orders can be found in work by Wu and Shang[17].

Acknowledgement

This work is supported by the Natural Science Foundation of China under Grant Nos.10705048, 10605006, 10731080. Authors would like to thank Prof. X.Zhang and Dr. J.A.Valiente-Kroon for their helpful discussion.

References

1. H. Bondi, M. G. J. van der Burg and A. W. K. Metzner, Proc. Roy. Soc. Lond. **A 269** (1962) 21.
2. R. Penrose and R. Rindler, *Spinors and Space-Time* Vol.I and II, Cambridge University Press, 1986.
3. E. T. Newman and T. W. J. Unti, J. Math. Phys. **3** (1962) 891.
4. D. Kramer, H. Stephani, E. Herlt and M. MacCallum, *Exact Solutions of Einstein's Field Equations*, Cambridge University Press, 1980.
5. H. Müller zum Hagen, Proc. Camb. Phil. Soc. **68** (1970) 199.
6. T. Damour and B. Schmidt, J. Math. Phys. **31** (1990) 2441.
7. S. Dain, Class. Quantum Grav. **18** (2001) 4329.
8. H. Friedrich, Proc. R. Soc. Lond. **A 378** (1981) 169-184, 401-421.
9. J. Kánnár, Proc. Roy. Soc. Lond. **A 452** (1996) 945.
10. J. Winicour, "Characteristic Evolution and Matching", Living Rev. Relativity **8** (2005)10, http://www.livingreviews.org/irr-2005-10.
11. D. C. Robinson, Phys. Rev. Lett. **34** (1975) 905.
12. M. Heusler, *Black Hole Uniqueness Theorems*, Cambridge University Press, 1996.

13. R. Hansen, J. Math. Phys. **15** (1974) 46.
14. P. K. Kundu, J. Math. Phys. **29** (1988) 1866.
15. H. Friedrich, Annales Henri Poincare **8** (2007) 817.
16. E. T. Newman and R. Penrose, Proc. Roy. Soc. Lond. **A 305** (1968) 175.
17. X. Wu and Y. Shang, Class. Quant. Grav. **24** (2007) 679.
18. D. S. Chellone, J. Phys. **A** : Gen. Math. **8** (1975) 1.

PULSARS AND GRAVITATIONAL WAVES

K. J. LEE* R. X. XU and G. J. QIAO

School of Physics and State Key Laboratory of Nuclear Physics and Technology, Peking University, Beijing, 100871, P. R. China
**E-mail: kjlee007@gmail.com*

The relationship between pulsar-like compact stars and gravitational waves is briefly reviewed. Due to regular spins, pulsars could be useful tools for us to detect ∼nano-Hz low-frequency gravitational waves by pulsar-timing array technique; besides, they would also be ∼kilo-Hz high-frequency gravitational wave radiators because of their compactness. The wave strain of an isolated pulsar depends on the equation state of cold matter at supra-nuclear densities. Therefore, a real detection of gravitational wave should be very meaningful in gravity physics, micro-theory of elementary strong interaction, and astronomy.

Keywords: Pulsars; Gravitational waves; Neutron stars

1. Introduction

Since the human mind first wakened from slumber, it has never ceased to feel the profound nature of space-time, especially the time-consciousness, in both philosophy and physics. However, a physical concrete of space-time is clarified only after Einstein's first insight: the space-time is a four dimensional continuum and the rule of the motion is the pure geometrical constrain that free particles follow the geodesics of the space-time, while the response of the space-time continuum to the matter is determined by the Einstein's equation in which a linear function of space-time curvature is in proportion to the energy-momentum of the matter.

It is worth noting that the nature of space-time is a debating topic starting earlier than the general relativity and would not be terminated only by Einstein's pure geometrical arguments. Guided by different perceptions of space-time philosophies, many gravity theories are proposed,[1] with different interpretations of equivalence principle. Even in the general relativity, the equivalent principle is still a matter of debate. Therefore, experimental tests of gravity theories, including in strong field and with fast motion, are critical to differentiate or falsify the gravity theories. Such experimental environments are only available in astrophysics, especially related to compact stars known as white dwarfs, pulsars/neutron stars, and black holes.

Topics of gravitational waves relevant to the pulsar astronomy are focused on in this review. The binary pulsar tests for gravity theories are given in §2. Pulsars as tools of detecting and as sources of gravitational waves are presented in §3 and §4, respectively. Future prospects are discussed in §5.

2. Pulsars, binary pulsars, and tests of gravity theories

Soon after the discovery, pulsars are identified as a class of fast rotating rather than pulsating compact objects. It was used to believe that pulsars are neutron stars composed by hadronic matter, because of the very limited knowledge in sub-nucleon research at that time, but this view might not be true[2] since the lowest compact states could be of quark matters with strangeness rather than that of neutron liquid. Observationally, there are two main categories of pulsars: the millisecond pulsars (MSPs) and the normal pulsars. Normal pulsars have rotational period from a few tens of milliseconds to a few seconds, while the MSPs' periods range from about 1 millisecond to a few tens milliseconds.[3] Long term timing monitoring shows that the MSPs are more stable rotators compared to the normal pulsars, which might due to both observational reasons and pulsar intrinsic physics. For MSPs, the difference between model-expected time of arrival (TOA) of radio pulses and observed TOA is usually less than 10 percent of their periods, and the most stable MSPs can achieve ~ 100 ns level on the time scale of a few years.[4]

It turns out that most of the MSPs are in binary system. One particular interesting system is the recently discovered binary pulsar system, J0737-3039AB, where both of the two stars are radio pulsars.[5] The J0737 is also a highly relativistic celestial system. Binary pulsars with possible pulsar companions are listed in Table 1, obtained from ATNF pulsar catalog.[6]

Armed with such a kind of stable celestial clocks (i.e., pulsars) relativistically orbiting their companions, one can then test gravity theories in the case of strong gravitational fields, as illustrated in the classical system of PSR B1913+16.[7] In the J0737 system, two pulsars orbit each other with a period of 2.5 hours and a very low orbital eccentricity. Up to known, this J0737 system becomes the most relativistic binary system, the details of which can be found in the review by Kramer and Wex (2009).[8]

To test the gravity theories, one must compare the predicted TOA of

Table 1. Parameters for possible pulsar-neutron star systems

Name	P	P_b	a	e
J0737-3039A	0.022699	0.1023	1.4150	8.778e-02
J0737-3039B	2.773461	0.1023	1.5161	8.778e-02
J1518+4904	0.040935	8.6340	20.0440	2.495e-01
B1534+12	0.037904	0.4207	3.7295	2.737e-01
J1756-2251	0.028462	0.3196	2.7564	1.806e-01
J1811-1736	0.104182	18.7792	34.7827	8.280e-01
B1820-11	0.279829	357.7620	200.6720	7.946e-01
J1829+2456	0.041010	1.1760	7.2380	1.391e-01
J1906+0746	0.144072	0.1660	1.4202	8.530e-02
B1913+16	0.059030	0.3230	2.3418	6.171e-01
B2127+11C	0.030529	0.3353	2.5185	6.814e-01

Note: P is the pulsar period in unit of second, P_b is the orbit period in unit of days, a is the projected semi major axis in unit of light seconds, and e is the orbital eccentricity.

a theoretical model with the observation. In this way, one needs thus the binary motion dynamical models, in which we put the gravity theory in. One can then calculate the theoretical pulse TOA at the solar barycenter. Note that, from a pulsar to the barycenter, various processes set in, including the photon propagation effects due to the gravitational field of both the binary system and solar system, the dispersion of pulsar radio signal due to interstellar medium and solar wind, and so on. One needs also the solar system ephemeris to convert the pulsar TOA at the radio telescope to the barycenter. The modeled TOA will then be compared with observed ones to see if the gravity theories is able to account for the observation.

In reality, thanks to the phenomenological framework of post-Keplerian (PK) parameters, with which gravity theories can be approximated, we can independently measure these PK parameters by fitting the TOA data. A gravity test is then via checking the self-consistency of PK parameters for a particular gravity theory. There are 7 PK parameters which are possibly measurable in the near future: advance of periastron $\dot{\omega}$, gravitational redshift parameter γ, Shapiro delay parameters r and s, orbit period derivative $\dot{P}_{\rm orb}$, spin-orbital coupling induced precession $\Omega_{\rm so,p}$ and relativistic orbit deformation δ_θ. These PK parameters, except δ_θ, have been measured in the 0737 system.

All the 7 parameters measured are functions of two unknown parameters: pulsar masses, m_a and m_b. A double neutron star system is then overdetermined if one detects three or more PK parameters. It is worth noting that the double pulsar system of J0737 offers an extra Keplerian constrain, the mass ratio between two star, $R(m_a, m_b) = m_a/m_b = x_b/x_a$, with x the projected semimajor axes. Two recent reviews[8,9] are valuable in the topic of testing gravity with binary pulsars.

3. Detecting gravitational waves with pulsars

Directly detecting gravitational wave (GW) is the Holy Grail of present experimental researches, not only in gravity physics but also in astronomy. With the efforts since 1960s,[10] recent equipments (e.g., LIGO[11]) may finally allow us to directly detect GWs although there is no confirmed detection now yet. In this section, we will review the ability of detecting gravitational waves using *pulsar timing array* (PTA). Potential roles of testing gravity with PTA are also presented here.

GW is actually a perturbation of space-time, fully characterized by a wave-like metric perturbation. Detecting GW is thus identical to measure the wave-like metric perturbation[a] which can be performed by comparing geodesics of two test objects approaching to and departing from each other. Such experiments fall into four categories: 1. Tracing the motion of two free-falling test objects (e.g. LIGO, LISA, GEO, TAMA, and so on), 2. Detecting the deformation of finite extend solid body (e.g. Bar detector, Sphere detector, and so on), 3. Measuring the Doppler shift of electromagnetic signals from distance free-falling objects (e.g. Doppler tracking of satellite, pulsar timing array, laser ranging, LISA), 4. Checking the perturbation of a cosmological system (e.g. cosmic background B mode detection, weak lensing survey). Among all these possible ways, PTA is one of the promising techniques to directly detect gravitational waves, being unique to detect GW at nano-Hertz band.[12]

As we have shown, MSPs are very stable celestial clock in the Galaxy. GWs perturb the background space-time of the Galaxy, such that pulsar pulse signals get red or blue Doppler shift along the path from pulsar to earth. It turns out that such GW-induced frequency shift only involves the metric perturbation at the pulsar and that at the earth. The GW-induced frequency shift can be obtained to be[13]

$$\frac{\Delta\omega(t)}{\omega} = \frac{\hat{\mathbf{n}}^i\hat{\mathbf{n}}^j}{2\left(1+\mathbf{n_g}\cdot\hat{\mathbf{n}}\right)}\left[h_{ij}(t,0) - h_{ij}(t-D/c,\mathbf{D})\right], \tag{1}$$

[a]It should be born in mind that detecting of GW is not detecting any types of metric perturbation. GW detection focuses on detecting the oscillatory part of the metric perturbation with strain h decrease as r^{-1}, such that gravitational wave could carry energy and momentum to the infinity.

where ω is the pulsar angular frequency of spin, $\hat{\mathbf{n}}$ is the pulsar direction, $\mathbf{D} = D\hat{\mathbf{n}}$ with D the distance to the pulsar, $\mathbf{n}_{\rm g}$ is the GW propagation direction, and h is the perturbation of metric. The GW-induced timing residuals in pulsar TOA is therefore $R = \int \Delta\omega/\omega\, dt$.

Due to the intrinsic noises and possible non-modeled accelerations of pulsars, it is unlikely that one can use $R(t)$ of a single pulsar to detect GWs. Nevertheless, magic happens if we correlate the residuals of R_j and R_j of two pulsars. From Eq. (1), one can have a correlation of $\langle R_i R_j \rangle = C(\theta)\sigma^2$ for two different pulsars in general relativity, with σ the RMS (root-mean-square) of a single pulsar's residual. Note that the correlation $C(\theta)$ is a determined function only involving the angular, θ, between two pulsars,[14] and this correlation $C(\theta)$ certainly plays a vital role in detecting GW using a array of pulsar timing data (PTA) since the shape of $C(\theta)$ is uniquely determined by a gravity theory and there is no other physical processes to make the pulsar signal correlated for two pulsars widely separated with a distance of several thousand light years away from each other[b]. In the general relativity theory of gravity, fortunately, the correlation $C(\theta)$ has a very simple form of[13,14]

$$C(\theta) = \frac{3x \log x}{2} - \frac{x}{4} + \frac{1}{2}\left(1 + \delta(x)\right), \tag{2}$$

where $x = (1 - \cos\theta)/2$.

We may make sense of the $C(\theta)$-curves from simple symmetric reasons. If a monochromatic general relativistic GW is propagating along 'z-axis' direction (there will be 180° symmetry and 90° anti-symmetry in x-y plane), then correlation $C(\theta)$ between two pulsars with 0° or 180° angular separation is positive, while $C(\theta)$ for 90° will be negative. This make a U-shaped $C(\theta)$.[14] Note that $C(180°) \neq C(0)$ which will be explained later. One can then measure such multi-pulsar correlation to detect GWs. Jenet et al. (2005)[12] had investigated the statistical properties of such detection processes. Their results show that regular timing observations of 40 pulsars each with a timing accuracy of 100 ns will be able to make a direct detection of the predicted stochastic background from coalescing black holes within 5 years. We compare the detection abilities for GW detectors in Fig. 1, for GW background due to coalescing supermassive binary black holes (BBH).

---

[b]The imperfectness of terrestrial clock and un-modeled solar system dynamics may introduce also correlation between measured R_i of pulsars, however the angular dependence of such correlations is very different from that $C(\theta)$ presented in Eq.(2).

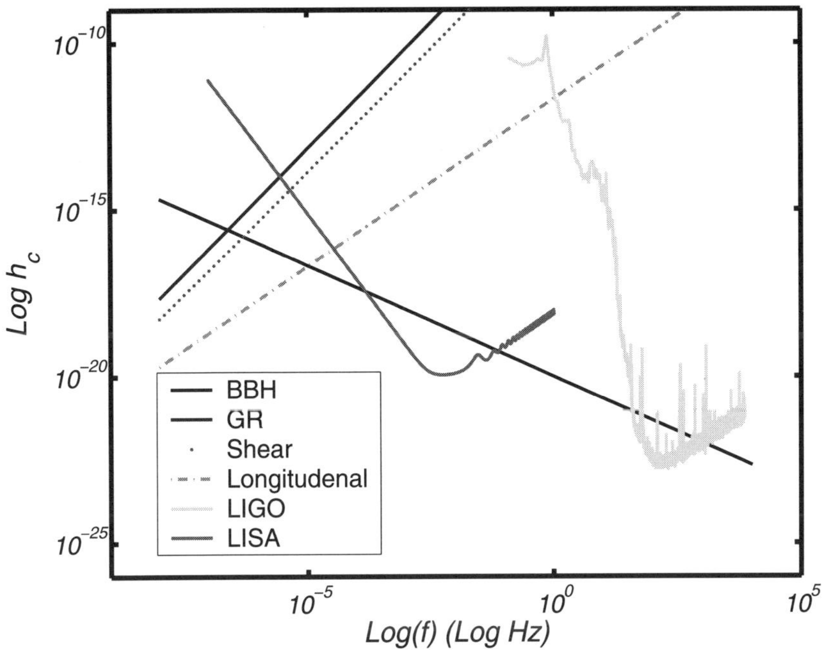

Fig. 1. The ability of detecting GWs for various GW detectors, where the x-axis is the frequency of GW, the y-axis is the characteristic strain of GW. The legends indicate the meaning of each curve, where GR is the possible pulsar timing array sensitivity for the two transverse-traceless polarized GW in general relativity, 'Longitudenal' is for GW with longitudinal polarization, 'Shear' is for GW with shear polarization. The pulsar timing sensitivity curves will be truncated at lower frequency due to finite observation length and at higher frequency due to finite duration between two successive observations.

From symmetry arguments above, it is clear that the shape of $C(\theta)$ depends on the polarization of GW (see Fig. 2). Einsteins theory of gravity predicts waves of the distortion of space-time with two degrees of polarization; alternative theories predict more polarizations, up to a maximum of six.[15] Lee et al. (2008)[13] analyzed such polarization effects and conclude that for biweekly observations made for five years with rms timing accuracy of 100 ns, detecting non-Einsteinian modes will require: 60 pulsars in the case of the longitudinal mode; 60 for the two spin-1 'shear' modes; and 40 for the spin-0 'breathing' mode. Further more, they showed that one can test gravity theories by checking GW polarization, i.e., to discriminate non-Einsteinian modes from Einsteinian modes, we need 40 pulsars for the breathing mode, 100 for the longitudinal mode, and 500 for the shear mode. These requirement is beyond present observation technology, but could be easily achieved using SKA or FAST telescope.[16,17]

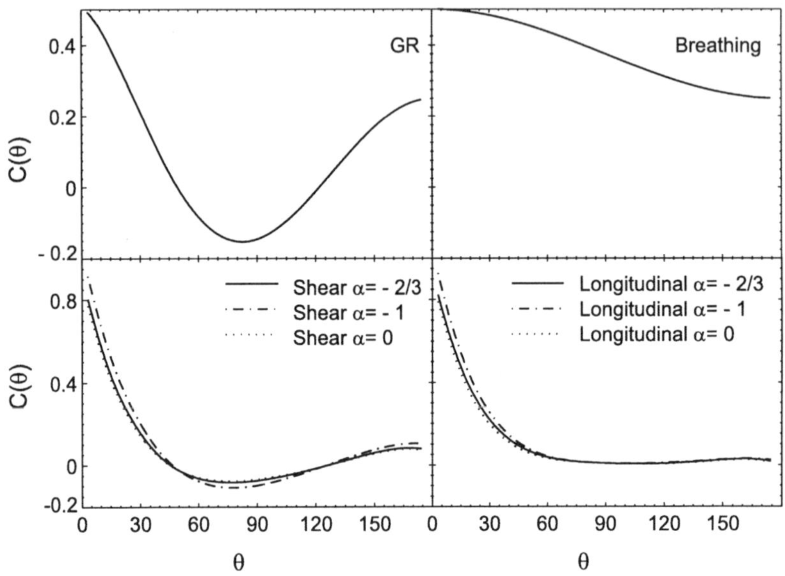

Fig. 2. The $C(\theta)$ curves for different kinds of GW polarization, with α denoting the power index of the GW background.

Another interesting topic on detecting GWs using PTA is about the dispersion relation of GW,[18] since the function of $C(\theta)$ and the detection statistics depends also on the mass of graviton. It is found[18] that $C(180°)$ increases to match the value of $C(0)$ as the graviton mass increases (see Fig. 3 for details). In the case of massless GW background, we know that the GW has 180° degree symmetry due to the polarization property, but why $C(0°) \neq C(180°)$? It turns out that GW propagation breaks up this 180° symmetry by the geometric factor in Eq. (1), which reads $1 + \hat{\mathbf{e}}_z \cdot \hat{\mathbf{n}}$ for the massless case. For the case of a massive GW background, the geometric factor reads $1 + \frac{c}{\omega_g}\mathbf{n}_g \cdot \hat{\mathbf{n}}$, where the graviton mass reduces the asymmetry. For the limiting case, where the GW frequency is just at the cut-off frequency, the dispersion relation tells us that such a GW is not propagative, then the 180° symmetry is restored. Therefore, we would expect that the correlation function are of 180° symmetry for very massive gravitons.

Lee et al. (2009) further find that it is possible to measure graviton mass

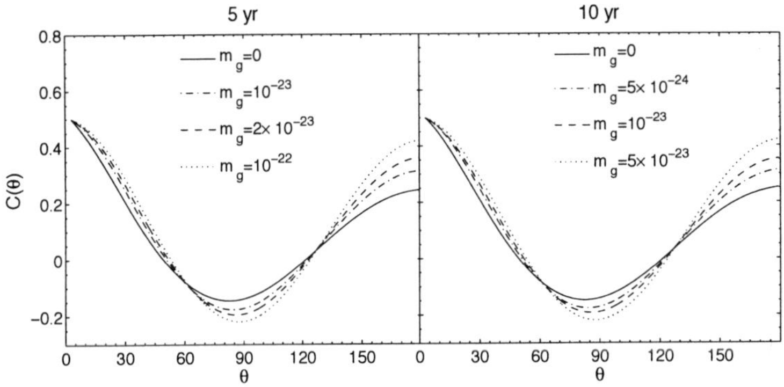

Fig. 3. The curves of $C(\theta)$ for different graviton masses.

using PTA and one will get 90% probability to differentiate between the results for massless graviton and that for graviton heavier than 3×10^{-22} eV, if biweekly observation of 60 pulsars are performed for 5 years with pulsar RMS timing accuracy of 100 ns in the future.

As we have shown that PTA can be constructed to measure the alternative polarization modes of GW and the GW dispersion relation. These measurements provide tests for gravity theory in the weak field/high velocity region, which are different from that of the solar system tests (i.e., the weak field/slow velocity case) and the binary pulsar tests (i.e., the strong field/slow velocity cases), because it is not completed to describing GW using post-Newtonian formalism and the scalar and tensor sectors of gravity theories are different.[19]

4. Gravitational wave radiation from pulsars

Pulsars are not only as tools to detect GWs, but also strong GW sources because of their compactness and the rapid mass changes. Indirect evidence for GW from binary pulsars has been discussed in §2, whereas a direct detection of GW from pulsars with ground-based facilities should be meaningful. It is recognized that the GW amplitudes of isolated pulsars depend on the equation of state (EoS) of cold matter at supra-nuclear density, which is strongly related to the understanding of QCD (quantum chromo-dynamics) at low energy scale, still another challenge for physicists today. A mixture of quantum (QCD) and gravity (relativity) makes this project more funny.

We are still not sure about the nature of pulsar-like compact stars though discovered since 1967. It is conventionally believed that these compact stars are normal neutron stars composed of hadronic matter, but one can not rule out the possibility that they are actually quark stars of quark matter[2] (see, e.g., a review[20]). Quark stars with strangeness are popularly discussed in literatures, which are called as strange (quark) stars. The EoS of realistic quark matter in compact stars, based on non-perturbative QCD, was supposed to be of Fermi gas or liquid, but could be of classical solid in order to understand different manifestations of pulsar-like stars.[21,22]

Besides QCD, that pulsars are quark stars should also be meaningful in GW physics.[23] (i). *GW being EoS-dependent.* Rotation (r) mode instability, which would result in GW radiation, may occur in fluid quark stars if the bulk and shear viscosities of quark matter is not sufficiently high, but no r-mode instability occurs in solid quark stars. Even in case of solid quark stars, the GW amplitude is relevant to the quadrupole deformations[27] (e.g., mountain building on stellar surface) sustained by elastic or magnetic forces on stellar surface. A quark or neutron star with quadrupole deformation would be a GW radiator if it has precession either free or torqued, and the precession amplitude (or the angle between spin axis and spindle of inertia ellipsoid) is determined by EoS[24] and determines GW strain. (ii). *GW being mass-dependent.* A very difference between quark and normal neutron stars is that the latter is gravity-bound while the former is confined additionally by self strong interaction, that results in the fact that quark stars could be very low massive[25] (even to be of $\sim 10^{-3} M_{\odot}$) but neutron star cannot. Low mass quark stars, either in liquid or solid states, are surely very weak GW emitters. This mass-depend nature makes it more complex to constrain EoS of target compact stars by negative results of LIGO GW detections. The points of above (i) and (ii) are certainly very useful for us to observationally distinguish quark stars from normal neutron stars in the future.

Pulsars spin usually at frequencies $> 10^0$ Hz, and we thus are interested in LIGO to detect their GWs, from Fig. 1. There are two kinds of GWs from pulsar-like compact stars: continuous GWs due to spin and bursting GWs due to stellar catastrophic events (e.g., star quake[26] or binary coalescence). It is worth noting that all the upper limits estimated from LIGO science runs depend on simulated waveform types of GWs (i.e., astrophysical GW radiative mechanisms). For continuous GWs, the waveforms could be better understood, and their searches are significantly more sensitive, especially when informed by observational photon astronomy and theoretical astrophysics.[28] The waveform of bursting GW is a matter of debate,[29] and such kind of GW searches is also focused by LIGO, especially on the super-flares of soft gamma-ray repeaters.[30]

5. Summary and Future prospects

Pulsars could be useful tools to detect GWs by PTA technique, they would also be strong GW radiators; a real detection of gravitational wave should be very meaningful in gravity physics, micro-theory of elementary strong interaction, and astronomy. A successful detection of GW by PTA may provide a test of GW polarization and measure the graviton mass. Thought indirect evidence for GWs from pulsar timing in binaries has been obtained, a direct GW detection of pulsar-like stars is also expected as persistent or transient sources. The strain of GW from an isolated compact star depends on the equation of state of cold matter at supra-nuclear densities.

Pulsar timing array projects are promising for detect GWs. We can achieve $h_c = 10^{-15}$ region for several year continuous pulsar timing monitoring. If bi-weekly observations are made for five years with RMS timing accuracy of 100 ns, then 40 pulsars are required for general relativistic modes, 60 for the longitudinal mode; 60 for the two spin-1 shear modes; and 40 for the spin 0 breathing mode. Additionally, we may measure the graviton mass through PTA techniques. With a 5-year observation of 100 or 300 pulsars, we can detect the graviton mass being higher than 2.5×10^{-22} and 10^{-22} eV, respectively. Ultimately, a 10-year observation of 300 pulsars allows us to probe the graviton mass at a level of 3×10^{-23} eV.

For the task of measuring the GW polarization and the graviton mass, there is one critical requirement: a large sample of stable pulsars. Thus the on-going and coming projects like the Parkes PTA,[31] the European PTA,[32] the Large European Array for Pulsars,[33] the FAST[16,17] and the SKA would offer unique opportunities to detect the GW background and to probe into the nature of GWs, both physical (the GW polarization and the graviton mass) and astronomy (the GW sources).

Other important requirements for a successful PTA include high stability of pulsar intrinsic noises and a low measurement noise. We need pulsar survey with better sky coverage as well as good observing system, especially with better band width[34] to get better signal to noise ratio and to subtract the interstellar medium effects. Better radio frequency interference filtering technology will also be very helpful such that we can use the full band data and reduce the terrestrial contamination. Better timing techniques (such as timing in full Stokes parameters[35]) could also be preferred.

Acknowledgments

We would like to acknowledge useful discussions at the pulsar group of PKU. This work is supported by NSFC (10833003, 10973002), the National Basic Research Program of China (grant 2009CB824800) and LCWR (LHXZ200602).

References

1. C. Will, *Living Reviews in Relativity* **9**, 3 (2006).
2. J. M. Lattimer, M. Prakash, *Science* **304**, 536 (2004).
3. D. R. Lorimer, *Living Reviews in Relativity* **8**, 7 (2005).
4. J. P. W. Verbiest, M. Bailes, W. A. Coles, et. al., *MNRAS* **400**, 951 (2009).
5. A. G. Lyne, M. Burgay, M. Kramer, et al., *Science* **303**, 1153 (2004).
6. R. N. Manchester, G. B. Hobbs, A. Teoh, et al., *AJ* **129**, 1993 (2005).
7. T. Damour and J. H. Taylor, *Phy. Rev. D* **45**, 1840 (1992).
8. M. Kramer and N. Wex, *Classical and Quantum Gravity* **26**, 073001 (2009).
9. I. H. Stairs, *Living Reviews in Relativity* **6**, 5 (2003).
10. R. L. Forward, D. Zipoy and J. Weber, *Nature* **189**, 473 (1961).
11. B. P. Abbott, et al., *Reports on Progress in Physics* **72**, 076901 (2009).
12. F. A. Jenet, G. B. Hobbs, K. J. Lee, et al., *ApJ* **625**, L123 (2005).
13. K. J. Lee, F. A. Jenet and R. H. Price, *ApJ* **685**, 1304 (2008).
14. R. W. Hellings and G. S. Downs, *ApJ* **265**, L39 (1983).
15. D. M. Eardley, D. L. Lee and A. P. Lightman, *Phys. Rev. D* **8**, 3308 (1973).
16. R. Nan, Q. Wang, L. Zhu, et al., *CJAA* **S6**, 020000 (2006).
17. R. Smits, D. R. Lorimer, M. Kramer, et al., *A&A* **505**, 919 (2009).
18. K. J. Lee, et al. (2010), in prepration.
19. M. Maggiore, *Gravitational waves* (Oxford: Oxford University Press, 2008).
20. R. X. Xu, *J. Phys. G: Nucl. Part. Phys.* **36**, 064010 (2009).
21. R. X. Xu, *ApJ*, **596** L59 (2003).
22. R. X. Xu, in *Compact stars in the QCD phase diagram II, May 20-24, 2009, KIAA-PKU, Beijing* (arXiv:0912.0349).
23. R. X. Xu, *Astroparticle Physics*, **25**, 212 (2006).
24. I. H. Stairs, A. G. Lyne, S. L. Shemar, *Nature*, **406** 484 (2000).
25. R. X. Xu, *Mon. Not. Roy. Astron. Soc.*, **356**, 359 (2005).
26. R. X. Xu, D. J. Tao, Y. Yang, *Mon. Not. Roy. Astron. Soc.* **373**, L85 (2006).
27. B. J. Owen, *Phys. Rev. Lett.*, **95** 211101 (2005).
28. B. J. Owen, preprint (arXiv:0904.4848).
29. P. Kalmus, PhD thesis submitted to Columbia Univeristy (arXiv:0904.4848).
30. J. Horvath, *Mod. Phys. Lett.* **A20** 2799 (2005).
31. G. B. Hobbs, M. Bailes, N. D. R. Bhat, et al.,, *PASA* **26**, 103 (2009).
32. B. W. Stappers, M. Kramer, A. G. Lyne, et al., *CJAA* **S6**, 020000 (2006).
33. B. Stappers, W. Vlemmings and M. Kramer, in *Proceedings of the 8th International e-VLBI Workshop. 22-26 June 2009. Madrid, Spain* (2009).
34. X. P. You, G. Hobbs, W. A. Coles, et al., *MNRAS* **378**, 493 (2007).
35. W. van Straten, *ApJ* **642**, 1004 (2006).

BRANEWORLD STARS: ANISOTROPY MINIMALLY PROJECTED ONTO THE BRANE

J. OVALLE

Departamento de Física,
Universidad Simón Bolívar,
AP 89000, Caracas 1080-A, Venezuela
E-mail: jovalle@usb.ve
http://www.usb.ve/

In the context of the Randall-Sundrum braneworld, an exhaustive and detailed description of the approach based in the minimal anisotropic consequence onto the brane, which has been successfully used to generate exact interior solutions to Einstein's field equations for static and non-uniform braneworld stars with local and non-local bulk terms, is carefully presented. It is shown that this approach allows the generation of a braneworld version for any known general relativistic solution.

Keywords: Braneworld

1. Introduction

Is well known that the non-closure of the braneworld equations represents an open problem in the study of braneworld stars.[1] A better understanding of the bulk geometry and proper boundary conditions is required to overcome this issue. Since the source of this problem is directly related with the projection $\mathcal{E}_{\mu\nu}$ of the bulk Weyl tensor on the brane, the first logical step to overcome this issue would be to discard the cause of the problem, namely, to impose the constraint $\mathcal{E}_{\mu\nu} = 0$ in the brane. However it was shown in[2] that this condition is incompatible with the Bianchi identity in the brane, thus a different and less radical restriction must be implemented. In this respect, a useful path that has been successful used consist in to discard the anisotropic stress associated to $\mathcal{E}_{\mu\nu}$, that is, $\mathcal{P}_{\mu\nu} = 0$. However, in our opinion, this constraint, which is useful to overcome the non-closure problem,[3] represents a restriction too strong in the brane. The reason is that some anisotropic effects onto the brane should be expected as a consequence of the "deformation" undergone by the 4D geometry due to five dimensional gravity effects, as was clearly explained in.[4] As was already shown in,[5] there is a constraint in the brane which represents a condition of minimal anisotropy projected onto the brane. In this paper, it is shown that this condition not only ensures a correct low energy limit, but also it represents a condition that is satisfied for *any* known general relativistic solution. The principal goal of this paper is to show that demanding the minimal anisotropic effects onto the brane, it is possible to construct the braneworld version of every general relativistic solution.

2. Non locality and the general relativity limit problem.

The effective Einstein's field equation in the brane can be written as a modification of the standard field equation through an energy-momentum tensor carrying bulk effects onto the brane:

$$T_{\mu\nu} \to T_{\mu\nu}^{\;T} = T_{\mu\nu} + \frac{6}{\sigma}S_{\mu\nu} + \frac{1}{8\pi}\mathcal{E}_{\mu\nu}, \tag{1}$$

here σ is the brane tension, with $S_{\mu\nu}$ and $\mathcal{E}_{\mu\nu}$ the high-energy and non-local corrections respectively. Using the line element in Schwarzschild-like coordinates $ds^2 = e^{\nu(r)}dt^2 - e^{\lambda(r)}dr^2 - r^2\left(d\theta^2 + \sin^2\theta d\phi^2\right)$ in the case of a static distribution having Weyl stresses in the interior, the effective equations can be written as

$$e^{-\lambda} = 1 - \frac{8\pi}{r}\int_0^r r^2\left[\rho + \frac{1}{\sigma}\left(\frac{\rho^2}{2} + \frac{6}{k^4}\mathcal{U}\right)\right]dr, \tag{2}$$

$$\frac{8\pi}{k^4}\frac{\mathcal{P}}{\sigma} = \frac{1}{6}\left(G_1^1 - G_2^2\right), \tag{3}$$

$$\frac{6}{k^4}\frac{\mathcal{U}}{\sigma} = -\frac{3}{\sigma}\left(\frac{\rho^2}{2} + \rho p\right) + \frac{1}{8\pi}\left(2G_2^2 + G_1^1\right) - 3p \tag{4}$$

$$p_1 = -\frac{\nu_1}{2}(\rho + p), \tag{5}$$

with

$$G_1^1 = -\frac{1}{r^2} + e^{-\lambda}\left(\frac{1}{r^2} + \frac{\nu_1}{r}\right), \tag{6}$$

$$G_2^2 = \frac{1}{4}e^{-\lambda}\left[2\nu_{11} + \nu_1^2 - \lambda_1\nu_1 + 2\frac{(\nu_1 - \lambda_1)}{r}\right]. \tag{7}$$

where $f_1 \equiv df/dr$ and $k^2 = 8\pi$. The general relativity is regained when $\sigma^{-1} \to 0$ and (5) becomes a lineal combination of (2)-(4).

As it is clearly shown through the Eqs. (2)-(5), in the case of a non-uniform static distribution with local and non-local bulk terms, we have an indefinite system of equations in the brane represented by the set of three unknown functions $\{p(r), \rho(r), \nu(r)\}$ satisfying one equation, that is, the conservation equation (5) . Hence to obtain a solution we must add additional information. However it is not clear what kind of restriction should be considered to close the system. To clarify the way to obtain some criterion which helps in searching a solution of this problem, let us start with the apparent simplest way: using (4) in the original form of the equation (2), which is the field equation

$$-8\pi\left(\rho+\frac{1}{\sigma}\left(\frac{\rho^2}{2}+\frac{6}{k^4}\mathcal{U}\right)\right)=-\frac{1}{r^2}+e^{-\lambda}\left(\frac{1}{r^2}-\frac{\lambda_1}{r}\right), \tag{8}$$

we have a first order linear differential equation to the geometric function $e^{-\lambda}$, given by

$$-\lambda_1 e^{-\lambda}+e^{-\lambda}\left(\frac{\nu_{11}+\nu_1^2/2+2\nu_1/r+2/r^2}{\nu_1/2+2/r}\right)=\frac{2}{r^2(\nu_1/2+2/r)}$$
$$-8\pi\frac{(\rho-3p-\frac{1}{\sigma}\rho(\rho+3p))}{(\nu_1/2+2/r)}, \tag{9}$$

which formal solution is

$$e^{-\lambda}=e^{-I}\left(\int_0^r\frac{e^I}{(\frac{\nu_1}{2}+\frac{2}{r})}\left[\frac{2}{r^2}-8\pi(\rho-3p-\frac{1}{\sigma}\left(\rho^2+3\rho p\right))\right]dr+c\right) \tag{10}$$

with

$$I\equiv\int\frac{(\nu_{11}+\frac{\nu_1^2}{2}+\frac{2\nu_1}{r}+\frac{2}{r^2})}{(\frac{\nu_1}{2}+\frac{2}{r})}dr. \tag{11}$$

Then when a solution $\{p,\ \rho,\ \nu\}$ to (5) is found, we would be able to find λ, $\mathcal{P}$ and $\mathcal{U}$ by (10), (3) and (4) respectively. Therefore, from the point of view of a brane observer, finding a solution in the brane, at least from the mathematical point of view, seems not very complicated. However it was shown in[5] that finding a consistent solution by starting from any arbitrary solution $\{p, \rho, \nu\}$ to the conservation equation (5), in general does not lead to a solution for $e^{-\lambda}$ having the expected form, which is

$$e^{-\lambda}=1-\frac{8\pi}{r}\int_0^r r^2\rho dr+\frac{1}{\sigma}(Bulk\ \ effects). \tag{12}$$

If the solution found to $e^{-\lambda}$ cannot be written by the way given by (12), then the general relativity limit, given through $\frac{1}{\sigma}\rightarrow 0$, will not be regained. Unfortunately this happen when we start from any arbitrary solution $\{p, \rho, \nu\}$

to (5). The source of this problem has to do with the formal solution to $e^{-\lambda}$ given through (10). Such a solution has mixed general relativity terms with non-local bulk terms in such a way that makes impossible to regain general relativity. A different way to explain why the formal solution (10) leads to the so-called "general relativity limit problem" is detailed next. First of all, it seems make sense to consider a solution to the geometric function λ as a generalization of the standard general relativity solution through

$$e^{-\lambda} = 1 - \frac{8\pi}{r}\int_0^r r^2\tilde{\rho}dr, \tag{13}$$

where

$$\tilde{\rho} \equiv \rho + \frac{1}{\sigma}\left(\frac{\rho^2}{2} + \frac{6}{k^4}\mathcal{U}\right) \tag{14}$$

is the effective density having local and non-local bulk effects on the brane. It can be seen that taking the limit $\frac{1}{\sigma} \to 0$ the well known general relativity solution is regained. Thus the lost of the general relativity seems not being a problem anymore. However we will see that the naive solution (13) is not a true solution at all. Let us start using (4) in (14) to obtain

$$\tilde{\rho} = \rho - \frac{1}{\sigma}\left(\rho^2 + 3\rho p\right) + \frac{1}{8\pi}\left(2G_2^2 + G_1^1\right) - 3p, \tag{15}$$

thus (13) is written as

$$e^{-\lambda} = 1 - \frac{8\pi}{r}\int_0^r r^2\rho dr + \frac{8\pi}{r}\int_0^r r^2\left[-\frac{1}{\sigma}\left(\rho^2 + 3\rho p\right) + \frac{1}{8\pi}\left(2G_2^2 + G_1^1\right) - 3p\right] dr. \tag{16}$$

Thus we can see that the "solution" (13) depends itself on $\lambda(r)$ and $\lambda_1(r)$ through G_1^1 and G_2^2, hence it represents an integral differential equation for the function $\lambda(r)$, something completely different from the general relativity case, and a direct consequence of the non-locality of the braneworld equations. A way to get it over will be explained in the next section.

3. Generating a constraint in the brane

So far we have two problems closed related making difficult the study of non-uniform braneworld stars, which any braneworld observer has to face, namely, the non possibility of regaining general relativity when the formal solution (10) is implemented or the existence of an integral differential equation when the standard solution (13) is enforced. A method already presented in[5] to overcome these two problems is explained in detail next.

First of all let us split the differential equation (9) as

$$\left[\frac{-\lambda_1 e^{-\lambda}}{r}+\frac{e^{-\lambda}}{r^2}-\frac{1}{r^2}+8\pi\rho\right]+\left[-\lambda_1 e^{-\lambda}(\frac{\nu_1}{2}+\frac{1}{r})+\right.$$

$$\left. e^{-\lambda}(\nu_{11}+\frac{\nu_1^2}{2}+2\frac{\nu_1}{r}+\frac{1}{r^2})-\frac{1}{r^2}-8\pi 3p-\frac{8\pi}{\sigma}\rho\left(\rho+3p\right)\right]=0, \quad (17)$$

here the left bracket has the standard general relativistic terms and the right bracket the bulk effects which modify the general relativistic equation. It can be seen that not all terms in the right bracket are manifestly bulk contributions, that is, not all of them are proportional to $\frac{1}{\sigma}$. Indeed only high energy terms are manifestly bulk contributions, while terms whose source is Weyl curvature remain as not explicit bulk contribution. This non-local terms, which are non explicit bulk contributions, are easily mixed with general relativistic terms, as is shown in the differential equation (9). Hence the solution eventually found to $e^{-\lambda}$, when (9) is solved, is written in such a way that will never have the form shown in (12). Therefore it will not be possible to regain the general relativistic solution when the $\frac{1}{\sigma}\to 0$ limit is taken. Keeping in mind general relativity as a limit, and the fact that a braneworld observer should see a geometric deformation due to five dimensional gravity effects, the following solution is proposed for the radial metric component:

$$e^{-\lambda}=\mu+Geometric\ \ Deformation, \quad (18)$$

where the μ function is the known general relativistic solution, given by

$$\mu=1-\frac{8\pi}{r}\int_0^r r^2\rho dr. \quad (19)$$

The unknown *geometric deformation* in (18) should have two sources: extrinsic curvature and five dimensional Weyl curvature, hence it can be written as a generic f function

$$e^{-\lambda}=\mu+f \quad (20)$$

which at the end will have the form

$$f=\frac{1}{\sigma}(\ high\ energy\ terms)\ +\ non\ local\ terms, \quad (21)$$

where, according to (16), the non local terms in (21) must be related with the anisotropy projected onto the brane. Demanding that the proposed solution (20) satisfies (17), a first order differential equation to f is obtained, given by

$$f_1 + \left[\frac{\nu_{11} + \frac{\nu_1^2}{2} + \frac{2\nu_1}{r} + \frac{2}{r^2}}{\frac{\nu_1}{2} + \frac{2}{r}}\right] f = \frac{\frac{8\pi}{\sigma}\rho(\rho + 3p) - H(p,\rho,\nu)}{\frac{\nu_1}{2} + \frac{2}{r}}, \tag{22}$$

where the function $H(p,\rho,\nu)$ is defined as

$$H(p,\rho,\nu) \equiv \left[\mu_1(\frac{\nu_1}{2} + \frac{1}{r}) + \mu(\nu_{11} + \frac{\nu_1^2}{2} + \frac{2\nu_1}{r} + \frac{1}{r^2}) - \frac{1}{r^2}\right] - 8\pi 3p. \tag{23}$$

Solving (22) the f function is written as[a]

$$f = e^{-I}\int_0^r \frac{e^I}{(\frac{\nu_1}{2} + \frac{2}{r})}\left[H(p,\rho,\nu) + \frac{8\pi}{\sigma}\left(\rho^2 + 3\rho p\right)\right] dr, \tag{24}$$

where local and non-local bulk effects can be seen. In (24) the function I is given again by the expression shown in (11). It is easy to see through the equations (2)-(4) evaluated at $\sigma^{-1} = 0$ (general relativity) that the non-local function $H(p,\rho,\nu)$ can be written as

$$H(p,\rho,\nu) = \left(2G_2^2 + G_1^1\right)|_{\sigma^{-1}=0} - 8\pi 3p, \tag{25}$$

which clearly correspond to an anisotropic term. In the general relativity case, that is, perfect fluid solution, the function $H(p,\rho,\nu)$ vanishes as a consequence of the isotropy of the solution. However, in the braneworld case, the general relativity isotropic condition $H(p,\rho,\nu) = 0$, in general should not be satisfied anymore. There is not reason to believe that the modifications undergone by p, ρ and ν, due to the bulk effects on the brane, do not modify the isotropic condition $H(p,\rho,\nu) = 0$. Therefore in general we have $H(p,\rho,\nu) \neq 0$. Thus the solution to the geometric function $\lambda(r)$ is finally written by

$$e^{-\lambda} = 1 - \frac{8\pi}{r}\int_0^r r^2 \rho dr + e^{-I}\int_0^r \frac{e^I}{(\frac{\nu_1}{2} + \frac{2}{r})}\left[H(p,\rho,\nu) + \frac{8\pi}{\sigma}\left(\rho^2 + 3\rho p\right)\right] dr. \tag{26}$$

In order to recover general relativity, the following condition must be satisfied

$$lim_{\sigma^{-1}\rightarrow\ 0} \int_0^r \frac{e^I}{(\frac{\nu_1}{2} + \frac{2}{r})} H(p,\rho,\nu) dr = 0. \tag{27}$$

[a]The constant of integration is put equal to zero to avoid a singular solution.

The expression (27) can be interpreted as a constraint whose physical meaning is nothing but the necessary condition to regain general relativity. A simple solution to (27) which has a direct physical interpretation is

$$H(p, \rho, \nu) = 0. \tag{28}$$

The constraint (28) explicitly ensure the general relativity limit[b] through the solution (26), and has been proven to be useful in finding solutions which posses general relativity as a limit.[5] It was clearly established through the equation (25) that the constraint (28) represents a condition of isotropy in general relativity. Thus in the context of the braneworld the constraint (28) has a direct physical interpretation: *eventual bulk corrections to p, ρ and ν will not produce anisotropic effects on the brane.* A clearer physical meaning of the constraint (28) will be illustrated in the next section.

4. Generating the braneworld version of any known general relativistic solution

After the constraint (28) is imposed, the problem on the brane is reduced to finding a solution $\{p(r), \rho(r), \nu(r)\}$ satisfying (5) and (28). The author found that the following general expression for the geometric variable $\nu(r)$

$$e^{\nu} = A(1 + Cr^m)^n \tag{29}$$

produces an analytic expression for the integral shown in (11) which has to be used in (26), and a complicated integral equation for ρ when (29) is used in (5) and (28). It is difficult to figure out appropriates values for the set of constants $\{A, C, m, n\}$ leading to exact expressions for p and ρ, and even more in the searching of exact and physically acceptable solutions. Nevertheless exact solutions for $\{p(r), \rho(r), \nu(r)\}$ where found in Ref.[5] and an exact solution for the complete system $\{p(r), \rho(r), \nu(r), \lambda(r), \mathcal{U}(r), \mathcal{P}(r)\}$ is reported in Ref.[4] It is worth noticing that in the case of a uniform distribution the system is *closed*, thus it is not necessary to impose any additional restriction except the constraint (28), which will produce a non lineal differential equation for the geometric function ν.

The way described in the previous paragraph was successfully used by the author. However now we become aware of an important issue regarding the constraint (28), which was not previously mentioned: *every solution for a perfect fluid satisfies the minimal anisotropic condition represented through the constraint H=0.* Therefore, the constraint (28) represents a

[b]Indeed, every perfect fluid solution satisfies H=0 (see next section)

natural way to generalize perfect fluid solutions (general relativity) in the context of the braneworld. The method consist in taking a known perfect fluid solution $\{p(r), \rho(r), \nu(r)\}$, then $\lambda(r)$, $\mathcal{P}(r)$ and $\mathcal{U}(r)$ are found through (26), (3) and (4) respectively. Of course the fact that the constraint (28) be satisfied by every known perfect fluid solution does not mean that any of these solutions can be directly used to find an exact braneworld solution. In order to investigate if one particular general relativistic solution has an exact braneworld version, the first step is to analyze the temporal component of the metric. If $\nu(r)$ is not simple enough, then hardly it will provide an analytic expression for (11). Even in the case where (11) be analytic, finding an exact expression for $\lambda(r)$ using (26) is very difficult. Nevertheless, this approach, which is entirely based in the constraint $H = 0$, always can be used to obtain the braneworld version of *any* general relativistic solution by numerical methods. Next it is explained a direct physical meaning of this constraint. Using the geometric deformation f as shown in Eq. (20) in the expression for $\mathcal{P}$ given in (3), it is found that the anisotropy induced by five dimensional effects may be written in term of the geometric deformation as

$$\frac{48\pi}{k^4}\mathcal{P} = -\frac{1}{r^2} + \mu(\frac{1}{r^2} + \frac{\nu_1}{r}) - \frac{1}{4}\mu(2\nu_{11} + \nu_1^2 + 2\frac{\nu_1}{r}) - \frac{1}{4}\mu_1(\nu_1 + \frac{2}{r})$$
$$+f(\frac{1}{r^2} + \frac{\nu_1}{r}) - \frac{1}{4}f(2\nu_{11} + \nu_1^2 + 2\frac{\nu_1}{r}) - \frac{1}{4}f_1(\nu_1 + \frac{2}{r}). \quad (30)$$

When the constraint (28) is imposed the anisotropy induced shown in (30) is written in terms of the geometric deformation by

$$\frac{48\pi}{k^4}\mathcal{P} = \frac{3}{2}\left[-\frac{1}{r^2} + \mu(\frac{1}{r^2} + \frac{\nu_1}{r}) - 8\pi p\right]$$
$$+f^*(\frac{1}{r^2} + \frac{\nu_1}{r}) - \frac{1}{4}f^*(2\nu_{11} + \nu_1^2 + 2\frac{\nu_1}{r}) - \frac{1}{4}f_1^*(\nu_1 + \frac{2}{r}), \quad (31)$$

where

$$f^* = \frac{1}{\sigma}(\textit{ high energy terms}) + \underbrace{\textit{non local terms}}_{=0} \quad (32)$$

is the *minimal geometric deformation*, whose explicit form may be seen by (24) as following

$$f^* = \frac{8\pi}{\sigma}e^{-I}\int_0^r \frac{e^I}{(\frac{\nu_1}{2} + \frac{2}{r})}(\rho^2 + 3\rho p)\, dr. \quad (33)$$

The expression (33) represents a minimal geometric deformation in the sense that all sources of the geometric deformation f have been removed except those produced by the density and pressure, which are always present in a stellar distribution[c]. It is clear that the geometric deformation represented by the f^* function is a source of anisotropy, as may be seen in (31). However there is another source for $\mathcal{P}$, which is represented by the bracket shown in (31). Nevertheless, when the constraint (28) is imposed the bracket shown in (31) should be zero, since eventual bulk corrections to p, ρ and ν do not produce anisotropic effects on the brane. Indeed, every general relativistic solution produces

$$\frac{48\pi}{k^4}\mathcal{P} = \frac{3}{2}\underbrace{\left[-\frac{1}{r^2}+\mu(\frac{1}{r^2}+\frac{\nu_1}{r})-8\pi p\right]}_{-8\pi T_1^1-G_1^1=0}$$
$$+f^*(\frac{1}{r^2}+\frac{\nu_1}{r})-\frac{1}{4}f^*(2\nu_{11}+\nu_1^2+2\frac{\nu_1}{r})-\frac{1}{4}f_1^*(\nu_1+\frac{2}{r}), \quad (34)$$

thus leaving the minimal geometric deformation f^* as the only source for the anisotropy induced inside the stellar distribution. Therefore every general relativistic solution satisfies the minimal anisotropic effect onto the brane. In consequence any general relativistic solution may be used to obtain a consistent braneworld stellar solution. Here below is shown the basic steps to use this approach.

- Step 1: Impose the constraint $H(p,\rho,\nu)=0$ to make sure we have a solution for the geometric function $\lambda(r)$ with the correct limit:

$$e^{-\lambda(r)} = 1-\frac{8\pi}{r}\int_0^r r^2\rho dr+\frac{8\pi}{\sigma}e^{-I}\int_0^r\frac{e^I}{(\frac{\nu'}{2}+\frac{2}{r})}(\rho^2+3\rho p)\,dr.$$

- Step 2: Pick a known general relativistic solution (p,ρ,ν) to the conservation equation $p'=-\frac{\nu'}{2}(\rho+p)$.
- Step 3: Find $\mathcal{P}$ and $\mathcal{U}$ by equations shown in (3) and (4).
- Step 4: Drop out the condition of vanishing pressure at the surface to obtain the bulk effect on any constant $C\to C(\sigma)$. Then we are able to find the bulk effect on pressure and density.

5. Conclusions and outlook

In the context of the Randall-Sundrum braneworld, a detailed description of the approach based in the minimal anisotropic consequence onto the brane was carefully presented. The explicit form of the anisotropic stress was obtained in terms of the geometric deformation undergone by the radial metric component, thus showing the role played by this deformation

[c]An even minimal deformation is obtained for a dust cloud, where $p=0$.

as a source of anisotropy inside the stellar distribution. It was shown that this geometric deformation is minimal when a general relativistic solution is considered, therefore any general relativistic solution belongs to a subset of braneworld solutions producing a minimal anisotropic consequence onto the brane. It was found that through this approach, it is possible to generate the braneworld version of any known general relativistic solution, thus overcoming the non-closure problem of the braneworld equations. This approach might be extended in the case of braneworld theories without Z_2 symmetry or any junction conditions, as those introduced in[6] and.[7] Another subjects of interest is the use of this approach in brane theories with variable tension, as introduced by[8] in the cosmological context, and the study of codimension-2 braneworld theories, as those developed in[9] and.[10] A possible extension of this approach in all these theories is currently being investigated.

Acknowledgments

This work was supported by **DID, USB**. Grant: S1-IN-CB-002-09, and by **FONACIT**. Grant: S2-2009000298.

References

1. R. Maartens, Brane-world gravity, Living Rev.Rel. 7 (2004).
2. K. Koyama and R. Maartens, Structure formation in the DGP cosmological model, JCAP 0601, 016(2006).
3. A. Viznyuk and Y. Shtanov, Spherically symmetric problem on the brane and galactic rotation curves, Phys.Rev.D,76 064009 (2007)
4. J. Ovalle, Non-uniform Braneworld Stars: an Exact Solution, Int.J.Mod.Phys.D18,837,(2009).
5. J. Ovalle, Searching Exact Solutions for Compact Stars in Braneworld: a conjecture, Mod.Phys.Lett.A23,3247(2008).
6. M. D. Maia, E. M. Monte and J. M. F. Maia, The accelerating universe in brane-world cosmology, Phys. Lett. B585, 11 (2004)
7. M. D. Maia, E. M. Monte, J. M. F. Maia and J. S. Alcaniz, On the geometry of dark energy, Class. Quant. Grav.22, 1623(2005)
8. László Á. Gergely, Friedmann branes with variable tension, Phys.Rev.D78:084006 (2008)
9. Bertha Cuadros-Melgar, Eleftherios Papantonopoulos, Minas Tsoukalas, Vassilios Zamarias, Black Holes on Thin 3-branes of Codimension-2 and their Extension into the Bulk, Nucl.Phys.B810:246-265(2009).
10. Eleftherios Papantonopoulos, Black Holes and Black String-like Solutions in Codimension-2 Braneworlds, Int.J.Mod.Phys.A24:1489-1496(2009).

QUANTUM YANG-MILLS GRAVITY: THE GHOST PARTICLE AND ITS INTERACTIONS

JONG-PING HSU

Department of Physics, University of Massachusetts Dartmouth
North Dartmouth, MA 02747-2300, USA
E-mail: jhsu@umassd.edu

The classical Yang-Mills gravity with translation gauge symmetry in flat space-time was shown to be consistent with experiments in previous papers. We summarize the main features and then discuss the massless ghost particle and its interaction in pure gravity. It is convenient to formulate quantum Yang-Mills gravity by using the Lagrange multiplier method. We discuss the propagator and interactions of ghost vector particles. These results are necessary for the unitarity and gauge invariance of the S-matrix in quantum Yang-Mills gravity.

1. Introduction

A satisfactory theory in physics should have experimental support and theoretical coherence. Quantum electrodynamics (QED) is a well-known example. Theoretically, the fundamental reasons for quantum electrodynamics to reach such a high level of aesthetical satisfaction are (i) QED is formulated within the quantum-mechanical framework and (ii) QED is endowed with U(1) gauge symmetry and Lorentz-Poincaré invariance. One marvelous feature of QED is that the U(1) symmetry guarantees the conservation of charge, which is the source for generating the electromagnetic field. It is arguably that Einstein's gravity has also reached a high level of theoretical coherence in the classical aspect. Nevertheless, its quantum aspect has a long-standing problem. We know that the space-time translation symmetry assures the conservation of energy-momentum tensor, which turns out to be also the source for generating the gravitational field. Thus, it is tempting to investigate whether one can use Yang-Mills theory to combine the space-time translation symmetry and the Lorentz-Poincaré invariance in flat space-time. If such a Yang-Mills theory of gravity can be accomplished, then one probably can bring gravity into the framework of quantum mechanics and quantum fields. Such an investigation is non-trivial because we are dealing with an external gauge group, whose group generators have the dimension of 1/length in natural units and do not have constant matrix representations.

In previous papers, the classical Yang-Mills gravity with translation gauge symmetry (or T(4) group) in flat 4-dimensional space-time is shown to be consistent with experiments.[1,2] The reason is that wave equations of physical fields with translation gauge symmetry lead to an 'effective Riemannian metric tensor' in the limit of geometric optics of wave equations. These fields include usual scalar, fermion and electromagnetic fields. Therefore, it appears as if light rays and classical particles in Yang-Mills gravity move in a 'curved space-time' described by Riemannian geometry. In other words, the theory of Yang-Mills gravity also has an effective Einstein-Grossman action, $S_{EG} = \int mds$, for a classical object, so that the object appears to move along the geodesic in an 'effective curved space-time.' However, the real underlying physical spacetime of gauge fields and quantum particles is inherently flat, i.e., has a vanishing Riemann-Christoffel curvature tensor.

Since Utiyama paved the way for a gauge theory of gravity, most people followed the usual approach of general relativity to formulate their gauge theories of gravity within the framework of curved space-time.[3] However, in order to overcome the difficulties of quantizing Einstein's gravity, it was speculated that a solution to these difficulties requires 'some quantum-mechanical analog of Riemannian geometry.' "Some analog of a metric tensor must be introduced in order to give a meaning to space-like separation."[4] Generally speaking, the mathematical framework of general coordinate invariance appears to be too general and has made it impossible to define a space-like relationship between two regions. Thus the necessary and important postulate of 'local commutativity' in quantum field theory becomes meaningless. On the other hand, the framework of flat spacetime has the advantages that

(a) field theory with T(4) gauge symmetry has an effective Riemannian metric tensor in the classical limit,

(b) the quantization of Yang-Mills gravity can be carried out, and

(c) T(4) symmetry in flat spacetime assures the conservation of a bona fide energy-momentum tensor.

Experimental supports of classical Yang-Mills gravity[1,2] and the interesting properties (a)-(c) motivate our further investigation of quantum Yang-Mills gravity.

In sharp contrast to the usual electrodynamics with Abelian group U(1),

Yang-Mills gravity based on Abelian group T(4) of spacetime translation symmetry in flat spacetime needs 'ghost particles' to preserve the unitarity and gauge invariance of the S-matrix,[5] similar to Yang-Mills theories with non-Abelian gauge groups. In contrast to the usual gauge theories with internal gauge groups, the T(4) gauge field in Yang-Mills gravity is a symmetric tensor field, $\phi_{\mu\nu} = \phi_{\nu\mu}$, rather than a (Lorentz) vector field. Such a tensor field $\phi_{\mu\nu}$ may be termed 'spacetime gauge field.'

The gauge theory with external translation group T(4) in flat spacetime leads to the following important and unique properties for gravity:

(A). Yang-Mills gravity has only one kind of 'charge' (i.e., mass $m \geq 0$ and only attractive force between all matter-matter, matter-antimatter and antimatter-antimatter interactions, because the generator $i\partial/\partial x^\mu$ $(c = \hbar = 1)$ of the spacetime translation group contains $i(= \sqrt{-1})$.[1] This is in sharp contrast with all other Yang-Mills theories with internal gauge groups that have two kinds of 'charges' and have both attractive and repulsive forces, just like the electromagnetic forces.

(B) The gravitational coupling constant g in Yang-Mills gravity has the dimension of 'length', in sharp contrast to the dimensionless coupling constant in the usual Yang-Mills theories with internal gauge groups. This is due to the fact that the generators $i\partial/\partial x^\mu$ of the translation group has a dimension of (1/length), contrary to the dimensionless generators (that also have constant matrix representations) of internal gauge groups.

2. The Gauge-Invariant Action and Gauge Conditions

In general, Yang-Mills gravity can be formulated in both inertial and non-inertial frames and in the presence of fermion fields.[1,2] It is complicated to discuss quantum field theory and particle physics even in a simple non-inertial frame with a constant linear acceleration, where the accelerated transformation of spacetime is smoothly connected to the Lorentz transformation in the limit of zero acceleration.[6,7,8]

For simplicity, let us consider the quantization of Yang-Mills gravity without involving fermions and only in inertial frames (with the Minkowski metric tensor $\eta^{\mu\nu} = (1, -1, -1, -1)$). Nevertheless, we shall use general and arbitrary coordinates in the discussions of the T(4) gauge identity and the derivation of Hilbert's symmetrized energy-momentum tensor in section 3.[9] Now let us first discuss Yang-Mills gravity in inertial frames. The action S_{pg} for pure gravity involves spacetime gauge field $\phi_{\mu\nu}$ and gauge-fixing terms L_ξ, and is assumed to be[1]

$$S_{pg} = \int L_{pg} d^4x, \quad L_{pg} = L_\phi + L_\xi \tag{1}$$

$$L_\phi = \frac{1}{4g^2}\left(C_{\mu\nu\alpha}C^{\mu\nu\alpha} - 2C_{\mu\alpha}{}^{\alpha}C^{\mu\beta}{}_{\beta}\right), \tag{2}$$

$$C^{\mu\nu\alpha} = J^{\mu\sigma}\partial_\sigma J^{\nu\alpha} - J^{\nu\sigma}\partial_\sigma J^{\mu\alpha}, \qquad J_{\mu\nu} = \eta_{\mu\nu} + g\phi_{\mu\nu} = J_{\nu\mu},$$

where $C^{\mu\alpha\beta}$ is the T(4) gauge curvature in inertial frames and L_ξ will be discussed below. We note that the Lagrangian L_ϕ changes only by a divergence under the translation gauge transformation, and the action functional $S_\phi = \int L_\phi d^4x$ is invariant under the spacetime translation gauge transformation.[1] For quantization of Yang-Mills gravity with T(4) gauge symmetry, it is necessary to include the gauge fixing terms L_ξ in the action functional (1), just like the quantization of a Yang-Mills theory with gauge symmetry. The gauge-fixing term L_ξ in (1) is assumed to be

$$L_\xi = \frac{\xi}{2g^2}\left[\partial^\mu J_{\mu\alpha} - \frac{1}{2}\partial_\alpha J\right]\left[\partial_\nu J^{\nu\alpha} - \frac{1}{2}\partial^\alpha J\right], \tag{3}$$

$$J = J^\lambda_\lambda = \delta^\lambda_\lambda - g\phi, \quad \phi = \phi^\lambda_\lambda, \qquad c = \hbar = 1, \tag{4}$$

where L_ξ involves an arbitrary gauge parameters ξ. We may remark that the gauge-fixing term L_ξ, take the same expression (3) in a general frame, so that the action involving L_ξ is not gauge invariant.[10,11] However, in a general frame, the constant metric $\eta_{\mu\nu}$ and ∂_μ in the (2) should be replaced by a space-time-dependent metric tensor $P_{\mu\nu}$ and the covariant derivative D_μ respectively. In this way, the Lagrangian $L_\phi\sqrt{-P}$ in a general frame also changes only by a divergence under T(4) gauge transformations.[1]

The Lagrangian in (3) corresponds to a class of gauge conditions,

$$\frac{1}{2}[\eta^{\mu\rho}\eta^{\nu\lambda} + \eta^{\nu\rho}\eta^{\mu\lambda} - \eta^{\mu\nu}\eta^{\rho\lambda}]\partial_\lambda J_{\mu\nu} = \partial_\lambda J^{\rho\lambda} - \frac{1}{2}\partial^\rho J \equiv Y^\rho, \tag{5}$$

where Y^ρ is a suitable function of spacetime. The Lagrangian for pure gravity L_{pg} in (1) can be expressed in terms of spacetime gauge fields $\phi_{\mu\nu}$:

$$L_{pg} = L_2 + L_\xi + L_{int}, \tag{6}$$

where

$$L_2 = \frac{1}{2}\left(\partial_\lambda\phi_{\alpha\beta}\partial^\lambda\phi^{\alpha\beta} - \partial_\lambda\phi_{\alpha\beta}\partial^\alpha\phi^{\lambda\beta} - \partial_\lambda\phi\partial^\lambda\phi\right. \tag{7}$$

$$\left. + 2\partial_\lambda\phi\partial^\beta\phi^\lambda_\beta - \partial_\lambda\phi^\lambda_\mu\partial^\beta\phi_{\mu\beta}\right),$$

$$L_\xi = \frac{\xi}{2}\left[(\partial_\lambda \phi^{\lambda\alpha})\partial^\rho \phi_{\rho\alpha} - (\partial_\lambda \phi^{\lambda\alpha})\partial_\alpha \phi + \frac{1}{4}(\partial^\alpha \phi)\partial_\alpha \phi\right]. \tag{8}$$

The Lagrangians L_2 and L_ξ involve quadratic tensor field and determine the propagator of the graviton in Yang-Mills gravity. Other terms L_{int} in (6) correspond to the interactions of 3 and 4 gravitons, which will not be discussed here.

3. Lagrange Multipliers and Ghost particles

The Lagrange multiplier method is a powerful tool to deal with a physical system with constraints.[10,11] For various gauge conditions, it is convenient to use the Lagrange multiplier χ^μ to discuss the corresponding S-matrix and unitarity. If one chooses the class of gauge conditions given in (5), the gauge-fixing terms L_ξ in (3) can be replaced by the following Lagrangian L_χ involving the Lagrange multiplier χ^μ

$$L_\chi = \frac{1}{g^2}\chi^\mu(\partial^\nu J_{\mu\nu} - \frac{1}{2}\partial_\mu J) - \frac{1}{2\xi g^2}\chi_\mu \chi^\mu. \tag{9}$$

The action functional $S_{\phi\chi}$ for pure gravity in inertial frames is

$$S_{\phi\chi} = \int d^4x(L_\phi + L_\chi), \tag{10}$$

which is equivalent to (1). It leads to spacetime gauge field equations and a constraint involving χ^μ:

$$H^{\mu\nu} + (\partial^\mu \chi^\nu - \frac{1}{2}\eta^{\mu\nu}\partial_\lambda \chi^\lambda) = 0, \tag{11}$$

$$\chi^\lambda = \xi(\partial_\sigma J^{\sigma\lambda} - \frac{1}{2}\partial^\lambda J), \tag{12}$$

$$H^{\mu\nu} \equiv Sym_{\mu\nu}\ [\partial_\lambda(J^\lambda_\rho C^{\rho\mu\nu} - J^\lambda_\alpha C^{\alpha\beta}{}_\beta \eta^{\mu\nu} + C^{\mu\beta}{}_\beta J^{\nu\lambda}) \tag{13}$$

$$-C^{\mu\alpha\beta}\partial^\nu J_{\alpha\beta} + C^{\mu\beta}{}_\beta \partial^\nu J^\alpha_\alpha - C^{\lambda\beta}{}_\beta \partial^\nu J^\mu_\lambda],$$

where the indices μ and ν in gauge field equations (11) and (13) should be made symmetric, e.g., $\partial^\mu \chi^\nu \to [\partial^\mu \chi^\nu + \partial^\nu \chi^\mu]/2$.

For discussions of the ghost particle and its interaction, it is useful to obtain the identity of the T(4) gauge invariance. The gauge curvature $C^{\mu\alpha\beta}$ in (2) and the action functional (1), are expressed in terms of the tensor $J_{\mu\nu}$, it is possible to consider the gauge variation of the invariant action functional S_ϕ with respect to $J_{\mu\nu}$. As we have shown in the paper I,[1] the gauge variation of the action functional involves the variation of both the physical tensor field $\phi_{\mu\nu}$ and the geometric metric tensor $P_{\mu\nu}$ of a general frame. In other words, the action (14) is a scalar, and therefore does not change for an infinitesimal coordinate transformation. In Yang-Mills gravity, the variations of $J_{\mu\nu}$ and $\phi_{\mu\nu}$ occur as a result of the infinitesimal transformation $x'^\mu = x^\mu + \Lambda^\mu$.[1,12] In our discussions, the variation of the 10 metric tensors, $\delta P_{\mu\nu}$, are not all independent because they are the result of a transformation of the 4-component coordinates.[12] Such a variation of the geometric metric tensor $P_{\mu\nu}$ can lead to the conservation of Hilbert's symmetrized energy-momentum tensor in flat space-time with arbitrary coordinates.[9,12,13]

To obtain the T(4) gauge identity and Hilbert's symmetrized energy-momentum tensor, let us write the action functional in a general frame

$$S_\phi = \int d^4x L_\phi \sqrt{-P}, \qquad P = det P_{\mu\nu}, \tag{14}$$

where L_ϕ in a general frame is given by (2) with $C^{\mu\nu\alpha} = J^{\mu\sigma} D_\sigma J^{\nu\alpha} - J^{\nu\sigma} D_\sigma J^{\mu\alpha}$, and $J^{\mu\nu} = P^{\mu\nu} + g\phi^{\mu\nu} = J^{\nu\mu}$. The action (14) is invariant under the complete variation of $\phi_{\mu\nu}$ and $P_{\mu\nu}$.[1,12] In Yang-Mills gravity, the Poincaré metric tensor $P_{\mu\nu}$ is not a physical field. Thus, the transition to arbitrary coordinates can be regarded as an intermediate step in the derivation of the energy-momentum tensor.[12,13] In such a calculation, we can expressed both $\delta P_{\mu\nu}$ and $\delta J^{\mu\nu}$ in terms of Λ^μ, e.g., $\delta P_{\mu\nu} = -\Lambda^\sigma \partial_\sigma P_{\mu\nu} - P_{\mu\sigma}\partial_\nu \Lambda^\sigma - P_{\sigma\nu}\partial_\mu \Lambda^\sigma$.[1] Because the gauge function Λ^μ is arbitrary, we obtain a gauge identity from the invariant action (14). In inertial frame with $P_{\mu\nu} = \eta_{\mu\nu}$, this T(4) gauge identity takes the form,

$$(\partial_\sigma J^{\mu\nu}) H_{\mu\nu} + 2\partial_\mu (H_{\sigma\nu} J^{\mu\nu}) + 2\partial_\nu (\eta_{\mu\sigma} T^{\mu\nu}) = 0, \tag{15}$$

where Hilbert's symmetrized energy-momentum tensor $T^{\mu\nu}$ is given by

$$T^{\mu\nu} = Sym_{\mu\nu} \, \{-(\partial^\nu J^{\beta\alpha}) C^\mu{}_{\beta\alpha} + \frac{1}{8}[2C^{\mu\alpha\beta} C^\nu{}_{\alpha\beta} + 2C^{\mu\alpha}_\beta C^{\beta\nu}{}_\alpha \tag{16}$$

$$+2C^{\alpha\beta\mu} C_{\alpha\beta}{}^\nu + \eta^{\mu\nu} C_{\alpha\beta\lambda} C^{\alpha\beta\lambda}]\} + V^{\mu\nu},$$

$$V^{\mu\nu} = Sym_{\mu\nu} \left(-\frac{1}{2}\right) \partial_\lambda [J^\lambda_\alpha f^{\nu\beta} C^{\alpha\mu}_\beta + J^\lambda_\alpha f^{\nu\beta} C^{\alpha\mu}_\beta - J^\lambda_\alpha f^{\nu\beta} C^{\alpha\mu}_\beta$$

$$+J^{\alpha\lambda} f^{\beta\nu} C_{\alpha\beta}{}^\mu + J^{\alpha\lambda} f^{\beta\nu} C_{\alpha\beta}{}^\mu - J^{\alpha\lambda} f^{\beta\nu} C_{\alpha\beta}{}^\mu],$$

where $f^{\mu\nu} = J^{\mu\nu} - \eta^{\mu\nu}$. In equation (15), the indices μ and ν in $H_{\mu\nu}$ and $T^{\mu\nu}$ should be symmetrized. We use "$Sym_{\mu\nu}$" in (16) explicitly to indicate that μ and ν should be made symmetric.

We note that the energy-momentum tensor calculated from the Noether theorem is generally not symmetric. Yet it is always possible to symmetrized it by adding suitable terms of the form $\partial^\lambda F_{\mu\nu\lambda}$, where $F_{\mu\nu\lambda}$ is anti-symmetric in the indices ν and λ. The advantage of the method of derivation in steps (14)-(16) is that the resultant energy-momentum tensor $T_{\mu\nu}$ is automatically symmetrized and is conserved, $\partial^\mu T_{\mu\sigma} = 0$ in flat space-time.[13]

Since (i) the energy-momentum tensor $T_{\mu\nu}$ is conserved by Noether's theorem (or by Hilbert's method),[9,12,13] (ii) the action functional S_ϕ is gauge invariant and (iii) the gauge function Λ^σ is arbitrary, one obtains an important formula for the spacetime gauge field $\phi_{\mu\nu}$:

$$[(\partial_\sigma J^{\mu\nu}) + 2\delta^\mu_\sigma(\partial_\lambda J^{\lambda\nu}) + 2\delta^\mu_\sigma J^{\lambda\nu}\partial_\lambda]H_{\mu\nu} = 0. \tag{17}$$

This T(4) gauge formula is useful for deriving an effective Lagrangian for the ghost particles in quantum Yang-Mills gravity.

4. Propagator and Interaction Vertex of the Ghost Particle

Let us discuss the general propagator (involving an arbitrary gauge parameter ζ) for the ghost particle and the interaction vertices of the ghost vector fields. From field equations (11) and the identity (17), we obtain the field equation for the Lagrange multiplier χ^μ. This equation can be derivative from an effective action S_{eff}:

$$S_{eff} = \int \{-\partial^\nu \overline{V}^\mu \partial_\nu V_\mu + \frac{g}{2}\overline{V}^\sigma(\partial_\sigma \phi_{\mu\nu})(\partial^\mu V^\nu + \partial^\nu V^\mu - \eta^{\mu\nu}\partial_\lambda V^\lambda) \tag{18}$$
$$-g(\partial^\sigma \overline{V}_\mu)\phi_{\sigma\nu}(\partial^\mu V^\nu + \partial^\nu V^\mu - \eta^{\mu\nu}\partial_\lambda V^\lambda)\} d^4x.$$

Note that we have replace χ^μ by the vector field V^μ for the ghost particle. The field $\overline{V}^\mu$ in (18) is considered as an independent field. The quanta of the fields V^μ and $\overline{V}^\mu$ in the effective Lagrangian L_{eff} are the ghost (or fictitious) particles in Yang-Mills gravity. An equivalent effective Lagrangian for ghost particle can also be obtained by using the gauge condition (5) and the Faddeev-Popov method.[14,15] A similar situation also occurs in Yang-Mills theories with internal gauge groups.[11]

The propagator of the ghost vector particle can be derived from the effective Lagrangian (18) for the ghost field. In (9), the gauge condition is $\partial^\mu J_{\mu\nu} - 1/2\partial_\nu J \equiv Y_1^\nu$ and the corresponding ghost propagator can be derived from (18), we find the following Feynman propagator,

$$G^{\mu\nu} = \frac{-i\eta^{\mu\nu}}{k^2}. \tag{19}$$

This propagator is consistent with those obtained under the same gauge condition in Einstein's theory of gravity.[10,16,17] If one chooses a different gauge condition, the ghost propagator (19) will be changed. To wit, suppose the first term $g^{-2}\chi^\mu(\partial^\nu J_{\mu\nu} - \frac{1}{2}\partial_\mu J)$ in (9) is replaced by $g^{-2}\chi^\mu\partial^\nu J_{\mu\nu}$, then the effective action (18) will be modified and the ghost propagator is

$$G^{\mu\nu} = \frac{-i}{k^2}\left(\eta^{\mu\nu} - \frac{k^\mu k^\nu}{2k^2}\right). \tag{20}$$

The $i\epsilon$ prescription for Feynman propagators in (19) and (20) is understood.

Let us consider the ghost-ghost-graviton vertex, which is denoted by

$$\overline{V}^\nu(p)V^\mu(q)\phi^{\alpha\beta}(k), \qquad p_\lambda + q_\lambda + k_\lambda = 0, \tag{21}$$

where we have used the convention that all momenta are incoming to the vertices. The Feynman rule for such a vertex is given by

$$Sym_{\alpha\beta}\ ig[-q^\alpha k^\nu \eta^{\mu\beta} + p^\alpha q^\nu \eta^{\mu\beta} + p^\alpha q^\beta \eta^{\mu\nu} \tag{22}$$

$$+\frac{1}{2}q^\mu k^\nu \eta^{\alpha\beta} - p^\alpha q^\mu \eta^{\nu\beta}].$$

The ghost particle may be considered as a 'vector-fermion.' All vector-fermion vertices are bilinear in the vector-fermions, as shown in the effective Lagrangian (18). The vector-fermion appears, by definition of the physical subspace for the S-matrix, only in closed loops in the intermediate states of a physical process, and there is a factor of -1 for each vector-fermion loop.[5,11]

In general, the graviton and ghost propagators in Feynman rules may depend on the specific form of gauge condition and the gauge parameters. In the action (9), we choose the gauge condition $\partial_\mu J^{\mu\nu} - 1/2\partial^\nu J \equiv Y_1^\nu$ with an arbitrary ξ, the ghost propagator is given by (19) which is independent of ξ. But the corresponding graviton propagator depends on ξ and takes the form:

$$G_{\alpha\beta,\rho\sigma} = i\left[\frac{1}{k^2}(\eta_{\alpha\rho}\eta_{\beta\sigma} + \eta_{\beta\rho}\eta_{\alpha\sigma 0} - \eta_{\alpha\beta}\eta_{\rho\sigma})\right.$$

$$\left. - \frac{1}{k^4}\left(1 + \frac{2}{\xi}\right)(\eta_{\rho\alpha}k_\sigma k_\beta + \eta_{\rho\beta}k_\sigma k_\alpha + \eta_{\sigma\alpha}k_\rho k_\beta + \eta_{\sigma\beta}k_\rho k_\alpha)\right]. \tag{23}$$

This graviton propagator is consistent with that in Einstein gravity obtained by Fradkin and Tyutin. The reason is that both Einstein gravity and Yang-Mills gravity have the same linearized field equations.[1,10]

5. Discussions

The translation gauge invariance of Yang-Mills gravity implies that the observable results such as those obtained from the S-matrix should be independent of the gauge parameter ξ in the gauge-fixing term L_ξ.

Suppose one considers quantum electrodynamics with a non-linear gauge condition[18] or a Yang-Mills theory with linear or non-linear gauge conditions, the equation for the Lagrange multiplier field has source terms and is not free.[10,11] This implies that the unitarity of the S-matrix will be upset. The field equation of the Lagrange Multiplier can lead to an effective action S_{eff} of a scalar ghost particle. Roughly speaking, the unphysical components of the gauge field in the original gauge invariant action produce some unphysical amplitudes.[19] These unwanted amplitudes are cancelled by those produced by ghost particles described by the action S_{eff}. By the definition of the physical subspace for the S-matrix, the ghost particles can only appear in the intermediate states of a physical process.[11]

In contrast to the previous works of Utiyama and Fukuyama,[20] and Cho[21] formulated in curved space-time, Ning Wu recently attempted to give Einstein's gravity a new interpretation based on a T(4) gauge symmetry in flat spacetime.[22] Namely, the Lagrangian in the Hilbert-Einstein action can be expressed in terms of the T(4) gauge curvature $C^{\mu\nu\alpha}$ and the T(4) tensor field within the framework conventional field theory. He gave a formal proof that the quantum gravity based on the usual Hilbert-Einstein action with the T(4) gauge symmetry in flat spacetime is renormalizable.[23] In view of the profound difficulties in the quantization of gravity in curved spacetime, it is desirable to have explicit calculations to substantiate a formal proof of renormalizability, because one has a non-compact external gauge group.[5]

We stress that there is a significant difference in the Feynman rules for interaction vertices between the Yang-Mills gravity in flat spacetime and all other theories of gravity in curved spacetime (including the formulations of gravity in flat space-time by Wu and Logunov).[22] Namely, in Yang-Mills gravity, the maximum number of graviton interaction in a vertex is 4 in Feynman rules, as one can see in the Lagrangian L_ϕ (2). In contrast, there exists vertices involving an arbitrary large number of gravitons in all other theories of gravity and in formulations of gravity by Wu and Logunov. This maximum 4-vertex for graviton interaction together with the T(4) gauge symmetry may be a gateway to the renormalizable quantum Yang-Mills gravity.

References

1. J. P. Hsu, Int. J. Mod. Phys. A, **21**, 5119 (2006).
2. J. P. Hsu, Int. J. Mod. Phys. A. (to be published, 2009)
3. R. Utiyama, Phys. Rev. **101**, 1597 (1956). See also ref. 20.
4. See, for example, *100 Years of Gravity and Accelerated Frames, The deepest Insights of Einstein and Yang-Mills* (Ed. J. P. Hsu and D. Fine, World Scientific, 2005), ch. 7, and pp. 347-352. Dyson said that "The most glaring incompatibility of concepts in contemporary physics is that between Einstein's principle of general coordinate invariance and all the modern schemes for a quantum-mechanical description of nature."
5. J. P. Hsu and M. D. Xin, Phys. Rev. D**24**, 471 (1981). See also ref. 11.
6. J. P. Hsu, Chin. J. Phys. **40**, 265 (2002).
7. D. Schmidt and J. P. Hsu, Int. J. Mod. Phys. A **20**, 5989 (2005).
8. J. P. Hsu and D. Fine, Int. J. Mod. Phys. A **20** 7485 (2005).
9. D. Hilbert, ref. 4, pp. 120-131; S. Deser, Gen. Rel. Grav. **1**, 9 (1970); A. A. Logunov, *The Theory of Gravity* (Moscow, NAUKA, 2001) p. 28.
10. F. S. Fradkin and I. V. Tyutin, Phys. Rev. D**2**, 2841 (1970).
11. J. P. Hsu and J. A. Underwood, Phys. Rev. D**15**, 1668 (1977) and ref. therein.
12. V. Fock, *The Theory of Space Time and Gravitation*(trans. N. Kemmer, Pergamon Press, 1958) pp. 158-166.
13. L. Landau and E. Lifshitz, *The Classical Theory of Fields* (trans. M. Hamermesh, Addison-Wesley, Cambridge, MA. 1951) pp. 293-296.
14. V. N. Popov, *Functional Integrals in Quantum Field Theory and Statistical Physics* (trans. from Russian by J. Niederle and L. Hlavaty, D. Reidel Publishing Comp., Boston, 1983) chapters 2 and 3.
15. L. D. Faddeev and V. N. Popov, in *100 Years of Gravity and Accelerated Frames, The deepest Insights of Einstein and Yang-Mills* (Ed. J. P. Hsu and D. Fine, World Scientific, 2005) p. 325.
16. S. Mandelstam, Phys. Rev. **175**, 1604 (1968).
17. B. S. DeWitt, Phys. Rev. **162** 1239 (1967).
18. J. P. Hsu, Phys. Rev. D**8**, 2609 (1973).
19. X. S. Chen and B. C. Zhu,'Physical decomposition of the gauge and gravitational fields' (preprint, HUST Wuhan; KITPC, Chinese Aca. Sci., 2009.)
20. R. Utiyama and T. Fukuyama, Prog. Theor. Phys. **45**, 612 (1971).
21. Y. M. Cho, Phys. Rev. D, **14**, 2521 (1976).
22. Ning Wu, Commun. Theor. Phys. (Beijing) **42** 543 (2004); gr-qc/0309041. See also A. A. Logunov, ref. 9.
23. Ning Wu, "Quantum Gauge Theory of Gravity", talk given at Meeting of the Division of Particle and fields of Am. Phys. Soc. at College of William & Mary, May 24-28, 2002, Williamsburg, Virginia, USA.

GRAVITATIONAL ENERGY

JAMES M. NESTER*
Department of Physics, Institute of Astronomy, and
Center for Mathematics and Theoretical Physics
National Central University,
Chungli, 320, Taiwan
**E-mail: nester@phy.ncu.edu.tw*

Gravity is the unique interaction which is universal and attractive, distinctive properties directly connected with energy. Identifying a good expression which describes the (quasi-)local energy-momentum of gravitating systems is still an outstanding fundamental puzzle. The traditional pseudotensor approach is considered here along with the more modern quasi-local idea. Using a covariant Hamiltonian boundary-term approach clarifies the geometric and physical ambiguities. Certain criteria can be used as theoretical tests of any proposed quasi-local energy-momentum expression, including positivity, the spatial and null asymptotic limit, and the small region limit. The argument for the positive energy requirement is recalled and some positive energy proofs are noted. Positivity in general is a very strong criterion, but it is not so easy to prove or disprove. Positivity in the small vacuum region limit is simpler, and is also a quite strong test. In particular none of the traditional pseudotensors passes this test. Two natural quasi-local expressions and some other contrived ones do satisfy this small region requirement. The natural expressions have a positive energy proof for finite regions. Conversely, circumstances in which it is appropriate for the energy to be negative are noted. Our covariant Hamiltonian boundary term quasi-local gravitational energy-momentum expression requires, on the boundary, a choice of a displacement vector and field reference values. We have proposed obtaining the reference values from an energy-optimized isometric embedding of the 2-boundary into Minkowski space. This gives reasonable results at least for spherically symmetric regions.

Keywords: gravitational energy, Hamiltonian, pseudotensors, quasilocal energy

1. Two Special properties of gravity

The presently known physical interactions: strong, weak, electromagnetic and gravity can all be formulated as local gauge theories (we cite just a few treatments[1–3]). Nevertheless gravity is quite special: it is the only interaction that is *universal* and *purely attractive*. These two *unique* properties of gravity have deep significance in the scheme of things; they are not at all accidental but rather are essential consequences of fundamental principles. They naturally play key roles in many situations.

1.1. *Attraction*

The *purely attractive* property of gravity is a natural requirement following from the fundamental physical principles of thermodynamics and stability (e.g., no perpetual motion, no infinite source of energy) as we briefly argue.

Consider a gravitating system: *repulsion* would be caused by a *negative* mass. Can a negative mass exist? Suppose there were a positive and negative mass pair. Then they would attract and repel each other, self accelerating to a high speed, gaining unlimited kinetic energy that could be extracted—contrary to a fundamental thermodynamic principle. Thus physically there should only be positive masses which attract.

Attraction is associated with *positive energy*. An isolated gravitating system is expected to settle down to its equilibrium state. The gravity field should then asymptotically be Newtonian and not dynamic. The energy of such a system can be determined from its effective Newtonian mass, found from the Kepler orbit of a distant test particle using the 1-2-3 law: $(GM)^1 = \omega^2 a^3$. A bound orbit means attractive gravity, a positive mass, and, via Einstein's famous relation $E = Mc^2$, a positive energy.

Moreover, suppose an isolated system could have negative energy, we could then combine such systems to make one with a negative energy of arbitrarily large magnitude. However, since physical systems will naturally spontaneously radiate energy until they reach their lowest energy state, if unlimited negative energy states were allowed that would then permit systems to radiate an infinite amount of energy. Physical systems are unstable unless there is a non-negative lower bound to energy. Consequently (under appropriate conditions) energy must be positive and gravity is fundamentally purely attractive.

This property should be satisfied by any acceptable gravity theory. Hence it can be used as a test to evaluate proposed gravity theories.

1.1.1. *Test and proof*

Positive energy is a strong test: it is not easy to create a relativistic theory which satisfies this requirement under all conditions.[4–6] For an acceptable gravity theory it is important to have a proof that the energy is positive for every appropriate solution; for a given theory finding such a proof may not be so simple. For Einstein's General Relativity (GR), people worked seriously on trying to prove positive energy for at least 20 years before Schoen and Yau[7] succeeded (by an indirect geometric argument). Soon thereafter Witten (with the aid of a spinor field) presented his much more direct proof,[8–10] other proofs followed later, including some by the present author.[11–14] Thus GR passes this important test.

1.1.2. *Dark energy*

Now a dark cloud has been cast on our thesis. About 12 years ago it was announced that observations indicate that, contrary to all expectations, the expansion of the universe is not slowing down. The scale of the universe is expanding faster and faster, accelerating. There is some kind of global repulsion—a challenge to our thesis that gravity is purely attractive. The cause of this acceleration has been given a name: *dark energy*. Dark energy could be a (quite small) positive cosmological constant or perhaps some unusual type of matter with an effective negative pressure which causes the repulsion (in the usual cosmological model in GR the acceleration is given by $\ddot{a}/a = -(4\pi G/3)(\rho + 3P) + \Lambda$). However—unless perhaps the dark energy comes from gravity itself (which can happen in some alternatives to GR theory[15,16])—gravity itself still would have positive energy and would still be regarded as fundamentally attractive.

1.2. *Universal*

Gravity is *universal.* The source of gravity is all matter and all interaction fields—including itself. In Newton's theory mass density produced gravity. From Einstein we know that "energy" is equivalent to mass. Hence energy density should also produce gravity. Moreover, from relativity we learn that energy is a part of a 4-vector of energy-momentum. Thus, relativistically, the source of gravity should be the *total energy-momentum density*.

Energy-momentum should be conserved. When sources interact with the gravity they can exchange (and this happens *locally*) energy-momentum with the gravitational field (e.g., recall how the space probes Voyager and Pioneer acquired the energy to travel so far from the sun). Note also that the binary pulsar provides indirect evidence for both gravitational waves and gravitational energy. Moreover Bondi[17,18] has presented a compelling theoretical argument that (even in Newtonian theory) energy can be transferred through empty space via the gravitational field. A dramatic illustration of this is Io's volcanoes being powered by Jupiter's tidal heating.[19–21]

2. Gravitational local energy-momentum density

Thus gravity itself should have some kind of local energy-momentum density—which should also produce gravity. Hence gravity should be inherently non-linear, and truly universal: it affects and is affected by everything, including itself. (Note that both of the special properties of the gravitational interaction: universal and attractive, are associated with energy.)

Einstein included energy conservation in his search for his gravitational equations. In fact he had his expression for gravitational energy even before he found the correct field equations for his GR theory.[22] However this expression for the energy-momentum density—as well as many others proposed later—has a problematical feature: it is not a true tensor, it is inherently reference frame dependent. Energy-momentum *pseudotensors* can be obtained via the Lagrangian using Noether symmetry (and are thus subject to the usual Noether current ambiguity) or via rearranging field equations (which naturally includes a similar ambiguity).[23]

2.1. *Canonical energy momentum tensor*

Let us briefly review a simple construction of the canonical energy momentum tensor. From the Lagrangian density $\mathcal{L} = \mathcal{L}(\varphi^A, \partial_\mu \varphi^A)$ one obtains the key variational formula

$$\delta\mathcal{L} = \partial_\mu(\delta\varphi^A P^\mu{}_A) + \delta\varphi^A \frac{\delta\mathcal{L}}{\delta\varphi^A}, \tag{1}$$

which implicitly determines the Euler-Lagrange expression and the canonical momentum:

$$\frac{\delta\mathcal{L}}{\delta\varphi^A} := \frac{\partial\mathcal{L}}{\partial\varphi^A} - \partial_\mu P^\mu{}_A, \qquad P^\mu{}_A := \frac{\partial\mathcal{L}}{\partial\partial_\mu\varphi^A}. \tag{2}$$

When (1) is integrated over a region with fixed variations on the boundary, from Hamilton's principle for extreme action we get our field equations: the vanishing of the Euler-Lagrange expressions. For invariance under translation along the coordinate directions (assuming that $\mathcal{L}$ depends on position only through the fields) the relation (1) should identically hold with $\delta \to \partial_\nu$:

$$\partial_\nu\mathcal{L} \equiv \partial_\mu(\partial_\nu\varphi^A P^\mu{}_A) + \partial_\nu\varphi^A \frac{\delta\mathcal{L}}{\delta\varphi^A}. \tag{3}$$

This, in turn, shows that the *canonical energy-momentum tensor*,

$$T^\mu{}_\nu := P^\mu{}_A \partial_\nu\varphi^A - \delta^\mu_\nu\mathcal{L}, \tag{4}$$

satisfies

$$\partial_\mu T^\mu{}_\nu \equiv -\frac{\delta \mathcal{L}}{\delta \varphi^A}\partial_\nu \varphi^A. \tag{5}$$

Consequently, "on shell" (i.e., when the field equations are satisfied) the energy-momentum "current" density $T^\mu{}_\nu$ has vanishing divergence, and hence the energy-momentum within a region,

$$p_\nu(V) := \int_V T^\mu{}_\nu d\Sigma_\mu, \tag{6}$$

is *conserved* (i.e., its change is determined by a flux through the boundary).

It is important to note, however, that this energy-momentum density is not the unique quantity of this kind, since

$$T'^\mu{}_\nu := T^\mu{}_\nu + \partial_\lambda V_\nu{}^{[\mu\lambda]} \tag{7}$$

is also likewise likewise divergence free—and thus defines a (generally different) conserved energy-momentum value—for any choice of $V_\nu{}^{[\mu\lambda]}$. As we shall see later (for electodynamics as a specific example), one cannot just always settle for the canonical energy-momentum tensor that follows directly from the Lagrangian; one may want to exploit this freedom to obtain within this class of objects a more suitable energy-momentum density.

2.2. *Einstein and Møller pseudotensors*

Turning to gravity, the Hilbert action $\mathcal{L}_{\rm H} := R\sqrt{-g}$ is not suitable for immediate application of the above procedure, as it depends on second derivatives of the dynamical field. Using the metric as the dynamical variable, from $\mathcal{L}_{\rm H}(g, \partial g, \partial\partial g)$, one can remove a total divergence to obtain the Einstein Lagrangian:

$$\mathcal{L}_{\rm E}(g, \partial g) := \mathcal{L}_{\rm H} - \text{total divergence}, \tag{8}$$

which gives the same field equations. The associated canonical energy-momentum tensor is the *Einstein pseudotensor*.[24]

Instead one may use as the dynamic variable an orthonormal frame $e^\alpha{}_j$, where $g_{ij} = \eta_{\alpha\beta} e^\alpha{}_i e^\beta{}_j$ and $\eta_{\alpha\beta} = \text{diag}(-1, +1, +1, +1)$ is the Minkowski metric. Removing a certain total divergence gives the Møller Lagrangian:

$$\mathcal{L}_{\rm M}(e, \partial e) := \mathcal{L}_{\rm H} - \text{total divergence}. \tag{9}$$

Now the canonical energy-momentum tensor construction using $e^\alpha{}_i$ as the dynamical variable gives the Møller 1961 tetrad-teleparallel pseudotensor.[25]

2.3. *Other famous pseudotensors*

All the classical pseudotensors can be obtained by suitably rearranging Einstein's equation:[23,24,26–30]

$$G_{\mu\nu} = \kappa T_{\mu\nu}, \tag{10}$$

(where $\kappa := 8\pi G/c^4$). One may choose a *superpotential* $U^{\mu\lambda}{}_\nu \equiv U^{[\mu\lambda]}{}_\nu$ and *define* the related gravitational energy-momentum density pseudotensor by

$$2\kappa t^\mu{}_\nu := \partial_\lambda U^{\mu\lambda}{}_\nu - 2|g|^{\frac{1}{2}} G^\mu{}_\nu. \tag{11}$$

Einstein's equation now takes the form

$$2\kappa \mathcal{T}^\mu{}_\nu := 2\kappa(|g|^{\frac{1}{2}} T^\mu{}_\nu + t^\mu{}_\nu) = \partial_\lambda U^{\mu\lambda}{}_\nu. \tag{12}$$

Because of the antisymmetry of the superpotential the total energy-momentum density complex is automatically conserved: $\partial_\mu \mathcal{T}^\mu{}_\nu \equiv 0$; this is a true conservation law, in contrast to

$$\nabla_\mu \sqrt{-g} T^\mu{}_\nu \equiv \partial_\mu(\sqrt{-g} T^\mu{}_\nu) - \Gamma^\lambda{}_{\nu\mu} T^\mu{}_\lambda = 0, \tag{13}$$

which shows the change of material energy-momentum due to the gravitational intereaction.

There are some variations on the above formulation. The classical pseudotensorial total energy-momentum density complexes all follow from superpotentials according to one of the patterns

$$2\kappa \mathcal{T}^\mu{}_\nu = \partial_\lambda U^{\mu\lambda}{}_\nu, \quad 2\kappa \mathcal{T}^{\mu\nu} = \partial_\lambda U^{\mu\lambda\nu}, \quad 2\kappa \mathcal{T}^{\mu\nu} = \partial_{\alpha\beta} H^{\alpha\mu\beta\nu}, \tag{14}$$

where the superpotentials have certain symmetries which automatically guarantee conservation: specifically $U^{\mu\lambda}{}_\nu \equiv U^{[\mu\lambda]}{}_\nu$, $U^{\mu\lambda\nu} \equiv U^{[\mu\lambda]\nu}$, while $H^{\alpha\mu\beta\nu}$ has the symmetries of the Riemann tensor (this latter form guarantees also a good formula for angular momentum). In particular the Einstein total energy-momentum density follows from the Freud superpotential[31]

$$U_{\rm F}^{\mu\lambda}{}_\nu := -|g|^{\frac{1}{2}} g^{\beta\sigma} \Gamma^\alpha{}_{\beta\gamma} \delta^{\mu\lambda\gamma}_{\alpha\sigma\nu}; \tag{15}$$

while the Bergmann-Thompson,[32] Landau-Lifshitz,[33] Papapetrou,[34] Weinberg[35,37] and Møller[38] expressions can be obtained respectively from

$$U_{\rm BT}^{\mu\lambda\nu} := g^{\nu\delta} U_{\rm F}^{\mu\lambda}{}_\delta, \tag{16}$$

$$U_{\rm LL}^{\mu\lambda\nu} := |g|^{\frac{1}{2}} U_{\rm BT}^{\mu\lambda\nu}, \quad \text{equivalently} \quad H_{\rm LL}^{\alpha\mu\beta\nu} := |g| \delta^{\mu\alpha}_{ma} g^{a\beta} g^{m\nu}, \tag{17}$$

$$H_{\rm P}^{\alpha\mu\beta\nu} := \delta^{\mu\alpha}_{ma} \delta^{\nu\beta}_{nb} \bar{g}^{ab}(|g|^{\frac{1}{2}} g^{mn}), \tag{18}$$

$$H_{\rm W}^{\alpha\mu\beta\nu} := \delta^{\mu\alpha}_{ma} \delta^{\nu\beta}_{nb} |\bar{g}|^{\frac{1}{2}} \bar{g}^{ab}(-\bar{g}^{mc}\bar{g}^{nd} + \frac{1}{2}\bar{g}^{mn}\bar{g}^{cd}) g_{cd}, \tag{19}$$

$$U_{\rm M58}^{\mu\lambda}{}_\nu := -|g|^{\frac{1}{2}} g^{\beta\sigma} \Gamma^\alpha{}_{\beta\nu} \delta^{\mu\lambda}_{\alpha\sigma} \equiv |g|^{\frac{1}{2}} g^{\beta\mu} g^{\lambda\delta}(\partial_\beta g_{\delta\nu} - \partial_\delta g_{\beta\nu}). \tag{20}$$

(all indicies in these expressions refer to spacetime and range from 0 to 3, otherwise our conventions follow MTW.)

What criteria can one use to decide which, if any, of these expressions are satisfactory descriptions of energy-momentum? Perhaps the most basic property is that, for asymptotically flat space, the expression should give the desired ADM[36] and linearized theory values[37] for the energy and momentum at spatial infinity. It turns out that all of these famous pseudotensors—*except* for the Møller 1958 expression—are satisfactory at spatial infinity. (There is also a requirement concerning angular momentum; angular momentum will not be considered in detail in this report.)

2.4. *The small sphere limit*

Another requirement concerns the limiting value assigned to a small region. Taylor expand the pseudotensor obtained from the superpotential. According to the *equivalence principle* one should get— to zeroth order—the matter energy-momentum. This requirement turns out to be even weaker than getting good asymptotic values; once again all the famous pseudotensors except the Møller 1958 expression satisfy this property.

Continuing the expansion to higher order, *in vacuum* the first non-vanishing contribution should appear at second order. It has been argued[39] that to this order the expression should be proportional to the Bel-Robinson tensor:[37,39,40]

$$\mathcal{T}^{\mu\nu} \propto B^{\mu\nu}{}_{\alpha\beta}x^\alpha x^\beta. \tag{21}$$

Why? Because the Bel-Robinson tensor assures *positive energy* (and in fact the *dominant energy condition* for the small region energy-momentum vector). This requirement is a strong constraint.

For holonomic frames, using Riemann normal coordinates, *none* of the traditional pseudotensors satisfies this requirement[29,30] (it is satisfied by certain contrived linear combinations of the classical pseudotensors[29,30] as well as by many new pseudotensors which we constructed.[29,41]). For *orthonormal frames*, using Riemann normal coordinates and *normal frames*:[42]

$$\vartheta^\alpha_i = \delta^\alpha_i, \qquad \partial_j\vartheta^\alpha{}_k = 0, \tag{22}$$

$$\Gamma^\alpha{}_{\beta k} = 0, \qquad \partial_j\Gamma^\alpha{}_{\beta k} = \frac{1}{2}R^\alpha{}_{\beta jk}, \tag{23}$$

we found that the Møller 1961 tetrad/teleparallel energy-momentum pseudotensor[25] *does not* satisfy this requirement. However, both our favorite *quasi-local* expression (to be discussed below) as well as the *translational gauge current* associated with the *teleparallel* formulation of GR[43] do satisfy this important small vacuum region "positivity" requirement.[44]

2.5. *Doubts*

Some have questioned the whole idea of trying to find a local energy density for gravitating systems. We note just two examples: (1) Cooperstock,[45] has argued that the gravitational energy-momentum density vanishes outside of matter. (2) A very influential textbook states: *Anyone who looks for a magic formula for "local gravitational energy-momentum" is looking for the right answer to the wrong question. Unhappily, enormous time and effort were devoted it the past to trying to "answer this question" before investigators realized the futility of the enterprise."*[37]

Certainly, the serious ambiguities of the pseudotensor approach must be recognized. First, given all the above classical pseudotensors along with the infinite number of possible ones constructed from all the possible choices of superpotentials (which essentially corresponds to the freedom noted in Eq. (7) above), which, if any, gives the true energy-momentum localization? Second, since pseudotensors are inherently reference frame dependent objects, which reference frame gives the physically correct localization?

Below it will be argued that these pseudotensor ambiguities are much mitigated by the Hamiltonian approach.

3. Geometry and gravity: quasi-local energy-momentum

Gravity is necessarily connected with geometry. The source of gravity is energy-momentum. By Noether's first theorem associating physically conserved quantities with symmetry, energy-momentum is related to the translation symmetry of space-time geometry.

According to the *equivalence principle* gravity cannot be detected at a point. There is thus, when it comes to energy, a fundamental incompatibility between the requirements of the equivalence principle and the *principle of general covariance.* There is simply no suitable generally covariant expression which can completely describe the local energy-momentum density for gravity. A consequence is that *gravitational energy-momentum*—and hence the energy-momentum of gravitating systems—and hence the energy-momentum of *all physical systems*—is *fundamentally non-local.* The modern idea is that all physical energy-momentum is *quasi-local*, i.e., it is associated with a closed 2-surface (and not with a density at a point).

4. Pseudotensors and the Hamiltonian boundary term

Choose any superpotential $U^{\nu\lambda}{}_{\mu} \equiv U^{[\nu\lambda]}{}_{\mu}$, and use it to split the Einstein tensor, thereby defining the associated gravitational energy-momentum pseudotensor via Eq. (11). Einstein's equation, $G^{\mu}{}_{\nu} = \kappa T^{\mu}{}_{\nu}$, is thus transformed into (12), a form with the total effective energy-momentum pseudotensor as source.[23,26,46,47]

Fix a coordinate system and (for later convenience) multiply (12) by a vector field N^{μ} having constant components in this system. Now integrate over a finite spacetime 3-volume Σ to get the total energy-momentum associated with the chosen vector field:

$$\begin{aligned}-N^{\mu}p_{\mu} &:= -\int_{\Sigma} N^{\mu}\mathcal{T}^{\nu}{}_{\mu}\sqrt{-g}(d^3x)_{\nu} \\ &\equiv \int_{\Sigma}\left[N^{\mu}\sqrt{-g}\left(\frac{1}{\kappa}G^{\nu}{}_{\mu} - T^{\nu}{}_{\mu}\right) - \frac{1}{2\kappa}\partial_{\lambda}(N^{\mu}U^{\nu\lambda}{}_{\mu})\right](d^3x)_{\nu}. \quad (24)\end{aligned}$$

This expression has the form

$$H(N,\Sigma) = \int_{\Sigma} N^{\mu}\mathcal{H}_{\mu} + \oint_{S=\partial\Sigma}\mathcal{B}(N) = \oint_{S=\partial\Sigma}\mathcal{B}(N), \quad (25)$$

since $\mathcal{H}_{\mu}$ vanishes by the field equation; $\mathcal{B}(N)$ is a certain 2-boundary integrand which is here linear in the superpotential. In a later section it will be shown that the GR Hamiltonian always has this same form.

5. A first order geometric formulation

For a more efficient formulation, we will now use the notation of differential forms. Paralleling at first the treatment in Section 2.1, consider a *first order* Lagrangian 4-form for a form field φ (which might have suppressed indicies):

$$\mathcal{L} = d\varphi \wedge p - \Lambda(\varphi, p). \quad (26)$$

Its variation leads to the key relation

$$\delta\mathcal{L} = d(\delta\varphi \wedge p) + \delta\varphi \wedge \frac{\delta\mathcal{L}}{\delta\varphi} + \frac{\delta\mathcal{L}}{\delta p} \wedge \delta p. \quad (27)$$

On the one hand, this relation implicitly defines the two variational derivatives. Integrating (27) and invoking Hamilton's principle (with vanishing $\delta\varphi$ on the boundary) one finds that the vanishing of these two expressions constitute a *pair of first order equations.* (The explicit form, which contains the first differentials of the dynamic quantities φ, p will not be needed here.)

On the other hand, for geometric theories $\mathcal{L}$ should be diffeomorphic invariant. The changes induced by an infinitesimal diffeomorphism are given by the Lie derivative with respect to some vector field. On form fields $\pounds_N \equiv di_N + i_N d$ (here i_N is the interior product, aka contraction). Consequently, for diffeomorphism invariant systems, one can let $\delta \to \pounds_N$, then the above key relation (27) should be identically satisfied:

$$di_N\mathcal{L} \equiv \pounds_N\mathcal{L} \equiv d(\pounds_N\varphi \wedge p) + \pounds_N\varphi \wedge \frac{\delta\mathcal{L}}{\delta\varphi} + \frac{\delta\mathcal{L}}{\delta p} \wedge \pounds_N p. \tag{28}$$

This invites the definition of the *translational current* 3-form:

$$\mathcal{H}(N) := \pounds_N\varphi \wedge p - i_N\mathcal{L} =: N^\mu\mathcal{H}_\mu + d\mathcal{B}(N), \tag{29}$$

which is (on shell) a conserved "current" (Noether's first theorem):

$$-d\mathcal{H}(N) \equiv \pounds_N\varphi \wedge \frac{\delta\mathcal{L}}{\delta\varphi} + \frac{\delta\mathcal{L}}{\delta p} \wedge \pounds_N p. \tag{30}$$

Substituting the rhs of (29) into the lhs of (30) gives

$$-dN^\mu \wedge \mathcal{H}_\mu - N^\mu d\mathcal{H}_\mu. \tag{31}$$

Since N is locally a free choice, one can conclude (this is Noether's second theorem), by expanding out the Lie derivatives on the rhs of (30), that both terms of (31) must be separately proportional to certain field equation expressions. Hence, in particular, $\mathcal{H}_\mu$ must vanish "on shell".

The conserved quantity obtained from the translational current 3-form is the energy-momentum. The energy-momentum associated with a given spacetime 3-volume and a vector field is, since $\mathcal{H}_\mu$ vanishes (on shell), given by an integral of the boundary term:

$$E(N,\Sigma) := \int_\Sigma \mathcal{H}(N) = \int_\Sigma N^\mu\mathcal{H}_\mu + d\mathcal{B}(N) = \oint_{\partial\Sigma} \mathcal{B}(N). \tag{32}$$

However, it is clear that one can still freely modify the boundary term $\mathcal{B}(N)$ (and thereby the value of the conserved energy-momentum) without changing the conservation property. In the next section we will see that the Hamiltonian approach tames this ambiguity.

6. The Hamiltonian approach

The Hamiltonian 3-form associated with a given first order Lagrangian 4-form (26) can be obtained simply by contracting with the timelike spacetime displacement vector field that specifies the time evolution:

$$i_N\mathcal{L} = \pounds_N\varphi \wedge p - \mathcal{H}(N), \tag{33}$$

(analogous to $L = \dot{q}p - H$). This specifies the same 3-form earlier referred to as the translational gauge current (29). However its integral,

$$H(N,\Sigma) = \int_\Sigma \mathcal{H}(N) = \int_\Sigma N^\mu \mathcal{H}_\mu + \oint_{\partial\Sigma} \mathcal{B}(N), \tag{34}$$

is now seen as the generator of "time" evolution. The boundary term now has a two-fold role. One the one-hand, as we have already seen, it determines the quasi-local energy-momentum. This boundary term can (and, as we shall see below, often should) be modified. A modification to the boundary term does not change the evolution, yet it does change the value of the the conserved quantities. Here in this Hamiltonian formulation, however, this freedom is not arbitrary; it has a clear physical significance. Which brings us to the other important (often overlooked) role of the Hamiltonian boundary term. Namely, it *controls what should be held fixed on the boundary*, i.e., *the boundary conditions.*[48,49] Here we do not have the space to give a detailed discussion (which can be readily found in certain of our published works[23,46,47,50–53]). Let us illustrate this idea by two familiar examples.

6.1. *Example: thermodynamics*

With volume V, pressure P, temperature T and entropy S, a thermodynamic system can be described by various energies. The most common are

$$\begin{aligned} dU &= TdS - PdV, && \text{internal energy} \\ dF &= -SdT - PdV, && \text{Helmholtz free energy} \\ dH &= TdS + VdP, && \text{enthalpy} \\ dG &= -SdT + VdP, && \text{Gibbs free energy.} \end{aligned}$$

Note that there is not one unique physical energy. Rather there are several physically meaningful energies: $U(S,V)$, $F(T,V)$, $H(S,P)$ and $G(T,P)$, each associated with a certain distinct specific pair of independent variables that can be externally controlled.

6.2. *Application: electromagnetism*

Consider the following Hamiltonian for vacuum electrodynamics (using for convenience here ordinary vector notation):

$$H(\sigma) = \int \left[\frac{1}{2}(\mathbf{E}^2 + \mathbf{B}^2) + \phi\nabla\cdot\mathbf{E}\right] d^3x, \tag{35}$$

where $\mathbf{B} = \nabla \times \mathbf{A}$. Note the familiar energy density and the *gauge* term (which both generates the gauge transformation and gives the Gauss constraint). The variation of this Hamiltonian includes the boundary term

$$\delta H(\sigma) \sim \oint \phi \, \delta(\mathbf{E} \cdot \mathbf{n}) \, dS. \tag{36}$$

This will vanish—if we fix on the boundary the normal component of the electric field. Physically this means fixing σ, the surface charge density.

On the other hand one could use the alternative Hamiltonian

$$H(\phi) = \int \left[\frac{1}{2}(\mathbf{E}^2 + \mathbf{B}^2) - \mathbf{E} \cdot \nabla\phi \right] d^3x = H(\sigma) - \oint \phi \mathbf{E} \cdot \mathbf{n} \, dS; \tag{37}$$

this differs simply by a boundary term, which does not change the evolution equations. However the variation of the Hamiltonian now includes a different boundary term, namely,

$$\delta H \sim - \oint \delta\phi \, \mathbf{E} \cdot \mathbf{n} \, dS. \tag{38}$$

This will vanish—if we fix the scalar potential on the boundary.

Each of these Hamiltonians accurately describes a certain class of physically realizable systems. In particular, for (36) one could connect a battery to a parallel plate capacitor until it was charged up, and then disconnect the battery and measure the work required to insert/remove a dielectric. Conversely, if we leave the battery connected (this will allow current to flow) one can measure the work with the fixed potential boundary condition. This is an example of a physical system described by (37). As is well known the amount of work required is not the same in these two cases.

This is a good example of the point we wish to emphasize: typically we have a choice between a Dirichlet or Neumann type boundary condition. In fact the Hamiltonian (36) corresponds to the Hilbert (symmetric) energy-momentum tensor, while the Hamiltonian (37) is essentially the canonical energy-momentum tensor of electromagnetism. Although both of these have physical meaning, they are nevertheless not of equal value. One should naturally prefer to use the first Hamiltonian (36), because (since the Gauss constraint vanishes on shell) it is *gauge invariant*. Along with this comes a bonus: with vanishing Gauss constraint the energy is nonnegative and, moreover, vanishes only for vanishing field. On the other hand, the Hamiltonian (37) is not gauge invariant (of course, for the fixed potential boundary condition is gauge dependent) and, moreover, its value can have any sign or magnitude, it can even vanish for non-vanishing fields. For similar reasons, we generally favor the Hilbert energy-momentum tensor over the canonical—they are related by a transformation of the form (7)—for most physical applications. This resolves the classical ambiguity for the choice of energy-momentum for all material sources as well as for all the gauge interaction fields—except for gravity.

7. The quasi-local Hamiltonian boundary term for GR

We have developed a covariant Hamiltonian formalism for geometric gravity theories. In this formulation the ambiguity regarding the choice of energy expression is given a clear physical and geometric meaning. Briefly, there are an infinite number of possible energy-momentum expressions simply because there are an infinite number of possible types of boundary conditions. Each of the possible energy expressions (which includes all the classical pseudotensors) corresponds to a Hamiltonian with evolution satisfying some specific boundary condition. Naturally not all boundary conditions are equally nice or equally physically reasonable or meaningful.

For GR, using the orthonormal coframe ϑ^α and the connection one-form $\Gamma^\alpha{}_\beta$ as the basic variables, we identified several nice boundary terms which correspond to certain Dirichlet/Neumann physical boundary condition choices for these quantities. One stands out above all the others as having the nicest properties. Our preferred Hamiltonian boundary term for GR is

$$\mathcal{B}(N) = \frac{1}{2\kappa}(\Delta\Gamma^\alpha{}_\beta \wedge i_N \eta_\alpha{}^\beta + \bar{D}_\beta N^\alpha \Delta\eta_\alpha{}^\beta), \tag{39}$$

where $\eta^{\alpha\beta} := *(\vartheta^\alpha \wedge \vartheta^\beta)$ is the dual coframe, and for any quantity we define

$$\Delta\varphi := \varphi - \bar{\varphi}, \tag{40}$$

where $\bar{\varphi}$ is a *reference value.*

Reference values specify the choice of zero point, the "ground state" of the variable. They can be used with other fields (e.g., for electromagnetism to include a background field) but they are not essential, simply because for all other fields it is possible—and usually desirable—to take the ground state as vanishing field value. However, for gravity the ground state is not a vanishing metric but rather the non-vanishing Minkowski metric values. Thus for gravity non-trivial reference values are unavoidable. Physically, the effect of including the reference values in our quasi-local expressions is the following: if the fields take on the reference values on the boundary, then all the geometric quasi-local quantities (energy-momentum etc.) vanish, indicating that the interior of the region is flat empty Minkowski space.

Thus our Hamiltonian boundary term quasi-local energy-momentum expression still has an ambiguity: namely, the explicit choice of reference (it is not enough to say it is the Minkowski metric, one needs to effectively give a reference coordinate system on the boundary in which the metric takes its standard Minkowski value). This reference choice freedom here plays essentially the same role as the pseudotensor coordinate ambiguity. However, we now have a geometric formulation which clarifies the physical meaning of the choice of reference and clearly shows that we only need to make a choice on the boundary of the region. We will consider below the problem of how to choose the reference.

8. Some properties of our quasi-local expression

From the integral of our preferred choice of Hamiltonian boundary term (39) over the 2-boundary of any region one can obtain values for the quasi-local quantities. With a suitable choice of reference, one can get quasi-local energy, momentum, angular momentum and center-of-mass by choosing, respectively, the space time displacement vector field to be an appropriate infinitesimal timelike or spacelike translation, or a rotation or boost.

The expression reduces to some other well regarded expressions in appropriate limits. At spatial infinity it has good limits to the asymptotic weak field expressions[37] and to the ADM energy-momentum[36] and the angular momentum/center-of-mass, as given by the expressions of Regge and Teitelboim,[54] or better the refined expressions of Beig and Ó Murchadha[55] and the further refinement of Szabados.[56] At future null infinity the expression gives the Bondi energy and, via a remarkable variational identity, the Bondi energy flux.[52,57] In the small region vacuum limit our quasi-local expression is, as desired, proportional to the Bel-Robinson tensor.[53]

Our expression has two terms: one, linear in the vector field, is essentially the Freud superpotential[31] (which generates the Einstein pseudotenor), the other, linear in the derivative of the displacement, is rather like Komar's expression.[58] The second term is essential for the center-of-mass[59] and also can contribute to some angular momentum calculations.[60]

It should be noted that essentially the same boundary term expression was (using a quite different approach with holonomic methods) independently found by Katz, Bičák and Lynden-Bel,[61] who have worked out a number of nice applications. Under suitable conditions the first term in our expression reduces exactly to the famous Brown-York energy, momentum and angular momentum quasi-local expressions.[62]

There is a proof of positive total energy for asymptotically flat gravitating systems that applies to our quasi-local expression.[11] It can be adapted to give a positive quasi-local energy proof.

9. Reference choices

To give specific values for the quasi-local energy-momentum our Hamiltonian boundary term expression must be supplied with a choice of spacetime displacement vector field and reference values for the dynamical variables. The usual choice of reference geometry is Minkowski space. A reasonable choice of evolution vector could be a constant timelike vector in this Minkowski reference. What is needed is, effectively, an embedding of the 2-boundary surface from the dynamic geometry into Minkowski space. Locally this can be described by four functions of two variables. Finding a good embedding, satisfying appropriate conditions, is presently an active pursuit. It is usually presumed that one would like to embed the spatial 2-boundary isometrically. That imposes three conditions. The standard quasi-local criteria[40,63] are that one should have positive energy, with vanishing energy only for Minkowski space. These criteria have been used to select both the embedding and the energy expression.[63–66] A reasonable proposal is to regard the energy as a function of the embedding variable(s) and examine its extreme. For our Hamiltonian boundary term quasi-local expression we will next give a brief report of the results we have obtained for the special case of spherically symmetric spacetimes.

9.1. *The optimal reference choice for spherical systems*

For the Schwarzschild spacetime with the evolution vector as the timelike Killing field of the reference corresponding to a static observer, we find our optimal quasi-local energy to have the standard Brown-York value:[62]

$$E = r\left(1 - \sqrt{1 - \frac{2m}{r}}\right) = \frac{2m}{1 + \sqrt{1 - \frac{2m}{r}}}. \tag{41}$$

Note: at spatial infinity the value is m, at the horizon it is $2m$, and it is not defined inside the horizon (there is no static observer inside).

Similarly, for the static observer in Reissner-Nördstrom spacetime

$$E = r\left(1 - \sqrt{1 - \frac{2m}{r} + \frac{Q^2}{r^2}}\right) = \frac{2m - \frac{Q^2}{r}}{1 + \sqrt{1 - \frac{2m}{r} + \frac{Q^2}{r^2}}}. \tag{42}$$

Note that this energy is *negative* at $r < \frac{Q^2}{2m}$, which is is exactly the turn-around radius inside which the gravitational force is repulsive.

However if we consider a radial geodesic observer who falls initially with velocity v_0 from a constant distance $r = a > 2m$. Then our energy for Schwarzschild is

$$E = r\left(1 - \sqrt{\frac{1-2m/a}{1-v_0^2}}\right). \tag{43}$$

When the initial velocity v_0 is less, equal, or greater than the escape velocity $\sqrt{2m/a}$, the energy is positive, zero, or negative, respectively.

9.2. *For the FLRW spacetimes*

Our program also works well for dynamic spherically symmetric spacetimes. For the Friedmann-Lemaître-Robertson-Walker (FLRW) spacetime,

$$ds^2 = -dt^2 + \frac{a^2(t)}{1-kr^2}dr^2 + a^2(t)r^2 d\Omega^2, \tag{44}$$

for a freely falling co-moving observers the quasi-local energy is

$$E = \frac{kar^3}{1+\sqrt{1-kr^2}}, \tag{45}$$

which vanishes for $k = 0$, is positive for $k = +1$ (but vanishing for the whole universe) and *negative* for $k = -1$.

9.3. *Application to Bianchi cosmologies*

Our Hamiltonian-boundary-term quasi-local energy-momentum ideas have been applied to homogenous cosmological models.[68]

Cosmological models which are homogeneous, but not (in general) isotropic can be described in terms of a metric of the form

$$ds^2 = -dt^2 + \delta_{ab}\vartheta^a \otimes \vartheta^b, \qquad a, b = 1, 2, 3 \tag{46}$$

where the spatial coframe,

$$\vartheta^a = h^a{}_j(t)\sigma^j(x), \tag{47}$$

is spatially homogeneous, i.e.,

$$d\sigma^i = \frac{1}{2}C^i{}_{jk}\sigma^j \wedge \sigma^k, \qquad i, j, k = 1, 2, 3 \tag{48}$$

where $C^i{}_{jk}$ are certain constants. The distinct possibilities have been systematically classified. Briefly, there are nine Bianchi types of such frames, falling into two classes:

Class A: $A_j := C^k{}_{jk} = 0$ (Types I, II, VI_0, VII_0, VIII, IX),

Class B: $A_k \neq 0$ (Types III, IV, V, VI_h, VII_h).

Here we will not need any more detail, except to note that for Type I the spatial curvature vanishes, for Type IX it is positive, and for all other types the spatial curvature is negative.

Within this framework we examined the energy of all such models with completely general sources (e.g., matter, radiation, dark matter, dark energy, cosmological constant etc.) We used, as seems appropriate, the *comoving time evolution vector* and *homogeneous boundary conditions* and a *homogeneous reference.* With this specialization our favored Hamiltonian boundary term quasi-local energy coincides with some other respectable energy expressions, including the teleparallel gauge current and the Hamiltonian associated with the Witten positive energy proof. The value of this common energy for these models works out to be

$$E(V) = -\frac{1}{\kappa} A_k A_l g^{kl}(t) V(t) \leq 0. \tag{49}$$

There are two noteworthy features: (i) The energy vanishes for all regions for all class A models (this is reasonable as Class A models are compactifiable, and the energy must vanish for a closed universe). (ii) The energy is *negative* for all regions for all class B models.

Thus, according to this *reasonable measure* of quasi-local energy for these models, one can have (i) negative energy, and (ii) vanishing energy for a non-trivial dynamic geometry!

10. Negative energy

We have used *the same energy expressions that give positive energy* for asymptotically flat isolated gravitating system and found, for physically and geometrically reasonable choices of evolution vector and reference, in some cases *negative quasi-local energy.* However, it should be noted that for the cosmological models for which this happens the gravitating systems are not at all like asymptotically flat isolated gravitating systems approaching a static or stationary equilibrium (for which there are compelling arguments in favor of positivity). For these dynamic models the negative spatial curvature geometry acts like a concave lens causing null geodesics to be defocused, as if they were being repelled by a negative mass, so a negative quasi-local energy value may be appropriate.

Under certain appropriate circumstances our quasi-local Hamiltonian boundary term is expected to have positive values. In some other circumstances it seems reasonable to have a negative value. Deeper investigation is required to sharpen the criteria for when the value should be positive, for when it is acceptable or even appropriate to have a negative value, and under what conditions it is acceptable to have a non-trivial geometry with zero energy.

11. Concluding thoughts

To better understand our work it may help to note that our principle aim has not been to find a unique "best quasi-local energy". Rather it has aimed to find the best general choice for the Hamiltonian boundary term.

Going back to our opening theme, gravity is the universal attractive interaction, moreover it connects all of existence together and is the prime cause of the order in the cosmos. It seems that gravity is like love, something worthy of meditation.

Acknowledgement

This is a report of material presented at The Summer School on Theories and Experimental Tests of GR 2009-07-02 and ICGA9 at HUST at Wuhan, China, 2009-06-29. The kind and generous support, assistance, and patience of the organizers, especially Zebing Zhou, was much appreciated. This presentation was based in part on work with C. M. Chen, R. S. Tung, L. L. So, J. L. Liu, and M. F. Wu. This work was supported by the Taiwan National Science council under the project NSC 97-2112-M008-001 and by the National Center of Theoretical Sciences.

References

1. J. M. Nester, "Gravity, Torsion and Gauge Theory" in *An Introduction to Kaluza-Klein Theories*, ed. H. C. Lee (World Scientific, Singapore, 1984) pp 83–115.
2. M. Blagojević, *Gravitational and Gauge Symmetries*, (Institute of Physics, 2002).
3. F. Gronwald and F. W. Hehl, "On the gauge aspects of gravity", *Proc. 14th Course of the School of Cosmology and Gravitation on Quantum Gravity (Erice, Italy, 1995)* ed. P. G. Bergmann, V. de Sabatta and H. J. Treder (Singapore: World Scientific) p 148.
4. V. V. Zhytnikov and J. M. Nester, *Phys. Rev. Lett.* **73**, 2950–53 (1994).
5. H. H. Chen, D. C. Chern and J. M. Nester, *Chin. J. Phys.* **25**, 481 (1987).
6. D. C. Chern, J. M. Nester and H. J. Yo, *Int. J. Mod. Phys. A* **7**, 1993–2003 (1993).
7. R. Schoen and S.T. Yau, *Phys. Rev. Lett.* **43**, 1457–59 (1979).
8. E. Witten, *Comm. Math. Phys.* **30**, 381–402 (1981).

9. J. M. Nester, *Phys. Lett.* A **83**, 241 (1981).
10. J. M. Nester, The Gravitational Hamiltonian, in *Asymptotical Behavior of Mass and Space-time Geometry* ed. F. Flaherty *Springer Lecture Notes in Physics* **202** (Springer, 1984) pp 155–163.
11. J. M. Nester, *Int. J. Mod. Phys.* A **4**, 1755 (1989); *Phys. Lett.* **139A**, 112 (1989).
12. J. M. Nester and R. S. Tung, *Gen. Rel. Grav.* **27**, 115–119 (1995).
13. J. M. Nester, R.-S. Tung and Y. Z. Zhang, *Class. Quantum Grav.* **11**, 757–766 (1994).
14. J. M. Nester and R. S. Tung, *Phys. Rev.* D **49**, 3958–3962 (1994).
15. K.-F. Shie, J.M. Nester and H.-J. Yo, *Phys. Rev.* D **78**, 023522 (2008).
16. H. Chen, F.H. Ho, J.M. Nester, C.H. Wang and H.J. Yo, *JCAP* **10**(2009)027.
17. J. V. Narlikar, *The Lighter Side of Gravity*, (Cambridge Univ. Press, 1996) pp 100–104.
18. H. Bondi and W. H. McCrea, *Proc. Camb. Phil. Soc.* **556**, 410–413 (1960).
19. P. Purdue, *Phys. Rev.* D **60**, 104054 (1999)
20. I. S. Booth and J. D. E. Creighton, *Phys. Rev.* D **62**, 067503 (2000).
21. M. Favata *Phys. Rev.* D **63**, 067503 (2001).
22. M. Janssen and J. Renn, Untying the Knot: How Einstein Found His Way Back to Field Equations Discarded in the Zurich Notebook, in *The Genesis of General Relativity* Vol. 2, *Einstein's Zurich Notebook: Commentary and Essays* ed. J. Renn (Springer, 2006) pp 849–925.
23. J. M. Nester, *Class. Quantum Grav.* **21**, S261–S280 (2004).
24. A. Trautman, in *An introduction to Current Research*, ed. L. Witten (Wiley, New York, 1962) pp 169–198.
25. C. Møller, *Ann. Phys. (NY)* **12**, 118–33 (1961); *Mat. Fys. Dan. Vid. Selsk.* **1**, No. 10, 1–50 (1961).
26. C. C. Chang, J. M. Nester and C.-M. Chen, *Phys. Rev. Lett.* **83**, 1897–1901 (1999).
27. J. N. Goldberg, *Phys. Rev.* **111**, 315 (1958).
28. L. B. Szabados, *Class. Quantum Grav.* **9**, 2521–41 (1992).
29. L. L. So, J. M. Nester and H. Chen, in *Gravitation and Astrophysics: on the occasion of the 90th year of General Relativity*, eds. J. M. Nester, C.-M Chen and J. P. Hsu (World Scientific, 2007), pp 356–362 [ICGA7 proceedings].
30. L. L. So, J. M. Nester and H. Chen, *Class. Quantum Grav.* **26**, 085004 (2009).
31. Ph. Freud, *Ann. Math.* **40**, 417–9 (1939).
32. P. G. Bergmann and R. Thompson, *Phys. Rev.* **89**, 400 (1953).
33. L. D. Landau and E. M. Lifshitz, *The Classical Theory of Fields* 2nd Edition, (Addison-Wesley, Reading MA, 1962).
34. A. Papapetrou, *Proc. Roy. Irish Acad. A* **52**, 11–23 (1948).
35. S. Weinberg, *Gravitation and Cosmology*, (Wiley, New York, 1972) p 371.
36. R. Arnowitt, S. Deser and C. W. Misner, The dynamics of general relativity, in *Gravitation: An Introduction to Current Research*, (Wiley, New York; London, 1962) pp 227–265.
37. C. W. Misner, K. Thorne and J. A. Wheeler, *Gravitation* (Freeman, San Fransisco, 1973) §20.2 and p 467.

38. C. Møller, *Ann. Phys. (NY)* **4**, 347 (1958).
39. S. Deser, J. S. Franklin and D. Seminara, *Class. Quantum Grav.* **16**, 2815–2821 (1999).
40. L. B. Szabados, Quasilocal energy-momentum and angular momentum in GR: a review article, *Living Rev. Relativity* **12**, 4 (2009); http://www.livingreviews.org/lrr-2009-4.
41. L. L. So and J. M. Nester, *Phys. Rev.* D **79**, 084028 (2009).
42. J. M. Nester, *Ann. Phys. (Berlin)* **19**, 45–52 (2010).
43. V. C. de Andrade, L. C. T. Guillen and J. G. Pereira, *Phys. Rev. Lett.* **84**, 4533–4536 (2000).
44. L. L. So and J. M. Nester, *Chin. J. Phys.* **47**, 10 (2009).
45. F. I. Cooperstock, *Mod. Phys. Lett.* A**14**, 1531 (1999); *Ann. Phys.* **282** 115–137 (2000).
46. C.-C. Chang, J. M. Nester and C.-M. Chen, Energy-Momentum (Quasi-)Localization for Gravitating Systems, in *Gravitation and Astrophysics* eds Liao Liu, Jun Luo, X.-Z. Li and J. P. Hsu (World Scientific, Singapore, 2000) pp 163–73 [ICGA4 proceedings].
47. C.-M. Chen and J. M. Nester, *Gravitation & Cosmology* **6**, 257–70 (2000).
48. J. Kijowski, *Gen. Relativ. Gravit.* **29**, 307-343, (1997).
49. J. Kijowski and W. M. Tulczyjew, *A Symplectic Framework for Field Theories, Lecture Notes in Physics No.* **107** (Springer-Verlag, Berlin, 1979).
50. C.-M. Chen, J. M. Nester and R.-S. Tung, *Phys. Lett.* A **203**, 5–11 (1995).
51. C.-M. Chen and J. M. Nester, *Class. Quantum Grav.* **16**, 1279–1304 (1999).
52. C.-M. Chen, J. M. Nester and R.-S. Tung, *Phys. Rev.* D **72**, 104020 (2005).
53. J. M. Nester, *Prog. Theor. Phys. Suppl.* ***172***, 30–39 (2008), [ICGA8 proceedings].
54. T. Regge and C. Teitelboim, *Ann. Phys. (N.Y.)* **88**, 286–318 (1974).
55. R. Beig and N. Ó Murchadha, *Ann. Phys. (N.Y.)* **174**, 463–498 (1987).
56. L. B. Szabados, *Class. Quantum Grav.* **20**, 2627–2661 (2003).
57. X.-n. Wu, C.-M. Chen, and J. M. Nester, *Phys. Rev.* D **71**, 124010 (2005).
58. A. Komar, *Phys. Rev.* **113**, 934–936 (1959).
59. J. M. Nester, F. F. Meng and C.-M. Chen, *J. Korean Phys. Soc.* **45**, S22–S25 (2004) [ICGA6 Proceedings].
60. R. D. Hecht and J. M. Nester, *Phys. Lett.* A **180**, 324–331 (1993); **217**, 81–89 (1996).
61. J. Katz, J. Bičák and D. Lynden-Bell, *Phys. Rev.* D **55**, 5957 (1997).
62. J. D. Brown and J. W. York, Jr, *Phys. Rev.* D **47**, 1407–1419 (1993).
63. C. C. M. Liu and S. T. Yau, *J. Amer. Math. Soc.* **19**, 181–204 (2006).
64. C. C. M. Liu and S. T. Yau, *Phys. Rev. Lett.* **90**, 231102 (2003).
65. N. Ó Murchadha, L. B. Szabados and K. P. Tod, *Phys. Rev. Lett.* **92**, 259001 (2004).
66. M. T. Wang and S. T. Yau, *Phys. Rev. Lett.* **102**, 021101 (2009).
67. C.-M. Chen, J.-L. Liu, J. M. Nester and M.-F. Wu, Optimal choices of reference for quasi-local energy, arXiv:0909.2754 [gr-qc].
68. L. L. So, J. M. Nester and T. Vargas *Phys. Rev.* D **78**, 044035 (2008).

Astrophysics

INTERACTIONS OF DARK ENERGY WITH OTHER COMPONENTS

SUNG-WON KIM*

Department of Science Education, Ewha Womans University,
Seoul 120-750, Korea
** E-mail: sungwon@ewha.ac.kr*

YONG-YEON KEUM

Department of Physics and BK21 Initiative for Global Leaders in Physics,
Korea University,
Seoul 136-701, Korea

In this paper, we studied the interactions of dark energy with dark matter, black hole, and wormhole. It was shown that, in phantom case, the interaction terms make arise the new behaviors, such as avoidance from big rip, the decrease of black hole mass, and the increase of the wormhole throat size.

Keywords: dark energy; dark matter; black hole; wormhole; phantom energy.

1. Introduction

Recent cosmological observations provide the crucial evidences and information of the universe. The data by WMAP show that the critical density and the Hubble parameter today are $\rho_c \equiv 3H_0^2/8\pi G \sim 10^{-29}\text{g/cm}^3$, $H_0 \sim 73\text{km} \cdot s^{-1}\text{Mpc}^{-1}$, respectively.[1] The data of CMBR, SN, large scale structure also indicate that our universe is composed of dark energy (73%), dark matter (23%), and baryon (4%). The main component of our universe is the dark energy that are being suggested as scalar field, inflation, or quintessence, even generalized Chaplygin gas through various models.

Now we want to think about the interactions of dark energy with other matter, such as dark matter, black hole, and wormhole. The first issue in this paper is the interaction with dark matter among them. There is a big plausible reason to consider the interaction of dark energy with dark matter. Since two components are the major constituent of our universe, there is no objection against the interaction between two. It also gives the answers to the coincidence problem and to the Big Rip problem for phantom energy. Though there are several models for couplings, there are not so strong model as yet. The conventional coupling models of dark energy and dark matter treat the quintessence potential.[2,3] In this model, even when the phantom energy is used for dark energy, there is no Big Rip which appeared in usual case without any interaction.

We also may think about the couplings of dark energy with other astrophysical objects, such as black hole and wormhole. Recent works on accretion of dark energy into them are good examples of the interactions. The accretion model of dark energy into black hole[5] followed the simple accretion law by Miachel.[4] They found the mass rate by accretion through the component of the energy-momentum tensor, ${T_0}^r$. If the accreting dark energy is phantom, the mass of the black hole diminishes. For the case of wormhole,[6] wormhole shell of radius expands faster than the universe and ends up engulfing it and destroying global hyperbolicity. When a spherically symmetric thin shell wormhole in a spatially flat FLRW universe,[9] a wormhole shell expanding relative to the cosmic substratum accretes positive cosmic fluid energy.

However, they only thought about the thin shell in cosmological model, but did not consider the backreaction of the metric. Therefore, when we want the issue in more realistic spacetime, the accretion should be extended into the case of backreaction. We should think about the total geometry, because the dark energy is originated from Friedman-Robertson-Walker (FRW) universe in geometry. The black hole models in FRW universe were suggested in separated papers.[12,14,15] Thus we have to count on the problem of the wormhole in FRW metric. This metric is already calculated and found the solution to the scale factor.[10] Also the combined model with wormhole exotic matter and dark energy model is considered without any interaction between the two.[11] Here we calculate the solution to the dark energy accretion to wormhole as the interaction of dark energy with wormhole in more realistic background.

2. Coupling Model

2.1. *Noninteracting Model*

The Friedmann equation for the non-interacting dark energy with various components including dark matter is given by

$$H^2 = \frac{8\pi G}{3}\rho_{\rm tot} = \frac{8\pi G}{3}[\rho_\gamma + \rho_m + \rho_\nu + \rho_\varphi + \rho_b], \tag{1}$$

where $\rho_\gamma, \rho_m, \rho_\nu, \rho_\varphi, \rho_b$ are the densities of radiation, dark matter, neutrino, dark energy, and baryon, respectively. Here $H = \dot{a}/a$, where a is the scale factor of the cosmological model. The relevant conservation laws are

$$\dot{\rho}_\gamma + 4H\rho_\gamma = 0, \tag{2}$$

$$\dot{\rho}_\nu + 3H\rho_\nu = 0, \tag{3}$$

$$\dot{\rho}_b + 3H\rho_b = 0, \tag{4}$$

$$\dot{\rho}_\varphi + 3H(\rho_\varphi + P_\varphi) = \dot{\rho}_\varphi + 3H(1+\omega_\varphi)\rho_\varphi = 0, \tag{5}$$

$$\dot{\rho}_m + 3H\rho_m = 0. \tag{6}$$

Here ω is the constant equation of state as $P = \omega\rho$. In radiation dominant era, the density of radiation is presented in terms of a by using the conservation law,

$$\frac{\dot{\rho}_\gamma}{\rho_\gamma} = -4\frac{\dot{a}}{a} \quad \text{or} \quad \rho_\gamma = \rho_\gamma^0 a^{-4}. \tag{7}$$

Similarly, the scale factor dependencies of densities for massive neutrino, baryon, and dark matter are

$$\frac{\dot{\rho}}{\rho} = -3\frac{\dot{a}}{a} \quad \text{or} \quad \rho_{b,\nu,m} = \rho_{b,\nu,m}^0 a^{-3}. \tag{8}$$

For quintessence field which is one of the candidates of dark energy, we calculate the a-dependencies of ω and ρ as

$$\omega_\varphi(a) = \frac{P_\varphi}{\rho_\varphi} = -\frac{1}{3}\frac{d\ln\rho_\varphi}{d\ln a} - 1, \tag{9}$$

$$\rho_\varphi(a) = \rho_\varphi^0 \exp\left[-3\int_1^a \frac{da'}{a'}[1+\omega_\varphi(a)]\right]. \tag{10}$$

Thus if we combine the whole elements considered here, the Friedmann equation becomes

$$H^2(a) = \frac{8\pi G}{3}\rho_{\text{tot}}^0 \left[\Omega_\gamma^0 a^{-4} + \Omega_b^0 a^{-3} + \Omega_\nu^0 a^{-3} + \Omega_m^0 a^{-3} \right.$$
$$\left. + \Omega_\varphi^0 \exp\left[-3\int_1^a \frac{da'}{a'}[1+\omega_\varphi(a)]\right]\right]. \tag{11}$$

By this equation, we can obtain the time-dependence of a when Ω values and the detailed form of $\omega_\varphi(a)$ are known.

2.2. *Interacting Model of Dark Energy coupled to Dark Matter*

For simplicity, exponential type quintessence potential is given as

$$V(\varphi) = V_0 e^{\rho_\varphi / m_p}, \tag{12}$$

where

$$m_p = \frac{M_p}{\sqrt{8\pi}} = 2.436 \times 10^{18}\ \text{GeV}, \tag{13}$$

M_p is the Planck mass, and V_0 is the potential when $\rho_\varphi = 0$. The mass of dark matter is given as

$$M_m = M_{m0} e^{-\lambda(\varphi - \varphi_0)/m_p}, \tag{14}$$

where M_{m0} is the mass of dark matter when $\varphi = \varphi_0$, today.

Since dark matter must be stable, its number density obeys the usual relation as

$$\dot{n}_m + 3Hn_m = 0 \quad \text{or} \quad n_m(a) \sim a^{-3}. \tag{15}$$

Assuming that dark matter is non-relativistic, its energy density is $\rho_m = M_m n_m$. With the coupling term Q,The conservation laws are[2,3]

$$\dot{\rho}_m = -3H\rho_m - Q, \tag{16}$$

$$\dot{\rho}_\varphi = -3H(1 + \omega_\varphi)\rho_\varphi + Q. \tag{17}$$

Here, $Q > 0$ means that energy transfer from dark matter sector to dark energy. Of course, $Q < 0$ means the energy transfer from dark energy to dark matter.

There can be several models:[3]

(i) $Q \propto \rho_m \dot{\varphi}$,

(ii) $Q \propto H\rho_\varphi$,

(iii) $Q = \Gamma \rho_c$, with const Γ,

We choose the first case as

$$Q = \frac{\lambda \dot{\varphi}}{m_p} \rho_m. \tag{18}$$

Since total energy-momentum tensor has to be conserved, the conservation law is

$$\dot{\rho}_{\text{tot}} + 3H(\rho_{\text{tot}} + P_{\text{tot}}) = 0, \tag{19}$$

where $\rho_{\text{tot}} = \rho_m + \rho_\varphi$. When it is separated by two components, the conservation laws are

$$\dot{\rho}_\varphi + 3H(\rho_\varphi + P_\varphi) = -(\rho_m + 3H\rho_m) = Q = \frac{\lambda \dot{\varphi}}{m_p} \rho_m \tag{20}$$

or

$$\dot{\rho}_\varphi + 3H(\rho_\varphi + P_\varphi) = \frac{\lambda\dot{\varphi}}{m_p}\rho_m. \tag{21}$$

We redefine the effective equation of state constants ω_m^e and ω_φ^e as

$$\omega_m^e \equiv \frac{\lambda\dot{\varphi}}{3Hm_p} = \frac{\lambda}{3m_p}\frac{\dot{\varphi}}{\dot{a}/a} = \frac{\lambda}{3m_p}\frac{d\varphi}{d\ln a} = \frac{\lambda}{3m_p}\varphi', \tag{22}$$

$$\omega_\varphi^e \equiv \omega_\phi - \frac{\lambda\dot{\varphi}}{3Hm_p}\frac{\rho_m}{\rho_\varphi} = \omega_\varphi - \frac{\lambda}{3m_p}\varphi'\frac{\rho_m}{\rho_\varphi}. \tag{23}$$

Conservation laws for the dark energy and dark matter can be rewritten as

$$\dot{\rho}_\varphi + 3H(1+\omega_\varphi^e)\rho_\varphi = 0, \tag{24}$$

$$\dot{\rho}_m + 3H(1+\omega_m^e)\rho_m = 0. \tag{25}$$

The $a-$dependencies are given as

$$\rho_{\varphi,m}(a) = \rho_{\varphi,m}^0 e^{-3\int_1^3 \frac{da'}{a}[1+\omega^e(a')]}. \tag{26}$$

Thus the equation of motion for dark energy becomes

$$\ddot{\varphi} + 3H\dot{\varphi} = H^2\varphi'' + \frac{1}{3m_p^2}\left[\frac{3}{2}(\rho_b+\rho_m)+\rho_\gamma+3V\right]\varphi' = \frac{\lambda}{m_p}\rho_m - \frac{\beta}{m_p}V, \tag{27}$$

$$\begin{aligned} H^2 &= \frac{1}{3m_p^2}\left[\rho_b+\rho_m+\rho_\gamma+\frac{1}{2}\dot{\varphi}^2+V\right] \\ &= \frac{1}{3m_p^2}[\rho_b+\rho_m+\rho_\gamma+V] + \frac{1}{6m_p^2}H. \end{aligned} \tag{28}$$

Therefore

$$H^2\left(1-\frac{\varphi'^2}{6m_p^2}\right) = \frac{1}{3m_p^2}[\rho_b+\rho_m+\rho_\gamma+V] \tag{29}$$

or

$$H^2 = \frac{8\pi G}{3}\frac{[\rho_b+\rho_m+\rho_\gamma+V]}{\left(1-\frac{\varphi'^2}{6m_p^2}\right)}. \tag{30}$$

This means that there is no Big Rip for phantom energy, $\varphi'^2 < 0$. The results show that the interaction term prevent from Big Rip even we use phantom energy for accelerated expansion.

3. Accretion of dark energy

For the accretion problem, we start from the spherically symmetric space-time whose line element is given as

$$ds^2 = e^\nu dt^2 - e^\lambda dr^2 - r^2(d\theta^2 + \sin^2\theta d\varphi^2), \tag{31}$$

where $u_0 \equiv \frac{dt}{ds}e^\nu = (e^\nu + e^{\lambda+\nu}u^2)^{1/2}$ and $u \equiv u^r = \frac{dr}{ds}$. We adopt the perfect fluid for sources. From the energy-momentum conservation, $T^{\mu\nu}{}_{:\nu} = 0$,

$$ur^2 e^{(\lambda+\nu)/2} \exp\left[\int_{\rho_\infty}^{\rho} \frac{d\rho'}{\rho' + P(\rho')}\right] = -A, \tag{32}$$

where A is an integration constant. Also energy-momentum flux conservation law, $u_\mu T^{\mu\nu}{}_{:\nu} = 0$, is

$$(P+\rho)u_0 u x^2 e^{(\lambda+\nu)/2} = C_1, \tag{33}$$

where $x \equiv r/M$, M is ADM mass of the object, and C_1 is another constant. If we divide Eq. (33) by Eq. (32), then

$$C_2 = -\frac{C_1}{A} = (P+\rho)u_0 \exp\left[-\int_{\rho_\infty}^{\rho} \frac{d\rho'}{\rho' + P(\rho')}\right]. \tag{34}$$

Here, C_2 is also a constant. When we move the interesting point to infinity, Eq. (34) becomes

$$C_2 = [\rho_\infty + P(\rho_\infty)] = -C_1/A. \tag{35}$$

Therefore,

$$C_1 = (P+\rho)u_0 u x^2 e^{(\lambda+\nu)/2} = -A(\rho_\infty + P(\rho_\infty)). \tag{36}$$

The mass rate in first approximation is given as[4]

$$\begin{aligned}\dot{M} &= -4\pi r^2 T_0{}^r = -4\pi r^2[(\rho+p)U_0 U]\\ &= 4\pi M^2 A e^{-(\lambda+\nu)/2}[\rho_\infty + P(\rho_\infty)].\end{aligned} \tag{37}$$

When we apply it to the Schwarzschild case,[5] the mass rate is

$$\dot{M} = 4\pi M^2 A[\rho_\infty + P(\rho_\infty)] = 4\pi M^2 A\rho_\infty(1+\omega). \tag{38}$$

In this case, usual dark energy such as quintessence field ($\omega > -1$) increase mass. However, when the accreted matter have the behavior of phantom energy ($\omega < -1$), it shows the negative $\dot{M}$ which means that the black hole mass will diminish by accreting. Since the phantom energy is strictly exotic matter, it is natural that such an phenomenon happens.

Later, González-Díaz extended the problem into the case of dark energy accretion to wormhole.[6] He just thought the spherical thin shell of exotic matter with mass[7]

$$\mu = -\pi b_0/2, \tag{39}$$

where b_0 is the radius of the spherical wormhole throat. By analogy of Visser type wormhole (cut-and-paste operation)[8] with $M \simeq \mu$, the mass rate is

$$\begin{aligned} \dot{M} &\simeq -\frac{\pi}{2}\dot{b}_0 \\ &\simeq 4\pi\mu^2 D\rho(1+\omega) = 4\pi\left(\frac{\pi^2 b_0^2}{4}\right) D\rho(1+\omega). \end{aligned} \tag{40}$$

Thus the throat size change rate is

$$\dot{b}_0 = -2\pi^2 D b_0^2 \rho(1+\omega), \tag{41}$$

where $D \simeq A$ is a dimensionless constant. In case of a universe dominated by dark energy, the scale factor can be written as

$$\begin{aligned} a(t) &= \left[a_0^{3(1+\omega)/2} + \frac{3}{2}(1+\omega)\sqrt{\frac{8\pi\rho_0}{3}}(t-t_0)\right]^{2/[3(1+\omega)]} \\ &= T^{2/[3(1+\omega)]}, \end{aligned} \tag{42}$$

where a_0 and t_0 is the value at the onset of dark energy domination. Thus

$$\dot{b}_0 = -2\pi^2 D(1+\omega)\rho_0 b_0^2 T^{-2}. \tag{43}$$

The result shows that for all quintessence type models ($\omega > -1$), the radius of wormhole throat will gradually decrease with time, while it will gradually increase when the wormhole accretes phantom energy ($\omega < -1$).[6] The wormhole shell of radius $b_0(t)$ expands faster than the universe and ends up engulfing it and destroying global hyperbolicity.[6]

However, the model have a critical point, which is restricted in the case of fixed wormhole throat. The advanced study was published afterwards by Faraoni and Israel.[9] In their paper, the model was the implication of the comoving radius of the shell as the wormhole, such as

$$r = \frac{R(t)}{a(t)} \equiv e^{-\alpha(t)}, \tag{44}$$

in spatially flat FRW universe

$$ds^2 = -dt^2 + a^2(t)[dr^2 + r^2(d\theta^2 + \sin^2 d\phi^2)], \tag{45}$$

where $R(t)$ is comoving radius of the shell and $a(t)$ is the scale factor. The mass of the wormhole is defined as

$$M \equiv 4\pi R^2\sigma = -2\left(\beta R - \frac{\alpha_t R_t R^2}{\beta}\right), \tag{46}$$

where

$$\beta = \sqrt{1-v^2}, \quad u^\alpha n_\alpha = -v/\beta \tag{47}$$

and σ is the surface mass density. The rate of accretion of the cosmic fluid by the wormhole is

$$\dot{M} + P_{(\Sigma)}\dot{\mathcal{A}} = \frac{2}{\rho^2}(P+\rho)\mathcal{A}v, \tag{48}$$

where $\mathcal{A}$ is the area of a spherical surface of isotropic radius. In De Sitter background $P = -\rho$, there is no accretion on the shell and static solutions with both $\dot{M}$ and $\dot{R}$ vanishing become possible. If the strong energy condition is satisfied, $(P+\rho > 0)$, then a wormhole shell expanding relative to the cosmic substratum $(v > 0)$ accretes positive cosmic fluid energy.[9]

4. Wormhole in Friedman-Roberton-Waler Universe

The wormhole in FRW universe is defined by the spacetime[10,11]

$$ds^2 = -e^{2\Phi(r)}dt^2 + a(t)^2\left[\frac{dr^2}{1-\kappa r^2 - b(r)/r} + r^2(d\theta^2 + \sin^2\theta d\phi^2)\right]. \tag{49}$$

Here $\Phi(r)$ is the redshift function, $b(r)$ is the wormhole shape function, κ is the spatial curvature. It is the combined model of Morris-Thorne type wormhole[7] and FRW cosmological model.

Its solution for the equation of state of $P = \omega\rho$ can be obtained by ansatz $a^2(t)\rho(r,t) = a^2(t)\rho_c(t) + \rho_w(r)$. Here, ρ_c is the density of the cosmological part and ρ_w is that of wormhole part. The a-dependence of ρ_c, time-dependence of a, and time-dependence of ρ_c are

$$\rho_c(a) \propto a^{-3(1+\omega)} + a^{-2}, \tag{50}$$

$$a(t) \propto t^{2/[3(1+\omega)]}, \tag{51}$$

$$\rho_c(t) \propto t^{-2} + t^{-4/[3(1+\omega)]}. \tag{52}$$

Since the general formula of the mass rate for spherically symmetric metric is Eq. (37), we simply apply the Morris-Thorne type wormhole to this formula as

$$\dot{b}_0 = -2\pi^2 D b_0^2 e^{-(\lambda+\nu)/2}(\rho_\infty + P(\rho_\infty)) = -2\pi^2 D b_0^2\sqrt{1-\frac{b_0^2}{r^2}}(1+\omega)\rho_\infty. \tag{53}$$

In this solution, there is no accretion at $r = b_0$. Of course, the phantom energy increase the size of wormhole throat b_0.

The black hole models in FRW universe are suggested in separated papers.[12,14,15] Their results were not quite different from the formers. For example,[12] the accretion rate is

$$\dot{M} = -\frac{G}{2} a B^2 \sqrt{1 + a^2 A^4 u^2} (P + \rho) A u. \tag{54}$$

For the case of wormhole in FRW universe, the metric components are given as

$$e^{\nu} = e^{2\Phi(r)}, \quad e^{\lambda} = \frac{a^2(t)}{1 - \kappa r - b(r)/r^2} \tag{55}$$

With the special case of $\Phi = 0$ and $b(r) = b_0^2/r$, the change rate of wormhole throat size is

$$\begin{aligned} \dot{b}_0 &= -4\pi r^2 T_0{}^r = -2\pi^2 D b_0^2 \sqrt{1 - \frac{b_0^2}{r^2}} (1 + \omega) \rho_\infty \\ &= -(1 + \omega) C(r) \left[\alpha' T^{-\frac{2}{3}\left(\frac{4+3\omega}{1+\omega}\right)} + \delta' T^{-\frac{2}{1+\omega}} \right]. \end{aligned} \tag{56}$$

Here, $\rho_\infty \simeq \rho_c$, and we substitute it in Eq. (42). The simple integration gives

$$b_0 \propto \alpha T^{-\frac{5+3\omega}{3+3\omega}} + \delta T^{\frac{\omega-1}{\omega+1}}, \tag{57}$$

with proper constants α and δ. When we neglect the wormhole effect comparing to the dark sector ($\delta \ll \alpha$),

$$b_0 \propto T^m. \tag{58}$$

Here the sign of the exponent m and the divergencies of b_0 are determined by the value of ω as

$$\begin{array}{lll} 0 > \omega > -1, & m < 0 & \text{or } b_0 \text{ diverges at finite time,} \\ -1 > \omega > -5/3, & m > 0, & \\ -5/3 > \omega, & m < 0 & \text{or } b_0 \text{ diverges at finite time.} \end{array} \tag{59}$$

Even $\omega > -1$, b_0 diverges at finite time. The accretion missing positions are located at

$$\begin{array}{ll} r = b_0, & \text{for } \kappa = 0, \\ r^2 = \frac{1}{2} \pm \sqrt{\frac{1}{4} - b_0^2}, & \text{for } \kappa = 1, \\ r^2 = -\frac{1}{2} + \sqrt{\frac{1}{4} + b_0^2}, & \text{for } \kappa = -1. \end{array} \tag{60}$$

By comparing with the former result, its behaviors are more complicated and the Big Rip appears under non-phantom energy.

5. Conclusion

In this paper, we studied the interactions of dark energy with dark matter, black hole, and wormhole. For the model of interaction with dark matter, the solution without big rip was found even in the case of phantom energy. The interactions with black hole and wormhole are considered as the accretions of dark energy into them. We also generalized the accretion model into the cosmological model with wormhole.

Acknowledgments

This research was supported by Basic Science Research Program through the National Research Foundation of Korea (NRF) funded by the Ministry of Education, Science and Technology (2009-0073859).

References

1. W. L. Freedman, et al., *Ap. J.* **170**, 377 (2001).
2. Jian-Hua He and Bin Wang, *JCAP* 0806, 010 (2008).
3. C. G. Böhmer, G. Caldera-Cabral, R. Lazkoz, and R. Maartens, *Phys. Rev. D* **78**, 023505 (2008).
4. F. C. Michel, *Astrophys. Space Sci.* **15**, 153 (1972).
5. E. Babichev, V. Dokuchaev, and Yu. Eroshenko, *Phys. Rev. Lett.* **93**, 021102 (2004).
6. D. F. González-Díaz, *Phys. Rev. Lett.* **93**, 071301 (2004).
7. Morris and Thorne, *Am. J. Phys.* **56**, 395 (1988).
8. M. Visser, *Lorentzian Wormholes*, Springer-Verlag, New York, (1996).
9. V. Faraoni and W. Israel, *Phys. Rev. D* **71**, 064017 (2005).
10. Sung-Won Kim, *Phys. Rev. D* **53**, 6889 (1996).
11. Sung-Won Kim, *J. Kor. Phys. Soc.* **49**, 755 (2006).
12. V. Faraoni and A. Jacques, *Phys. Rev. D* **76**, 063510 (2007).
13. G. C. McVittie, *Mon. Not. R. Astron. Soc.* **93**, 325 (1933).
14. V. Faraoni, C. Gao, X. Chen, and Y.-G. Shen, *Phys. Lett. B* **671**, 7 (2009).
15. C. Gao, X. Chen, V. Faraoni, and Y.-G. Shen, *Phys. Rev. D*, **78**, 024008 (2008).

BRIEF INTRODUCTION OF YINGHUO-1 MARS ORBITER AND OPEN-LOOP TRACKING TECHNIQUES

JIN-SONG PING, KUN SHANG, NIAN-CHUAN JIAN, MING-YUAN WANG, SU-JUN ZHANG, XIAN SHI, TING-TING HAN, JING SUN, GUANG-LI WANG, JIN-LING LI, LEEWO FUNG

Shanghai Astronomical Observatory, CAS. Nandan Rd. 80
Shanghai, 200030, P.R. China

China and Russia are planning to launch a joint Mars mission in 2011. In the joint mission, the 1st Chinese Mars Probe, Yinghuo-1 will explore the space weather of the Mars, and will test the deep space navigation techniques. Different from the close-loop tracking methods in common deep space mission, the open-loop methods like DOR/DOD and 1-way Doppler, are developed and applied to determine the s/c orbit and position.

1. Introduction of joint YH-1 and FGSC missions

Since the beginning of the new century, Mars exploration has attracted the huge attention from space communities. A new race and cooperation in lunar and planetary exploration has started. Being a beginner in this area, China has launched her 1st lunar orbiter Chang'E-1 successfully, and has got many new scientific results from this exploration. Beyond this, a joint Russian-Chinese Martian mission, Yinghuo-1 (YH-1) & Phobos-Grunt (FGSC), has been developed and promoted solidly [1].

The "Chinese-Russian collaborative Mission for Mars" is planned in the government level, and the final agreement has been signed in 2007. Two probes will be launched together in November, 2011. Figure 1 shows the configuration of them.

Lavochkin Association and IKI of Russia side will be responsible for developing the FGSC; Shanghai Academy of Spaceflight Technology and CSSAR/CAS will be responsible for developing the YH-1 mission. They will coordinate the issues of joint launching and joint exploration.

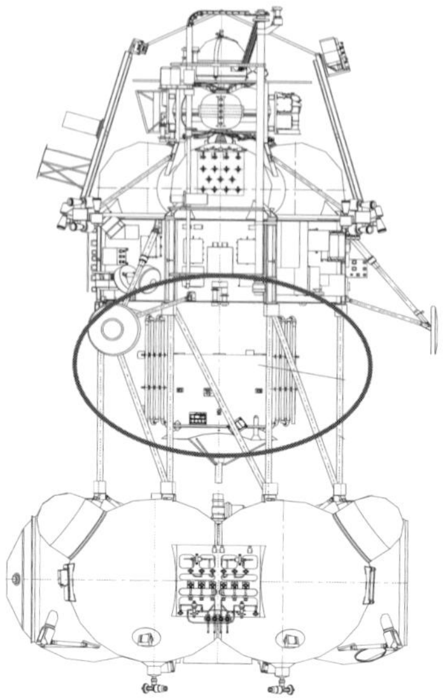

Figure 1, The configuration of launching package of YH-1 and FGSC

After a successful launching, the joint spacecraft YH-1 & FGSC will be sent to a transfer orbit flying to the Mars. After 10-11months, the joint craft will arrive the Martian system, and will be eject into an equatorial orbit of 800x80000km, with a period of ~72hours, inclination of 1^{O}.~5°. The joint craft will fly in this orbit for about 3 circles, then they will be separated, FGSC will change its orbit to find change of landing on Phobos, and YH-1 will free-fly in this orbit for 1year.

YH-1 is a small satellite focused on investigating the Martian space environment and the solar wind -Mars interaction. FGSC is a sampling return mission to land on the 1st Martian satellite, Phobos, and take some (0.1-0.2kg) soil back to the Earth. YH-1 and Phobos-Grunt forms a two-point measurement configuration in the Martian space environment. Equipped with similar plasma detecting payload on two spacecraft would give some coordinated exploration around Mars. The two S/Cs will also carry out satellite-to-satellite radio link, so as to study the Martian ionosphere by using radio occultation links at UHF. Some general characteristics for YH-1 are read as:

- Total mass of YH-1, 110kg
- Power supply 150W (average), and instant 180W。
- Data down link:

0.9m in diameter, HGA, directly to earth, 80bps---8kbps

- ➢ X-band Receiver and transmitter onboard s/c for communication.
- ➢ 3-axis stabilized attitude control, deployable solar panel.

The 6 kinds of payloads of YH-1 mission are list in table, where the main scientific objectives of YH-1 are read as:

- ➢ Martian space environmental structure, plasma distribution and characteristic in the regions;
- ➢ Solar wind-atmosphere coupling and energy deposition processes, and Martian ions escaping processes and possible mechanisms;
- ➢ Martian and Phobos surface imaging;
- ➢ Regional gravity field of the Mars.

Table.1 Payloads onboard YH-1 S/C

instrument	objective
Electron analyzer Proton analyzer Planetary ion analyzer	Electron, proton and planetary ion energy spectrum, ion identification e, 10eV---20 keV P, 10eV—20 keV I, 10eV--- 20 KeV, 2 π solid angle
416MHz receiver	Ionosphere occultation
Photo-imager	Sand storm and Mars imaging
FGM	Magnetic field, +/- 65000 nT, 8nT resolution +/- 256 nT, 0.01 resolution

2. Open Loop Tracking in YH-1 Mission

Considering that the Chinese deep space tracking system is still under construction, there will not be any uplink system in China can meet the power requirement of uplink communication for a distance about 2AU. To simplify and to minimize the designation, an X-band receiver and X-band transmitter system have been adopted for onboard communication. There is not a common PLL transponder used for tracking

Chinese VLBI network system has been playing important roles in lunar and planetary deep space tracking. To solve above tracking and OD problem, an Ultra-Stable-Oscillator (USO)-based 1-way open loop concept will be used for the s/c, and the ground astronomical Very Long Baseline Interferometry (VLBI) system [2] will be used to receive the radio signal, so as to retrieve the DOR/DOD and Doppler information. The open loop observable will be applied for S/C positioning and OD, by using Chinese VLBI network.

2.1. *Chinese VLBI network*

Chinese VLBI network system is composes 4 stations located at Beijing, Shanghai, Kunming and Urumqi, as well as a VLBI data analysis center at Shanghai. The antenna size and accomplish year are read as Shanghai 25m1987, Urumqi 25m1993, Beijing 50m2006, Kunming 40m 2006. See Figure 2.

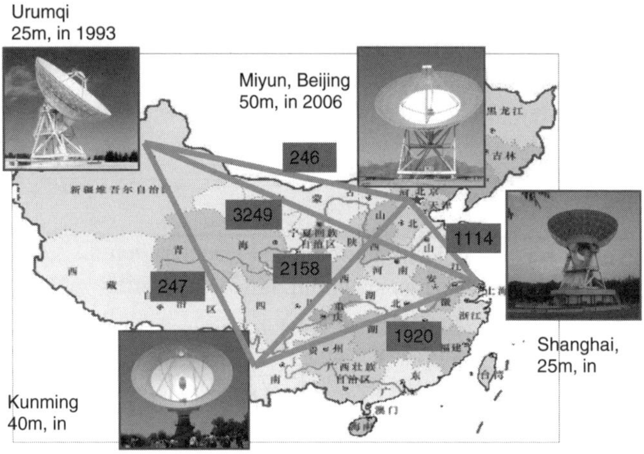

Figure 2. Distribution of Chinese Astronomical VLBI Network

This VLBI network can do radio observation at L/S/C/X/Ku band. The S/X dual-frequency mode covers the whole ITU satellite communication bandwidth, it can satisfy the observation requirement of Geodesy and satellite tracking. Since 1990s, this network has taken part in some international s/c VLBI tracking work, and some domestic s/c tracking work. In two recent lunar mission, the SELENE and Chinag'E-1 mission, this network realized the real-time s/c tracking and POD by VLBI method. Using its data together with unified S-Band tracking data, the lunar gravity field can be improved by either the common or the SBI VLBI mode.

In SELENE and Chang'E-1 VLBI tracking experiments, near by angle distance radio source, QUASARs, and lunar orbiters were tracked using switching mode. Using above method, the S/C signal was treated as white noise, the astronomical correlation method can be applied to get the radio signal delay between two stations located at the ends of a VLBI baseline. Where, the S/C VLBI observable, geometric group time delay $\tau_{s/c,g}$ can be obtained by removing the systematic errors of instruments and media from the observed time delay $\tau_{s/c,o}$ by using following algorithm.

$$\tau_{s/c,g} = \tau_{s/c,o} - (\tau_{q,o} - \tau_{q,g}) \quad (1)$$

where, the subscript q means the observable of QUASAR.

The tracking stations and the VLBI center are connected by Intelnet. In SELENE mission, the observation data can be sent to National Astronomical Observatory of Japan by using TCP/IP ftp data transferring mode. In Chang'E-1 mission, the VLBI tracking data can be sent to the Shanghai VLBI center using real-time IP data stream mode, to carry out real-time correlation. In YH-1 mission, the network will be slightly enhanced by including 2-3 Russian VLBI stations. After getting the VLBI observable, the VLBI center will also do the S/C orbit estimation and prediction for mission users.

Besides the common Delta-VLBI observation, different from the historical method, both of the YH-1 and FGSC have been equipped with the USO. Frequency instability of reference quartz oscillator is less than 1 or $2x10^{-12}$ under averaging time from 0.1 to 1000 seconds. The frequency characteristics of the USO are shown in Figure 3. In YH-1, the X-Band phase will be modulated by 90degree to get a pair of tone signal for DOR observation. The two missions will send VLBI signal at 8424 MHz and 8425 MHz, which can be used for SBI observation used in SELENE mission [3].

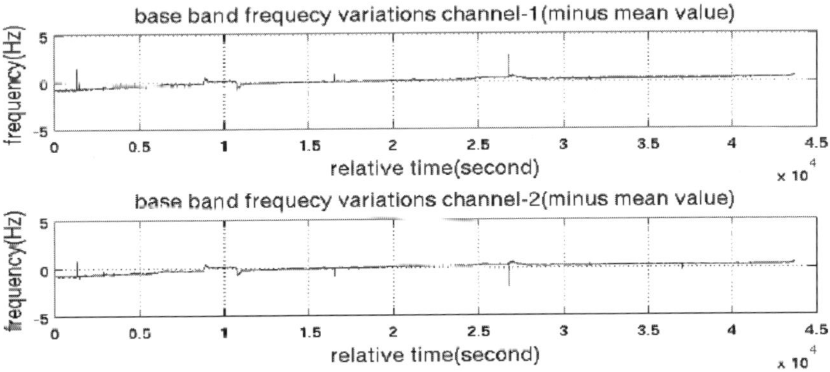

Figure3 The Frequency Drift of USO on YH-1 Proto-mode

Using the Delta-VLBI and SBI observation, many possible scientific results are expected to be reached from the VLBI tracking of the joint mission. These objectives are estimated as:

- To define more exactly Sun system's parameters (Astronomical constant, orbital parameters of Mars and Phobos);
- To evaluate experimentally Phobos life time on its orbit;
- To obtain mass distribution inside Phobos;
- To define more exactly the large asteroids masses from main belt;
- To define more exactly limits of variation of universal gravitational constant;

- To define more exactly geometrical tie of dynamical coordinate system having original in Sun system mass centre of with coordinate system based on quasar angular coordinate measurements.

To accomplish above joint tracking and experiments between China and Russia, a resolution of time scale and reference frame is coordinated and recommended to be adopted for manufacture, launch, telemetry, control and scientific application of YH-1 spacecraft in Mars exploration project. This resolution also provides the definitions of some concerned time scales and reference frames in the project, with the values of certain constants. The transformations between the different time scales are also provided [4].

2.2. *1-way Doppler Tracking*

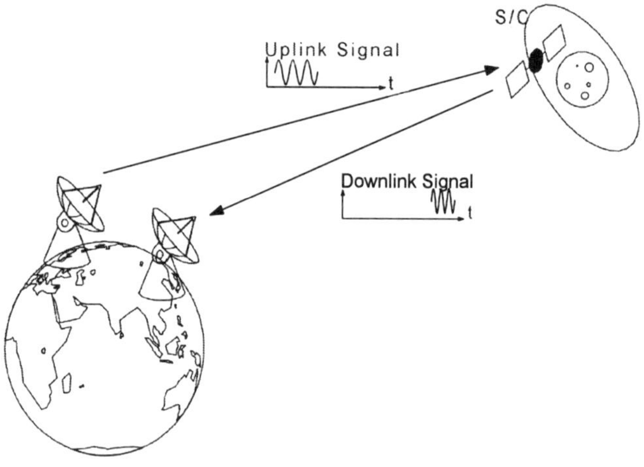

Figure 4. Open Loop TT&C Concept for YH-1 Mission.

In China, a domestic lunar and deep space tracking system is under construction. The designed technical specifications of this system can meet the tracking and navigation requirement of YH-1 mission. However, this deep space tracking system has less chance to be accomplished before the end of 2012, when the YH-1 will have arrived the Mars and start the exploration. To solve this problem, a few of international space agencies will support the uplink control and communication for YH-1 mission. And, to simplify and minimize the designation of the spacecraft, only an X-band receiver and transmitter system have been equipped for onboard communication. The open loop TT&C will be applied for S/C positioning and OD. Its concept is shown in Figure 4[5].

Using software radio method, we developed 3 kinds of 1-way Doppler receiver [5-7]: a software receiver can retrieving the Doppler signal from VLBI multi-bit ADC data using post-processing method; a DSP based digital Doppler counter can retrieve the Doppler signal in real-time way from VLBI analog base band output; a high speed ADC based local correlator can retrieve frequency signal in real-time way from VLBI analog base band output. Beyond above proto-type digital counter, we are developing the full dimensional digital Doppler counter based on software radio and analog IF output.

During the nominal mission period of Chang'E-1, we used 3-way method to test the open loop tracking ability. Which means, a tracking station sent Rb atomic clock generated uplink S-Band TT&C signal to Chang'E-1 S/C, the transponder of the S/C locked the uplink CW and sent it back to the Earth. The new developed Doppler counters were installed only in VLBI stations, using a local H-maser atomic clock, they can get 3way Doppler with rms of 3mm/s at 1Hz sampling rate. This kind of observable can also be used for S/C-Earth Martian atmospheric occultation [8] and other planetary radio scientific experiments. The main error is coming from the Rb clock. This result can satisfy the YH-1 mission requirement. Experiments have been done together with ESOC deep space tracking stations for tracking the MEX[9].

Based on above tracking condition, a simulation work has been carried out for OD of YH-1, it is found that a long term 1way tracking, 3circle or 10days, can meeting the orbiting requirement in YH-1 mission during its free flying period.

3. Summary

China will launch her 1st Martian mission YH-1 together with Russia in October, 2011. YH-1 mission will mainly investigate the space weather of the Mars and the coupling with the solar wind. Together with the Russian mission FGSC, they will study the dynamical specifications of Martian system.

To minimize the designation and to simply the tracking work, 1-way open loop tracking concept has been accepted by the mission system, based on using USO onboard S/C, and on using the sophisticated astronomical VLBI system. By developing the new digital Doppler counter and installing on the VLBI back-end, the VLBI system can be updated as an open-loop deep space tracking network. A successful tracking in the mission may benefit the future deep space exploration..

Acknowledgments

This work is supported by the Sino-Russian cooperation "YH-1" Mars exploration project, and by national high-tech projects 2008AA12A209 and 2008AA12A210 issues, as well as by CAS key direction research Program KJCX2-YW-T13-2 and NSFC No. 10973031.

References

[1]http://www.russianspaceweb.com/phobos_grunt.html; Proposal of Using VLBI for YH-1 Tracking and Orbit Determination, Shanghai Astronomical Observatory, 2007.04, in Chinese.

[2] O. J. Sovers and J. L. Fanselow, *Rev. of Modern Phys.* **70(4)**, 1393 (1998)

[3] Q. Liu, X. Shi, W. Wang, et al., S. *Physics*, **38(10)**, 712(2009), in Chinese;

[4] J. Li and J. Yan, YH-1 *Mars Exploration Project Resolutions on Time Scales and Reference Systems(Proposal)*, Shanghai Astronomical Observatory, CAS; Beijing Aerospace Control Center, March, 2009

[5]Catherine L T. Radiometric Tracking Techniques for Deep Space Navigation, Deep Space Communication and Navigation Series, 2002

[6] www2.nict.go.jp/w/w114/stsi/K5/VSSP/install_cor_e.html

[7] J. Ping, W. Frank, Yusuke K and H. Hideo, .*Journal of Planetary Geodesy*, **36**, 1 (2001).

[8] S. Zhang, J. Ping and Z. Hong, et al., *Physics*, **38(10)**, 722(2009), in Chinese;

[9] K. Shang, N. Jian and J. Ping, et al., *Physics,* **38(11)**, 799(2009), in Chinese.

APPLY MOVING PUNCTURE METHOD TO ADM FORMALISM

ZHOU-JIAN CAO* and CHEN-ZHOU LIU

Institute of Applied Mathematics, Academy of Mathematics and Systems Science, Chinese Academy of Sciences, Beijing 100190, China
**E-mail: zjcao@amt.ac.cn*

It is well known that moving puncture method and the specific gauge condition are critically important for successful simulation of binary black hole merging. On the contrary, the importance of formalism for numerical relativity is not very clear yet. Both generalized harmonic formalism and BSSN formalism work very well. So the simplicity of Einstein's equations in ADM formalism stimulates us to investigate a naive but interesting problem—can ADM formalism work as stably as BSSN formalism does with the moving puncture method and the advanced gauge condition, which were proved critical important for BSSN formalism's success. We apply this idea to Schwarzschild black hole simulation as a test example. Unfortunately, our result implies that ADM formalism has its intrinsic instable character which may be introduced by the property of the corresponding equations themselves instead of the gauge condition. And more, we find that the so called advanced gauge condition even make the situation worse. So we conclude that one gauge condition works well in one numerical formalism does not mean it works well also in other formalism. Through concrete examples, we give readers a primary sense on the roles that formalism, gauge condition, numerical method and other issues play in the problem of stability of numerical relativity.

Keywords: Numerical Relativity; ADM formalism; Moving puncture method; Gauge condition

1. Introduction

It was when numerical relativists discussed very actively the effects of different formalism of Einstein equations in numerical simulation that Pretorius made a breakthrough on stability problem of three dimensional numerical relativity.[1] He can successfully simulate the whole inspiraling, merging and ringing down process of binary black hole system stably and get some converged waveform of gravitational radiation. Pretorius's method includes many novel features; among them he uses generalized harmonic formalism, discretizes the four-dimensional Einstein equations directly, which is not a conventional approach so far. So the problem of formalism seemed very important to numerical relativity at that time. But interestingly, Campanelli *et al.*[2] and Baker *et al.*[3] can also get this success with more traditional formalism—BSSN formalism only half year about later. The new feature in their method is moving puncture method and careful choice of gauges. Then we can see that successful simulation of black hole system is critically undergirded by specific gauge conditions. But on the other hand, how important the formalism is to the stability is not very clear yet.

ADM formalism is the most simple and direct formalism based on $3+1$ decomposition of spacetime. And it has been successfully applied to spherical- and axi-symmetrical gravitational systems. But when numerical relativists applied it together with excision method and geodesic or maximal slicing gauge condition to general system without symmetry, all suffered instability.[5] There are many authors who discuss the stability of ADM formalism such as Refs. 6–9. Their results tell us the ADM formalism under those specific gauge conditions is unstable. From the viewpoint of well-posed property, many authors report that ADM formalism is not strongly hyperbolic.[7,10,11] But the concept of strongly hyperbolic is only applicable to first order in both time and space system. While ADM formalism is a system first order in time second order in space. So there is some ambiguity to tell the hyperbolic property of ADM formalism. And more, all of the published analysis are based on some specific gauge choice. In comparison, BSSN formalism is strongly hyperbolic in some gauge conditions while not in other gauge conditions. From the viewpoint of both hyperbolic property and numerical experiment, the success of BSSN formalism relies on the gauge condition and moving puncture method very much.[12] So we naively ask whether the previous failure of ADM formalism is due to the bad choice of gauge condition. Is there any possibility that ADM formalism can also work with some special gauge condition? Although no suitable gauge condition is found for ADM formalism, we are not very sure if there is no gauge can make ADM stable at all. Motivated by the success of BSSN formalism with moving puncture method and smart gauge condition choice (hereafter we call it advanced gauge condition), how this advanced gauge condition affects ADM formalism is an interesting problem. In this paper we apply moving puncture method and advanced gauge condition to ADM formalism and investigate these problems through numerical simulation of Schwarzschild black hole. But unfortunately our results shows that these new method and advanced gauge condition cannot help ADM formalism simulate black hole system stably. Given our results, we can not say there is no gauge condition can make ADM formalism stable yet. But we find unexpectedly that the advanced gauge condition even works worse than simpler gauge condition, which tells us that good behavior of one gauge condition in one formalism does not imply its good behavior in other formalism.

Courant factor is very important to compute hyperbolic partial differential equations numerically. In reference 13, the authors reported that the exponential instability of ADM formalism in simulation of perturbed Minkowski space resulted from Courant-Friedrichs-Lewy (CFL) violation. In order to avoid similar things happen in our case we take two Courant factors with large difference (0.25 and 0.025 respectively) to guarantee the convergence of Courant factor. In next section, we review the standard ADM formalism briefly to fix our notation and introduce our puncture method in ADM formalism. In section 3, we will describe our numerical code briefly and present our simulation setting. We present our numerical results in section 4 and give some discussions and conclusions in section 5.

2. Evolution equations and puncture method

We write the metric in the form

$$ds^2 = -\alpha^2 dt^2 + \gamma_{ij}(dx^i + \beta^i dt)(dx^j + \beta^j dt), \tag{1}$$

where α is the lapse function, β^i is the shift vector, and γ_{ij} is the spatial metric. Throughout this paper, Latin indices are spatial indices and run from 1 to 3. The extrinsic curvature K_{ij} is defined by the equation

$$K_{ij} = -\frac{1}{2\alpha}(\frac{\partial}{\partial t} - \mathcal{L}_\beta)\gamma_{ij}, \tag{2}$$

where $\mathcal{L}_\beta$ denotes the Lie derivative with respect to β. The Einstein equations can then be split into the Hamiltonian constraint

$$\mathcal{H} \equiv R + K^2 - K_{ij}K^{ij} = 16\pi\rho, \tag{3}$$

the momentum constraint

$$\mathcal{M}_i \equiv \nabla_j K^j{}_i - \nabla_i K = 8\pi s_i, \tag{4}$$

and the evolution equation for the extrinsic curvature. ρ and s_i are involved with matter. For vacuum case all of them vanish. In above equations and hereafter, R denotes the scalar curvature of γ_{ij} and K is the trace of the extrinsic curvature. Here our evolution equations of ADM formalism are standard

$$\begin{aligned}\partial_t\gamma_{ij} &= -2\alpha K_{ij} + D_i\beta_j + D_j\beta_i \\ &= -2\alpha K_{ij} + 2\gamma_{k(i}\beta^k{}_{,j)} + \beta^k\partial_k\gamma_{ij}, \qquad (5)\\ \partial_t K_{ij} &= \alpha R_{ij} + \alpha K K_{ij} - 2\alpha K_{ik}K^k{}_j - D_iD_j\alpha \\ &\quad +(D_i\beta^k)K_{kj} + (D_j\beta^k)K_{ki} - \beta^k\Gamma^l{}_{k(i}K_{j)l} + \beta^k\partial_k K_{ij}, \qquad (6)\end{aligned}$$

which are due to York.[14] These equations are the most widely adopted ADM system in numerical relativity. The basic variables for ADM formalism are three physical metric γ_{ij} and three physical extrinsic curvature K_{ij}. All the indices are raised and lowered by γ_{ij}. In order to rate the stability of different simulations we monitor the L2 norm of constraints $\int |\mathcal{H}|^2 d^3x$ and $\int |\mathcal{M}_i|^2 d^3x$.

For the evolution equations of gauge variables α and β^i, we put all the equations in the form of unification

$$\partial_t\alpha = f_\alpha, \qquad (7)$$

$$\partial_t\beta^i = f_{\beta^i}, \qquad (8)$$

$$\partial_t B^i = f_{B^i}, \qquad (9)$$

where B^i is an auxiliary variable to make the shift equation first order in time. This form of equations is adopted from popular gauge equations now used among numerical relativity community.[2,3]

Instead of excision, we borrow the idea from moving puncture method in BSSN formalism, we set puncture type initial data as following for Schwarzschild black hole.

$$\gamma_{ij} = (1 + \frac{m}{2r})^4\hat{\gamma}_{ij}, \qquad (10)$$

$$K_{ij} = 0, \qquad (11)$$

where $\hat{\gamma}_{ij}$ is the flat three metric, i.e., Euclidean metric. This initial data has only one nontrivial factor, $1 + \frac{m}{2r}$. But this nontrivial factor affects all coefficients of the three metric γ_{ij}. For BSSN formalism, the case is different. This nontrivial factor affects only one dynamical variable ϕ. In addition, the effect is reduced through logarithm operation. One disadvantage of ADM is that the value of dynamical variables is much larger than that of BSSN. The value of ADM variable is as large as several hundreds while that of BSSN variable is restricted at 1 around. But it is not clear how this disadvantage will affect the simulation of black hole system. In this paper we do not touch this problem because it needs change the form of evolution equation. This will be out of the scope of this paper.

3. Code description and simulation setting

We have developed a new 3D numerical code based on the above standard ADM formalism. This code is constructed on the same infrastructure as Ref. 4. The dynamical variables are gauge variables α, β^i, three metric γ_{ij} and extrinsic curvature K_{ij}. The time-integration is under the free-evolution scheme. Our time-integration method is the three-step iterative Crank-Nicholson method (ICN) with centered finite difference approximation of spacial derivatives except advection terms which is approximated with up wind method to deal with. As in Ref. 4, we use 4th order accurate difference approximation generally, but 4th order lopsided difference approximation for advection term such as $\beta^i\partial_i\gamma_{jk}$. We lower the difference order for near boundary region.

Convergence test is the necessary condition for correctness of code. So we introduce the Kerr-Schild coordinate form of Schwarzschild black hole to test our code's convergence. Since all variables in this form are nontrivial, testing with it is much more efficient than with isotropic coordinate. The specific Kerr-Schild coordinate form can refer to Ref. 15. The main part of code for ICN method is the computation of right hand side of equations (5) and (6). The test results are plotted in Fig.1. Since we use cell center grid structure, we interpolate the functions to y-axis (but marked with x in the figures). Due the symmetry, xx components are identical to zz components; xy components are identical to yz components. The convergence loses only when the value of functions touch the accuracy of machine. The convergent factor at least 4, which corresponds to our 4th order accuracy difference approximation, can be clearly seen. The nonsmooth behavior near boundary is due to our lowering order operation.

In this paper we only investigate Schwarzschild black hole, and we take advantage of the symmetry of this spacetime by using octant symmetry to evolve it. With octant symmetry we mean x, y and z all take positive value only and the boundary condition at $x = 0, y = 0, z = 0$ is planar symmetric boundary condition. In order to skip the influence of mesh refinement, we use unigrid to do all of the investigation. We set the computational domain $0 < x, y, z < 15M$ with grids $100 \times 100 \times 100$. Through our puncture method, there is no inner boundary in our simulation. And we take standard Sommerfeld boundary condition with 5th order polynomial interpolation at outer boundary.[16]

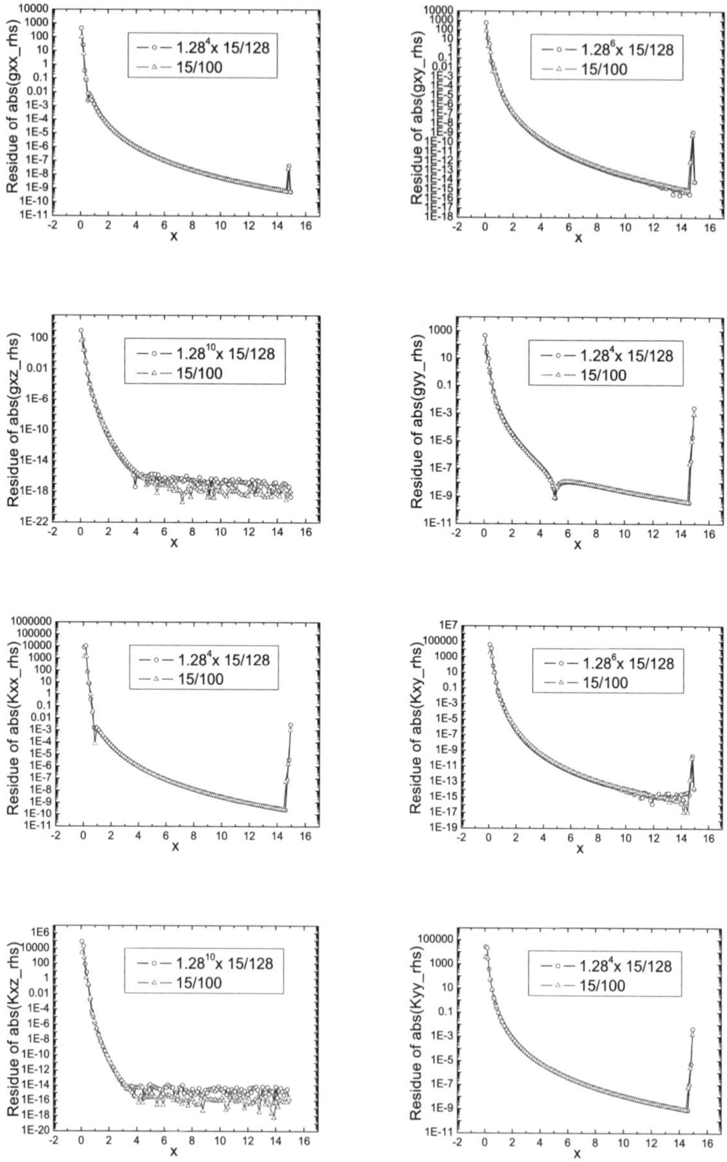

Fig. 1. Convergence test with Kerr-Schild form of Schwarzschild black hole. 15/128 and 15/100 mean two different resolutions.

4. Numerical result

In this section we will show our main numerical results. Our key point is to investigate the effect of the moving puncture method and the modified advanced gauge conditions[5,17,19] for ADM formalism. To do so, we adopt the gauge conditions ever been widely used before 2005 and test the effect of moving puncture method for ADM formalism. Then we combine the moving puncture method and the advanced gauge condition for ADM formalism. In the following subsections, we present these results one by one. In all of the following simulations, the mass of the black hole is set as $M = 1$.

4.1. *Geodesic gauge*

Our first test is geodesic gauge condition

$$\alpha = 1, \tag{12}$$

$$\beta^i = 0. \tag{13}$$

In our code this gauge is implemented through setting initial value of lapse function and shift vector as $\alpha = 1, \beta^i = 0$ and let $f_\alpha = 0, f_{\beta^i} = 0, f_{B^i} = 0$ in gauge equations (7),(8) and (9) during the simulation. Under this gauge condition, all observers represented by numerical points fall into black hole quickly. Theoretic prediction tells us the slices will touch the physical singularity in time of πM.[5,18] In our test, the code crashes at time much smaller than πM. The instability comes from the equation system itself instead of the physical singularity. In addition, the Courant factor does not bring any significant effects, that is to say the instability here does not result from CFL violation. So we only show the result with Courant factor 0.025 in Fig.2. Due to spherical symmetry, the three momentum constraints coincide with each other. We can see that the puncture method even makes the case worse compared with Fig.5 of Ref. 5. The simulation crashes at about $t = 0.45M$. And the abnormal behavior emerges from the region near the throat. The simulation in Ref. 5 used reflection boundary condition at the throat. So this instability is maybe hidden there.

4.2. *Zero shift plus Harmonic slice*

Our second test is zero shift plus Harmonic slice gauge condition

$$\partial_t \alpha = -2\alpha^2 K, \tag{14}$$

$$\beta^i = 0. \tag{15}$$

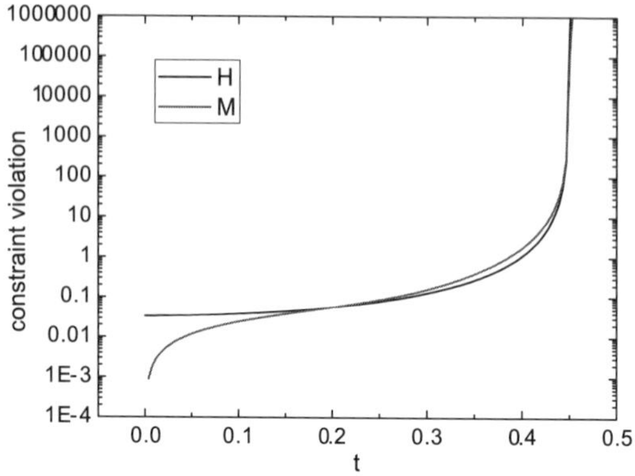

Fig. 2. Constraints violation behavior respect to time for geodesic gauge. H, M represent the violation of Hamiltonian constraint and linear momentum constraints respectively. All of the graphs in this paper adopt this same style. The code crashes at about $t = 0.45M$. The three momentum constraints coincide together, which is consistent with the spherical symmetry of Schwarzschild black hole. The crash time is much less than πM which indicates that the moving puncture method even makes ADM formalism worse.

Similar to geodesic gauge condition, we implement it through setting initial values of lapse function and shift vector as $\alpha = \frac{1}{(1+\frac{M}{2r})^2}, \beta^i = 0$, which follows closely Ref. 20 and let $f_{\beta^i} = 0, f_{B^i} = 0$ in gauge equations (8) and (9) during the simulation. Harmonic slice condition is widely used in hyperbolic formulations of Einstein's equation.[17,19,21] The result is plotted in Fig.3. Compared with Fig.2, we can see that this gauge condition is much better than geodesic gauge condition. Similarly, the Courant factor does not bring any significant effects. From this test, the importance of gauge choice for stability can be seen clearly. But even BSSN formalism can not work well with this gauge condition, so we do not expect this gauge condition together with moving puncture method can make ADM formalism stable. The code crashes at around $11M$. And the abnormal behavior emerges from the region near the throat. This is similar to the geodesic gauge case.

4.3. *Zero shift plus Bona-Masso slice*

In this subsection, we test the slice condition used in the advanced gauge condition while fix the shift to zero still. The slice condition is called Bona-Masso type slicing condition.[17,22] In all, Bona-Masso type slicing condition

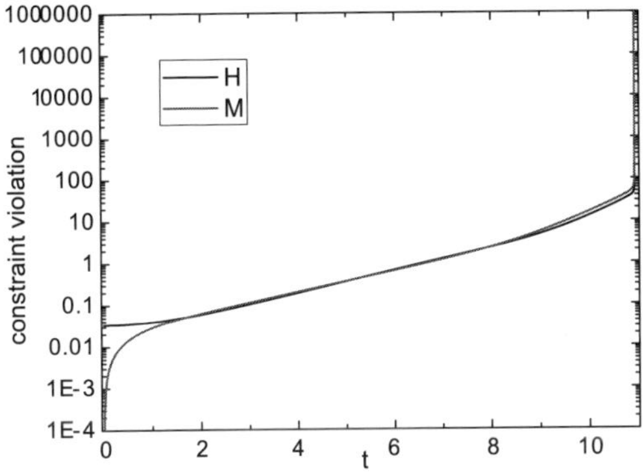

Fig. 3. As Fig.2, except with zero shift plus Harmonic slice gauge condition. The code crashes at around $t = 11M$. The evolution lasts much longer compared with geodesic gauge condition but crashes still.

and zero shift gauge condition read as

$$\partial_t \alpha = 2\alpha K + \beta^i \partial_i \alpha, \tag{16}$$

$$\beta^i = 0. \tag{17}$$

Similar to above subsection, we implemented it through setting initial values of lapse function and shift vector as $\alpha = \frac{1}{(1+\frac{M}{2r})^2}, \beta^i = 0$ and let $f_{\beta^i} = 0, f_{B^i} = 0$ in gauge equations (8) and (9) during the simulation. The numerical result is plotted in Fig.4. Bona-Masso slice condition is well known for its better ability of singularity avoiding than harmonic slice condition. Compared with above two kinds of gauge conditions, this gauge condition lasts the evolution much longer, but it still crashes at $t = 30M$ also. It does show the power of advanced slice condition and give us some hope to make ADM formalism stable if combining it with suitable shift condition. But one interesting issue is that the abnormal behavior emerges both near the throat of black hole and near the outer boundary. The abnormal behavior near outer boundary is consistent with the non-vanishing propagation speed of constraint violation for ADM equations[11] even with vanishing shift vector. In this case, it is the ill behavior near throat destroy the simulation in all. In the meanwhile, we did not find any significant difference the Courant factor can make.

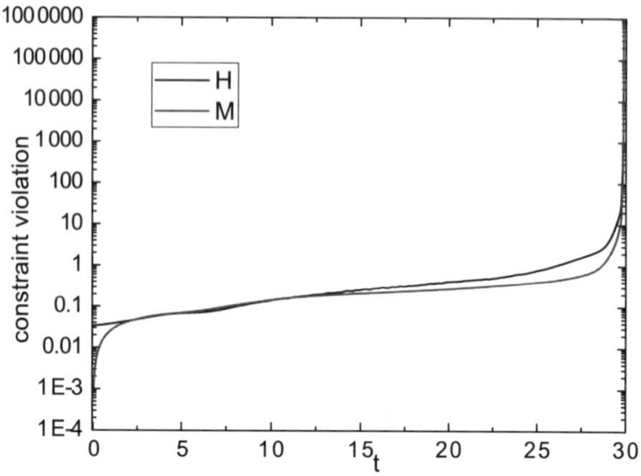

Fig. 4. Same as Fig.2, except with zero shift plus Bona-Masso slice gauge condition. Compared with Fig.2 and Fig.3, the evolution becomes much longer but still crashes at $t = 30M$ around.

4.4. *Modified popular gauge in BSSN formalism*

Since the leading work of Campanelli *et al.*[2] and Baker *et al.*,[3] the following gauge condition (or minor modification) almost becomes standard in numerical relativity community

$$\partial_t \alpha = -2\alpha K + \beta^i \partial_i \alpha, \tag{18}$$

$$\partial_t \beta^i = \frac{3}{4} B^i, \tag{19}$$

$$\partial_t B^i = \partial_t \tilde{\Gamma}^i - \eta B^i, \tag{20}$$

where $\tilde{\Gamma}^i$ is the dynamical variable in BSSN formalism. Naively, ones may expect this gauge condition may make ADM formalism stable also together with moving puncture method if they think the moving puncture and advanced gauge condition is the essence for the success of BSSN formalism instead of formalism itself. Following this idea, we test this gauge condition based on the moving puncture infrastructure of ADM formalism. But ADM formalism has different variables from BSSN formalism. Specifically, ADM formalism has no variables $\tilde{\Gamma}^i$. So before we adopt this gauge condition for ADM formalism, we must adjust it according to the character of ADM formalism. Since $\tilde{\Gamma}^i$ themselves are not independent variables, we can reconstruct them through other ADM variables. But we have some different methods to do so. In one method, we can reconstruct them through following procedures. Since

$$\gamma_{,t} = \gamma\gamma^{ij}\gamma_{ij,t} \tag{21}$$

$$\gamma^{ij}{}_{,t} = -\gamma^{il}\gamma^{jm}\gamma_{lm,t}, \tag{22}$$

we have

$$\tilde{\Gamma}^i_{,t} = -\tilde{\gamma}^{ij}{}_{,jt} \tag{23}$$

$$= -(\frac{\gamma^{ij}}{\gamma^{1/3}})_{,jt} \tag{24}$$

$$= -[\frac{\gamma^{ij}{}_{,t}}{\gamma^{1/3}} - \frac{\gamma^{ij}\gamma_{,t}}{3\gamma^{4/3}}]_{,j} \tag{25}$$

$$= [\frac{1}{\gamma^{1/3}}(\gamma^{ik}\gamma^{jl} + \frac{1}{3}\gamma^{ij}\gamma^{kl})\gamma_{kl,t}]_{,j} \tag{26}$$

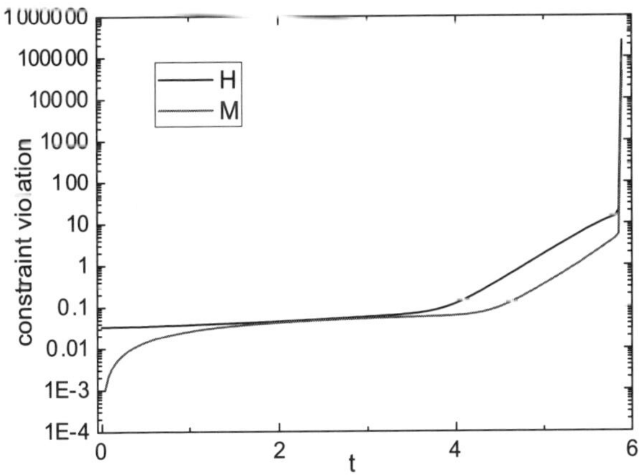

Fig. 5. As Fig.2, but with modified popular gauge in BSSN formalism. The code crashes at $t = 6M$ around. This result is worse than zero shift plus Bona-Masso slice gauge condition. Similarly, the Courant factor does not show any difference.

With this method, the result is even worse than zero shift plus Bona-Masso slice gauge condition. The result is plotted in Fig.5. The evolution can only last $6M$, which is much shorter than Fig.4 and even shorter than the harmonic slice condition. In another method to reconstruct $\tilde{\Gamma}^i$, we try to mimic the role of $\tilde{\Gamma}^i$ in BSSN formalism. Instead of constructing the term $\tilde{\Gamma}^i_{,t}$ in the gauge condition directly, we compute $\tilde{\Gamma}^i$ firstly on entire numerical grid according to

$$\tilde{\Gamma}^i = -\tilde{\gamma}^{ij}{}_{,j} \tag{27}$$

$$= -(\frac{\gamma^{ij}}{\gamma^{1/3}})_{,j} \tag{28}$$

$$= -[\frac{\gamma^{ij}{}_{,j}}{\gamma^{1/3}} - \frac{1}{3}\frac{\gamma^{ij}\gamma_{,j}}{\gamma^{4/3}}] \tag{29}$$

Then use the difference approximation to calculate the time derivative of $\tilde{\Gamma}^i$. Specifically we use the difference of two near time level over Δt to simulate $\tilde{\Gamma}^i{}_{,t}$. Or to say, we use second order accurate difference method to calculate the time derivative of $\tilde{\Gamma}^i$. Unfortunately, this method makes the situation even worse, the evolution can only last a little more than $1M$. With this gauge condition, the abnormal behavior comes from out boundary instead of throat of black hole. This is reduced from the non-zero shift vector effect on the propagation speed of constraint violation.[11] So our tests implies that the advanced gauge condition can not fix the stability problem of ADM formalism. And more interestingly, we find that the advanced gauge condition working well for BSSN formalism makes ADM formalism worse.

5. Discussion and conclusion

Both BSSN formalism and generalized harmonic formalism of Einstein's equation have been successfully implemented in three dimensional simulation of binary black hole systems. And years ago, ADM formalism has also been widely used for spherical- and axi-symmetrical spacetime simulation. In the mean time, we have many other formalisms of Einstein's equation. In fact, we are not very clear what's the advantages of different formalisms to numerical simulation of dynamical spacetime. But we do know the numerical method such as moving puncture method and gauge condition are fatal important to the stability of numerical calculation of Einstein's equation. Motivated by the successful simulation of binary black hole system with BSSN formalism under moving puncture method and advanced gauge condition, we investigate how these attracting methods affect ADM formalism. In addition, we test the Courant factor's influence on ADM formalism to see if the exponential instability of ADM formalism is resulted from too large Courant factor. Unfortunately, our result is some negative. We confirm again that the ADM formalism has intuitive unstable mode which is independent of gauge condition and Courant factor. Furthermore, our results imply that one numerical method or one gauge condition works well under some given formalism does not mean it can also work well under another formalism. Our explorations give those readers who are not familiar with the stability problem in numerical relativity a rough sense through some concrete examples.

Acknowledgments

This work is supported in part by the NSFC (Nos. 10671196 and 10731080) and the National 973 Project (No. 2006CB805905).

References

1. F. Pretorius, *Phys. Rev. Lett.* 95, 121101 (2005).
2. M. Campanelli, C. O. Lousto, P. Marronetti, and Y. Zlochower *Phys. Rev. Lett.* 96, 111101 (2006).
3. J. Baker, J. Centrella, D-I. Choi, M. Koppitz, and J. van Meter *Phys. Rev. Lett.* 96, 111102 (2006).
4. Z. Cao, H. Yo, and J. Yu, Phys. Rev. D 78, 124011 (2008).
5. P. Anninos, K. Camarda, J. Masso, E. Seidel, W-M. Suen, J. Towns *Phys. Rev. D* 52, 2059 (1995).
6. M. Alcubierre, G. Allen, B. Brugmann, E. Seidel and W. Suen *Phys. Rev. D* 62, 124011 (2000).
7. L. Kidder, M. Scheel and S. Toukolsky *Phys. Rev. D* 64, 064017 (2001).
8. B. Kelly, P. Laguna, K. Lockitch, J. Pullin, E. Schnetter, D. Shoemaker and M. Tiglio *Phys. Rev. D* 64, 084013 (2001).
9. H. Shinkai and G. Yoneda, *Class. Quantum Grav.* 19, 1027 (2002).
10. S. Frittelli, *Phys. Rev. D* 55, 5992 (1997).
11. G. Yoneda and H. Shinkai, *Phys. Rev. D* 63, 124019 (2001).
12. J. Meter, J. Baker, M. Koppitz and, D. Choi, *Phys. Rev. D* 73, 124011 (2006).
13. M. Babiuc, *et al.*,*Class. Quantum Grav.* 25, 125012 (2008).
14. J. York, Kinematics and dynamics of general relativity in *Sources of Gravitational Radiation* ed L. Smarr (Cambridge: Cambridge University Press,1979).
15. H.J. Yo, T.W. Baumgarte, and S.L. Shapiro, Phys. Rev. D **66**, 084026 (2001).
16. T. Baumgarte and S. Shapiro, *Phys. Rev. D* 59, 024007 (1998).
17. T.W. Baumgarte and S.L. Shapiro, Phys. Rept. 376, 41 (2003).
18. L. Smarr, and J. York, *Phys. Rev. D* 17, 2529 (1978).
19. M. Shibata and T. Nakamura, Phys. Rev. D **52**, 5428 (1995).
20. M. Alcubierre, B. Brugmann, P. Diener, M. Koppitz, D. Pollney, E. Seidel, and R. Takahashi, *Phys. Rev. D* 67, 084023 (2003).
21. G. Cook and M. Scheel, *Phys. Rev. D* 56, 4775 (1997).
22. C. Bona, J. Masso, E. Seidel and J. Stela, Phys. Rev. Lett. **75**, 600 (1995).

ANALYTIC SOLUTION FOR MATTER DENSITY FLUCTUATIONS IN $f(R)$ MODELS OF COSMIC ACCELERATION

HAYATO MOTOHASHI* [1,2], ALEXEI A. STAROBINSKY† [2,3] and JUN'ICHI YOKOYAMA‡ [2,4]

[1] *Department of Physics, Graduate School of Science, The University of Tokyo, Tokyo 113-0033, Japan*
[2] *Research Center for the Early Universe (RESCEU), Graduate School of Science, The University of Tokyo, Tokyo 113-0033, Japan*
[3] *L. D. Landau Institute for Theoretical Physics, Moscow 119334, Russia*
[4] *Institute for the Physics and Mathematics of the Universe(IPMU), The University of Tokyo, Kashiwa, Chiba, 277-8568, Japan*

We present an analytic solution for density perturbations in the matter component during the matter dominated stage in terms of Gauss' hypergeometric functions for a class of $f(R)$ gravity models of accelerated expansion. By using the analytic solution, we obtain an analytical expression for the matter transfer function at scales much less than the present Hubble scale.

Keywords: $f(R)$ gravity; cosmological models; density perturbations; analytic solution

1. Introduction

The origin of the accelerated cosmic expansion observed at the redshift $z \lesssim 0.7$ is undoubtedly the greatest mystery of contemporary cosmology. If interpreted in terms of the Einstein general theory of relativity, this acceleration requires the existence of some new component in the right-hand side of the Einstein equations, dubbed dark energy (DE), which remains practically unclustered at all scales at which gravitational clustering of baryonic and dark non-baryonic matter is seen, and which effective pressure p_{DE}

*motohashi@resceu.s.u-tokyo.ac.jp
†alstar@landau.ac.ru
‡yokoyama@resceu.s.u-tokyo.ac.jp

is approximately equal to minus its effective energy density ρ_{DE}. Thus, its properties are very close to those of a cosmological constant Λ (see Refs. SS00 - FTH08 for some reviews).

The simplest possibility of DE being exactly Λ combined with a non-relativistic non-baryonic dark matter (the standard spatially flat ΛCDM cosmological model) provides a good fit to all existing observational data[6]. In this case Λ acquires the status of a new fundamental physical constant. However, its required value is very small as compared to known atomic and elementary particle scales (not speaking about the Planck ones), so a firm theoretical prediction for this quantity from first principles is lacking currently (although it may arise due to some non-perturbative effects, see e.g. Ref. Yokoyama:2001ez).

On the other hand, in the second case when a component with qualitatively similar properties is assumed to exist – in the inflationary scenario of the early Universe, we are sure that this "primordial DE" may not be an exact cosmological constant since it should decay long time ago. That is why it is natural to assume by analogy that the present DE is not absolutely stable, too.

An interesting alternative to the standard ΛCDM model is provided by "geometric" DE models based on $f(R)$ gravity which modify and generalize the Einstein General Relativity by introducing a new function of Ricci scalar, $f(R)$, into the gravitational field action instead of $R - 2\Lambda$ (see Ref. SF08 for a recent review). $f(R)$ gravity, which in turn is a particular case of more general scalar-tensor gravity, contains an additional scalar degree of freedom or, in quantum language, a massive scalar particle. [a] Because of this, viable models of present DE in $f(R)$ gravity should satisfy a number of conditions which exclude many possible, in principle, forms of $f(R)$. In particular, in order to have the correct Newtonian limit for $R \gg R(t_0) \sim H_0^2$ where t_0 is the present moment and H_0 is the Hubble constant, as well as the standard matter-dominated FLRW stage with the scale factor behaviour $a(t) \propto t^{2/3}$ driven by cold dark matter and baryons,

[a] This particle was dubbed "scalaron" in Ref. S80 where the particular case $f(R) = R + R^2/6M^2$ (plus small additional terms) was used to construct the first internally self-consistent inflationary scenario of the early Universe having a graceful exit to the subsequent radiation-dominated Friedmann-Lemaître-Robertson-Walker (FLRW) stage through an intermediate matter-dominated reheating period. This inflationary model still remains viable since it predicts the slope of the primordial spectrum of scalar perturbations n_s and the tensor-to-scalar ratio r in agreement with the most recent observational data.

the following conditions should be fulfilled:

$$|f(R) - R| \ll R, \quad |f'(R) - 1| \ll 1, \quad Rf''(R) \ll 1, \quad R \gg R_0 \,, \tag{1}$$

where the prime denotes the derivative with respect to the argument R. In addition, the stability condition $f''(R) > 0$ has to be satisfied that guarantees that the standard matter-dominated FLRW stage remains an attractor with respect to an open set of neighboring isotropic cosmological solutions in $f(R)$ gravity (in quantum language, this condition means that scalaron is not a tachyon). [b] A number of functional forms has been proposed that can account for accelerated expansion without Λ while passing laboratory and astronomical tests[10,11].

The present paper considers the evolution of density perturbation in $f(R)$ model. After brief review of density perturbation in general $f(R)$ gravity, we choose a specific inverse power-law form of $f(R) - R$ which is the limiting form of the viable models proposed in Refs. Hu:2007nk,Starobinsky:2007hu for $R \gg R_0$ and derive an analytic solution for density perturbations. By using the analytic solution, we obtain the transfer function that allows us to compare the model with observational data.

2. Density perturbations in $f(R)$ gravity

We use the perturbed spatially flat FLRW metric in the longitudinal gauge,

$$ds^2 = -(1 + 2\Phi)dt^2 + a^2(t)(1 - 2\Psi)(dr^2 + r^2 d\theta + r^2 \sin^2 \theta d\phi). \tag{2}$$

During the matter-dominated era, non-zero components of the energy-momentum tensor are given by

$$T_0^0 = -\rho - \delta\rho, \quad T_i^0 = -\rho\partial_i v, \tag{3}$$

where ρ and $\delta\rho$ denote background matter density and its fluctuation, and v is the velocity potential of scalar perturbation.

From the Einstein equations, we can derive the differential equation for comoving density perturbation defined by

$$\delta = \frac{\delta\rho}{\rho} + 3Hav. \tag{4}$$

---

[b] The second stability condition $f'(R) > 0$ which means that gravity is attractive and graviton is not a ghost is automatically fulfilled in this regime.

We move on to Fourier space,

$$\delta_{\mathbf{k}}(t) = \int \frac{d^3x}{(2\pi)^{3/2}} \delta(t, \mathbf{x}) e^{i\mathbf{k}\cdot\mathbf{x}}, \tag{5}$$

where $\mathbf{k}$ denotes comoving wavenumber. In the following, we abbreviate $\delta_{\mathbf{k}}(t)$ just δ for simplicity.

We write the action in the following form:

$$S = \int d^4x\sqrt{-g}\left[\frac{1}{16\pi G} f(R) + \mathcal{L}^{(m)}\right], \tag{6}$$

where G is the Newtonian gravitational constant and $\mathcal{L}^{(m)}$ is matter Lagrangian. If we take $f(R) = R - 2\Lambda$, Eq. (6) reduces to the Einstein-Hilbert action for the ΛCDM model. Below we consider $f(R)$ which vanishes for $R = 0$, so no cosmological constant is introduced by hand in flat space-time.

In $f(R)$ gravity, modified Einstein equations have the form:

$$FR_{\mu\nu} - \frac{1}{2} f g_{\mu\nu} + (g_{\mu\nu}\Box - \nabla_\mu\nabla_\nu)F = 8\pi G T_{\mu\nu}, \tag{7}$$

where $F(R) \equiv df/dR$. For dust-like matter (3), the background equations take the form:

$$3FH^2 = \frac{1}{2}(FR - f) - 3H\dot{F} + 8\pi G\rho, \tag{8}$$

$$-2F\dot{H} = \ddot{F} - H\dot{F} + 8\pi G\rho, \tag{9}$$

$$\dot{\rho} + 3H\rho = 0. \tag{10}$$

When deviation from the Einstein gravity is small, namely, $f(R) \cong R$ and $F(R) \cong 1$, these equations yield the standard matter-dominated regime $a(t) = a_0(t/t_0)^{2/3}$.

The differential equation describing a density perturbation in the sub-horizon regime is: [12]

$$\ddot{\delta} + 2H\dot{\delta} - 4\pi G_{\rm eff}\rho\delta = 0, \tag{11}$$

where

$$G_{\rm eff} = \frac{G}{F} \frac{1 + 4\frac{k^2}{a^2}\frac{F_{,R}}{F}}{1 + 3\frac{k^2}{a^2}\frac{F_{,R}}{F}}. \tag{12}$$

From now on, we adopt a specific $f(R)$ model such that

$$F(R) \equiv f'(R) = 1 - \left(\frac{R_c}{R}\right)^{N+1}, \quad N > -1 \tag{13}$$

with $R_c \sim H_0^2$ (but still $R_c < R(t_0)$). This $f(R)$ corresponds to the models in Refs. Hu:2007nk,Starobinsky:2007hu in the regime $R \gg R_c$, and we shall use it in this regime only.[c] Eq. (11) then becomes

$$\ddot{\delta} + \frac{4}{3t}\dot{\delta} - \frac{2}{3t^2}\frac{1 + 4A(t/t_i)^{2N+8/3}}{1 + 3A(t/t_i)^{2N+8/3}}\delta = 0, \tag{14}$$

with

$$A(N, k) = \frac{(N+1)k^2}{a_i^2 R_c}\left(\frac{3R_c t_i^2}{4}\right)^{N+2}. \tag{15}$$

Here, we set t_i as an initial time in the matter dominated regime.

2.1. *Asymptotic behavior*

We can read asymptotic behavior from the differential equation Eq. (14).

(i) $t \to 0$

In this limit, we can neglect $A(t/t_i)^{2n+8/3}$ with respect to 1. It reproduces the same result of the matter dominated era,

$$\delta_{\mathbf{k}}(t) = \delta_{i\mathbf{k}}\left(\frac{t}{t_i}\right)^{\frac{2}{3}}. \tag{16}$$

(ii) $t \to \infty$

Then Eq. (14) reduces

$$\ddot{\delta} + \frac{4}{3t}\dot{\delta} - \frac{8}{9t^2}\delta = 0. \tag{17}$$

In this regime the growing mode is

$$\delta_{\mathbf{k}}(t) = \delta_{i\mathbf{k}}C(k)\left(\frac{t}{t_i}\right)^{\frac{-1+\sqrt{33}}{6}}. \tag{18}$$

The coefficient $C(k)$ is the transfer function for matter perturbations which will be derived from the analytic solution found below.

---

[c]The parameter N used here has the same sense as in Ref. Hu:2007nk, while it is equal to $2n$ in the notation of Ref. Starobinsky:2007hu.

2.2. *Analytic solution*

Now let us derive the analytic solution for general t. By changing the variable from t to $\tau = (t/t_i)^\alpha$ with $\alpha = 2N + 8/3$, Eq. (14) can be rewritten as

$$\delta'' + \left(1 + \frac{1}{3\alpha}\right)\frac{\delta'}{\tau} - \frac{2}{3\alpha^2}\frac{1+4A\tau}{1+3A\tau}\frac{\delta}{\tau^2} = 0. \qquad (19)$$

Here a prime denotes derivative respect to τ. In order to make the equation simpler form, we take $d = \tau^\beta \delta$.

$$d'' + (M - 2\beta)\frac{d'}{\tau} + \frac{\left[3\beta^2 - 3(M-1)\beta - 4L\right]A\tau + \beta^2 - (M-1)\beta - L}{1+3A\tau}\frac{d}{\tau^2} = 0, \qquad (20)$$

where $M = 1 + 1/3\alpha, L = 2/(3\alpha^2)$. We choose β to satisfy $\beta^2 - (M-1)\beta - L = 0$, that is,

$$\beta_\pm = -\frac{1 - M \pm \sqrt{P}}{2}, \quad P = (M-1)^2 + 4L = \frac{25}{9\alpha^2}. \qquad (21)$$

By substituting $\beta_\pm$,

$$d'' + (1 \pm \sqrt{P})\frac{d'}{\tau} - \frac{LA}{1+3A\tau}\frac{d}{\tau} = 0. \qquad (22)$$

Finally, by taking $z = -3A\tau$, we get

$$d''(z) + (1 \pm \sqrt{P})\frac{d'(z)}{z} + \frac{L}{3}\frac{1}{z(1-z)}d = 0 \qquad (23)$$

Clearly, this equation can be reduced the differential equation of Gauss' hypergeometric function ${}_2F_1(a, b, c; z)$,

$${}_2F_1(a, b, c; z) = \sum_{n=0}^{\infty} \frac{(a)_n (b)_n}{(c)_n}\frac{z^n}{n!}, \qquad (24)$$

$$(a)_n \equiv a(a+1)\cdots(a+n-1) = \frac{\Gamma(a+n)}{\Gamma(a)}. \qquad (25)$$

As a result,

$$d \propto {}_2F_1\left(\frac{\pm\sqrt{P} + \sqrt{P + \frac{4L}{3}}}{2}, \frac{\pm\sqrt{P} - \sqrt{P + \frac{4L}{3}}}{2}, 1 \pm \sqrt{P}; z\right) \qquad (26)$$

In terms of δ, the two independent solutions of $\delta_{\mathbf{k}}(t)$ are

$$\delta_{i\mathbf{k}}\left(\frac{t}{t_i}\right)^{\frac{-1\pm 5}{6}} {}_2F_1\left(\frac{\pm 5 - \sqrt{33}}{6\alpha}, \frac{\pm 5 + \sqrt{33}}{6\alpha}; 1 \pm \frac{5}{6\alpha}; -3A(N,k)\left(\frac{t}{t_i}\right)^\alpha\right). \qquad (27)$$

In the following discussion, we consider the upper sign case only, because the other solution corresponds to the decaying mode of perturbations and is singular at $t \to 0$.

Let us check the asymptotic behavior of the solution, Eq. (27).

(i) $t \to 0$

$$\delta_{\mathbf{k}}(t) \to \delta_{i\mathbf{k}} \left(\frac{t}{t_i}\right)^{\frac{2}{3}} . \tag{28}$$

(ii) $t \to \infty$

$$\begin{aligned}\delta_{\mathbf{k}}(t) \to & \delta_{i\mathbf{k}} \times \frac{\Gamma\left(1+\frac{5}{6\alpha}\right)\Gamma\left(\frac{\sqrt{33}}{3\alpha}\right)}{\Gamma\left(1+\frac{\sqrt{33}}{6\alpha}\right)\Gamma\left(\frac{5+\sqrt{33}}{6\alpha}\right)} \\ & \times \left[\frac{3(N+1)k^2}{a_i^2 R_c}\left(\frac{3R_c t_i^2}{4}\right)^{N+2}\right]^{\frac{-5+\sqrt{33}}{6\alpha}} \left(\frac{t}{t_i}\right)^{\frac{-1+\sqrt{33}}{6}} .\end{aligned} \tag{29}$$

In both limits, the asymptotic behavior agrees with that one given by Eq. (16) and Eq. (18), respectively. Furthermore, here we can read off the transfer function, $C(k)$, which appears in Eq. (18):

$$C(k) = \frac{\Gamma\left(1+\frac{5}{4(3N+4)}\right)\Gamma\left(\frac{\sqrt{33}}{2(3N+4)}\right)}{\Gamma\left(1+\frac{\sqrt{33}}{4(3N+4)}\right)\Gamma\left(\frac{5+\sqrt{33}}{4(3N+4)}\right)} \left[\frac{3(N+1)k^2}{a_i^2 R_c}\left(\frac{3R_c t_i^2}{4}\right)^{N+2}\right]^{\frac{-5+\sqrt{33}}{4(3N+4)}} . \tag{30}$$

3. Conclusions and discussion

We have obtained an analytic solution describing the growth of density perturbation at the matter-dominated stage for a specific class of viable cosmological models in $f(R)$ gravity. Initially, the solution behaves in the same way as in the ΛCDM model, while it experiences an anomalous growth at late times (redshifts of the order of a few). We also find an analytic expression for the matter transfer function which shows that an initial perturbation power spectrum acquires the additional power-law factor $\propto k^{\Delta n_s}$ with

$$\Delta n_s = \frac{-5+\sqrt{33}}{3N+4} \tag{31}$$

at scales much less than the present Hubble scale, as originally shown in Ref. Starobinsky:2007hu. [d] Clearly, this additional factor is absent in the matter power spectrum at the recombination time. So, by comparing the form of the primordial matter power spectrum derived from CMB data and from galaxy surveys separately, it is possible to obtain an important constraint on the parameter n characterizing this class of cosmological models in $f(R)$ gravity, although we do not have much stringent constraints on it at present[26]. If we take an upper limit on Δn_s as $\Delta n_s^{\max} = 0.05$, which is a conservative bound for now[11], and assume that R_c is not much less than H_0^2 (if otherwise $R_c \ll H_0^2$, deviation of the background FLRW model from the ΛCDM one is very small), we obtain a constraint

$$N > 4.96 \left(\frac{\Delta n_s^{\max}}{0.05} \right)^{-1} - 1.33. \tag{32}$$

Future observational data together with a more detailed theoretical analysis may well yield a more stringent bound on N.

Of course, the $f(R)$ gravity model (13) is viable for a finite range of R only, in particular, for $R \gg R_c$. For $R \sim R(t_0) \sim H_0^2$, it has to be substituted by a more complicated expression admitting a stable (or, at least metastable) de Sitter solution, e.g. by the models presented in Refs. Hu:2007nk,Starobinsky:2007hu. As a result, the equation for matter density perturbations has to be solved numerically for recent redshifts $z \lesssim 1$. However, evolution in this region may add only a k-independent factor to the total matter transfer function. Therefore, the k exponent in Eqs. (30) and (31) does not depend on a concrete form of $f(R)$ for $R \sim R(t_0)$.

Also, the model (13) should not be used for too large values of R for several reasons. First, the effective scalaron mass squared $m_s^2 = 1/3F_{,R}$ (in the regime $F_{,R}H^2 \ll 1$) grows quickly with R to the past and may even exceed the Planck mass making copious production of primordial black holes possible. As was noted in Ref. Starobinsky:2007hu, this problem may be avoided by adding the $R^2/6M^2$ term to $f(R)$ that bounds the scalaron mass from above just by M. Simultaneously, such a change of $f(R)$ at large R removes the "Big Boost" singularity (in terminology of Ref. BDK08 where such a singularity appeared in a different context), which generic appearance in the model (13) was shown in Ref. F08. However, the value of M

[d] Note the difference in the notation for N between Ref. Starobinsky:2007hu and our paper which was mentioned above.

should be sufficiently large in order not to destroy the standard cosmology of the present and early Universe. In particular, its values considered in Refs. D08,KM09 seem not to be high enough for this purpose. Indeed, the simplest way to solve one more problem of this $f(R)$ cosmological scenario (also noted in Ref. Starobinsky:2007hu) – overproduction of scalarons in the early Universe – is, as usual, to have an inflationary stage preceding the radiation-dominated one. Then M should not be smaller than H during its last 60 e-folds, and for $M \approx 3 \times 10^{13}$ GeV the scalaron will be the inflation itself according to the scenario[9].

Note that for sufficiently large N the corresponding correction to $F(R)$ may become more important than the second term in the right-hand side of Eq. (13) already at the matter-dominated stage. For example, even for M as large as 3×10^{13} GeV, the term $R/3M^2$ becomes larger than $(R_c/R)^{N+1}$ for $R = 3 \times 10^{10} R(t_0)$ (corresponding to the matter-radiation equality) if $N \geq 10$. However, this does not affect the exact solution obtained in the previous section since at this moment $F_{,R}/F \ll a^2/k^2$ for all scales of interest, so $G_{\rm eff} = G$ in Eq. (11) irrespective of an actual structure of $F(R) - 1$. In turn, if $F_{,R}/F \geq a^2/k^2$, the R^2 correction is negligible for all scales of interest at the matter-dominated stage. Therefore, this high-R correction to the model (13) needed to obtain a viable cosmological model of the early Universe does not change our results.

Acknowledgments

AS acknowledges RESCEU hospitality as a visiting professor. He was also partially supported by the grant RFBR 08-02-00923 and by the Scientific Programme "Astronomy" of the Russian Academy of Sciences. This work was supported in part by JSPS Grant-in-Aid for Scientific Research No. 19340054(JY), JSPS Core-to-Core program "International Research Network on Dark Energy", and Global COE Program "the Physical Sciences Frontier", MEXT, Japan.

References

1. V. Sahni and A. A. Starobinsky, *Int. J. Mod. Phys. D* **9**, 373 (2000) [arXiv:astro-ph/9904398].
2. T. Padmanabhan, *Phys. Rept.* **380**, 235 (2003) [arXiv:hep-th/0212290].
3. E. J. Copeland, M. Sami and S. Tsujikawa, *Int. J. Mod. Phys. D* **15**, 1753 (2006) [arXiv:hep-th/0603057].
4. V. Sahni and A. A. Starobinsky, *Int. J. Mod. Phys. D* **15**, 2105 (2006) [arXiv:astro-ph/0610026].

5. J. A. Frieman, M. S. Turner and D. Huterer, *Ann. Rev. Astron. Astroph.* **46**, 385 (2008) [arXiv:astro-ph/0803.0982].
6. E. Komatsu *et al.* [WMAP Collaboration], *Astrophys. J. Suppl.* **180**, 330 (2009) [arXiv:astro-ph/0803.0547].
7. J. Yokoyama, *Phys. Rev. Lett.* **88**, 151302 (2002) [arXiv:hep-th/0110137].
8. T. P. Sotiriou and V. Faraoni [arXiv:gr-qc/0805.1726].
9. A. A. Starobinsky, *Phys. Lett. B* **91**, 99 (1980).
10. W. Hu and I. Sawicki, *Phys. Rev. D* **76**, 064004 (2007) [arXiv:astro-ph/0705.1158].
11. A. A. Starobinsky, *JETP Lett.* **86**, 157 (2007) [arXiv:astro-ph/0706.2041].
12. S. Tsujikawa, *Phys. Rev. D* **76**, 023514 (2007) [arXiv:astro-ph/0705.1032].
13. Y.-S. Song, W. Hu and I. Sawicki, *Phys. Rev. D* **75**, 044004 (2007) [arXiv:astro-ph/0610532].
14. B. Boisseau, G. Esposito Farese, D. Polarski and A. A. Starobinsky, *Phys. Rev. Lett.* **85**, 2236 (2000) [arXiv:gr-qc/0001066].
15. Y.-S. Song, H. Peiris and W. Hu, *Phys. Rev. D* **76**, 063517 (2007) [arXiv:astro-ph/0706.2399].
16. W. Hu and I. Sawicki, *Phys. Rev. D* **76**, 104043 (2007) [arXiv:astro-ph/0708.1190].
17. S. Tsujikawa, K. Uddin, S. Mizuno, R. Tavakol and J. Yokoyama, *Phys. Rev. D* **77**, 103009 (2008) [arXiv:astro-ph/0803.1106].
18. R. Gannouji, B. Moraes amd D. Polarski, *J. Cosm. Astroph. Phys.* **0902**, 034 (2009) [arXiv:astro-ph/0809.3374].
19. T. Tatekawa and S. Tsujikawa, *J. Cosm. Astroph. Phys.* **0809**, 009 (2008) [arXiv:astro-ph/0807.2017].
20. H. Oyaizu, *Phys. Rev. D* **78**, 123523 (2008) [arXiv:astro-ph/0807.2449].
21. H. Oyaizu, M. Lima and W. Hu, *Phys. Rev. D* **78**, 123524 (2008) [arXiv:astro-ph/0807.2462].
22. F. Schmidt, M. Lima, H. Oyaizu and W. Hu, *Phys. Rev. D* **79**, 083518 (2009) [arXiv:astro-ph/0812.0545].
23. A. de la Cruz-Dombriz, A. Dobado and A. L. Maroto, *Phys. Rev. D* **77**, 123515 (2008) [arXiv:astro-ph/0802.2999].
24. K. N. Ananda, S. Carloni and P. K. S. Dunsby [arXiv:astro-ph/0809.3673].
25. K. N. Ananda, S. Carloni and P. K. S. Dunsby [arXiv:astro-ph/0812.2028].
26. M. Tegmark *et al.* [SDSS Collaboration], *Phys. Rev. D* **74**, 123507 (2006) [arXiv:astro-ph/0608632].
27. A. O. Barvinsky, C. Deffayet and A. Yu. Kamenshchik, *J. Cosm. Astroph. Phys.* **0805**, 020 (2008) [arXiv:astro-ph/0801.2063].
28. A. V. Frolov, *Phys. Rev. Lett.* **101**, 061103 (2008) [arXiv:astro-ph/0803.2500].
29. A. Dev *et al.*, *Phys. Rev. D* **78**, 083515 (2008) [arXiv:hep-th/0807.3445].
30. T. Kobayashi and K. Maeda, *Phys. Rev. D* **79**, 024009 (2009) [arXiv:astro-ph/0810.5664].

NORMAL MODES, ZERO MODES AND SUPER-RADIANT MODES FOR SCALAR FIELDS IN ROTATING BLACK HOLE SPACETIME

M. KENMOKU

Department of Physics, Nara Women's University,
Nara, 630-8506 Japan
**E-mail: kenmoku@asuka.phys.nara-wu.ac.jp*
http://asuka.phys.nara-wu.ac.jp/ kenmoku/

Normal modes, zero modes and super-radiant modes for scalar fields are studied in (2+1)-dimensional BTZ spacetime and in (3+1)-dimensional Kerr-anti-de Sitter spacetime. For BTZ spacetime, normal modes are obtained in solving the eigenvalue equation in numerical and analytical methods. All physical normal modes shown to lie above the zero mode line: 0=frequency - angular velocity × azimuthal angular momentum. For Kerr-anti-de Sitter spacetime, non-existence of zero modes is shown rigorously for the Dirichlet boundary condition. Non-existence of zero modes indicates that super-radiant instability modes are unphysical but super-radiant stability modes are physical.

Keywords: Rotating black hole; Normal mode; Zero and super-radiant mode.

1. Introduction

Black holes are interesting in theory and observation. Especially, super-massive black holes are observed in almost all galaxies. They may be well described by exact solutions of Einstein field equations. New exact solutions of Einstein equations are discovered for multi-dimensional rotating black holes and multi-dimensional Kerr-NUT black holes and others.

Interactions of black holes (BH) with matter fields are important. Especially normal modes and quasi-normal modes of matter fields are important with respect to super-radiant instability problem, BH thermodynamics and others.

To derive BH thermodynamics from microscopic statistical mechanics is interesting. One possible approach is the Brick Wall Model by 't Hooft,[1] which is constructed in the way: (1) standard statistical mechanics of scalar field around BH, (2) built the brick wall at horizon, (3) the Dirichlet boundary condition determines normal modes of scalar fields. (4) sum of normal modes drives partition function and entropy.

However, problems in the brick wall model for rotating BH cases are discussed because the statistical sum of normal modes can't be taken and the Boltzmann factor is ill-defined due to the super-radiant instability modes.[2]

Problems of super-radiant instability or stability are extensively discussed: The repeated instability causes the BH bomb by Press and Teukolsky,[3] Large Kerr anti-de Sitter (K-AdS) BH are stable, Hawking and Reall,[4] Small K-AdS BH are unstable by Cardoso et al.,[5] and many others.

The purpose of this note is to make clear the super-radiant instability problem. For this purpose, we study normal modes, zero modes for the scalar fields around rotating black hole spacetime in numerical method as well as analytical method based on our recent papers for (2+1)-dimensional BTZ black hole spacetime[6] and for (3+1)-dimensional K-AdS black hole spacetime.[7]

2. Scalar fields in BTZ spacetime

The eigenstate problem for scalar fields in BTZ spacetime is studied in our previous papers.[6]

2.1. *Metric and field equations*

The metric of BTZ spacetime with $\Lambda = -1/\ell^2$ is given through the line element:[8]

$$ds^2 = g_{tt}dt^2 + g_{\phi\phi}d\phi^2 + 2g_{t\phi}dtd\phi + g_{rr}dr^2 \,,$$
$$g_{tt} = M - \frac{r^2}{\ell^2}\,, \quad g_{t\phi} = -\frac{J}{2}\,, \quad g_{\phi\phi} = r^2\,, \quad g_{rr} = \left(-M + \frac{J^2}{4r^2} + \frac{r^2}{\ell^2}\right)^{-1} \tag{2.1}$$

where M and J denote the BH mass and the rotation parameter respectively. The event horizon is given by $r_+ = \ell(M/2(1+\sqrt{1-J^2M^{-2}\ell^{-2}}))^{1/2}$. The action of the complex scalar field $\Phi(x)$ with mass μ in BTZ spacetime is given by

$$I_{\rm scalar} = -\int dtdrd\phi\sqrt{-g}\left(g^{\mu\nu}\partial_\mu\Phi^*(x)\partial_\nu\Phi(x) + \frac{\mu}{\ell^2}\Phi^*(x)\Phi(x)\right)\,. \tag{2.2}$$

The scalar field is written in the form of separation of variables corresponding to two Killing vectors for t and ϕ as

$$\Phi = e^{-i\omega t}e^{im\phi}R(r)\,, \tag{2.3}$$

where ω and m denote the frequency and the azimuthal angular momentum of scalar fields respectively. The radial equation for $R(r)$ is expressed as

$$\left(g_{rr}(\omega - \frac{J}{2r^2}m)^2 - \frac{m^2}{r^2} + \frac{1}{r}\partial_r \frac{r}{g_{rr}}\partial_r - \frac{\mu}{\ell^2}\right) R(r) = 0\,. \tag{2.4}$$

Introducing the new variable z and the new radial function $F(z)$

$$z = \frac{r^2 - r_+^2}{r^2 - r_-^2}\,, \quad R(r) = z^{-i\alpha}(1-z)^{\beta} F(z)\,, \tag{2.5}$$

the radial equation reduces to hyper-geometric differential equation:

$$z(1-z)\frac{d^2F}{dz^2} + (c - (1+a+b)z)\frac{dF}{dz} - abF = 0\,, \tag{2.6}$$

where parameters are defined as

$$\begin{aligned} a &= \beta - i\frac{\ell^2}{2(r_+ + r_-)}\left(\omega + \frac{m}{\ell}\right)\,, \; b = \beta - i\frac{\ell^2}{2(r_+ - r_-)}\left(\omega - \frac{m}{\ell}\right)\,,\\ c &= 1 - 2i\frac{\ell^2 r_+}{2(r_+^2 - r_-^2)}(\omega - \Omega_{\rm H} m)\,, \; \beta = \frac{1-\sqrt{1+\mu}}{2}\,, \end{aligned} \tag{2.7}$$

with the angular velocity on the horizon $\Omega_{\rm H} = J/2r_+^2$.

2.2. *Boundary conditions*

The radial function is expressed by the hyper-geometric function imposing to converge at infinity, which is rewritten by incoming and outgoing waves near the horizon as

$$\begin{aligned} R_\infty &= \frac{z^{-i\alpha}(1-z)^{\beta+c-a-b}}{\Gamma(c-a-b+1)} F(c-a, c-b, c-a-b+1; 1-z)\,,\\ &= \frac{\Gamma(1-c)}{\Gamma(1-a)\Gamma(1-b)} R_{r_+,\rm in} + \frac{\Gamma(c-1)}{\Gamma(c-a)\Gamma(c-b)} R_{r_+,\rm out}\,, \end{aligned} \tag{2.8}$$

with

$$\begin{aligned} R_{r_+,\rm in} &= z^{-i\alpha}(1-z)^{\beta} F(a,b,c;z)\,,\\ R_{r_+,\rm out} &= z^{i\alpha}(1-z)^{\beta} F(1+b-c, 1+a-c, 2-c; z)\,. \end{aligned} \tag{2.9}$$

On the horizon, the Dirichlet and the Neumann boundary conditions (B.C.) are imposed on the radial function to obtain eigenvalue equations:

(i) The Dirichlet B.C.

$$\left[\frac{\Gamma(1-c)}{\Gamma(1-a)\Gamma(1-b)} R_{r_+,\rm in} + \frac{\Gamma(c-1)}{\Gamma(c-a)\Gamma(c-b)} R_{r_+,\rm out}\right]\Bigg|_{r_+ + \epsilon} = 0 \tag{2.10}$$

determines the eigenvalue equation for integer n:

$$(\omega - \Omega_{\rm H} m) r_{*,{\rm H}} + \alpha_0(\omega) + \beta_0(\omega) = -\pi \left(n + \frac{1}{2} \right) , \qquad (2.11)$$

(ii) The Neumann B.C.

$$\left[\frac{\Gamma(1-c)}{\Gamma(1-a)\Gamma(1-b)} R_{r_+,{\rm in}} - \frac{\Gamma(c-1)}{\Gamma(c-a)\Gamma(c-b)} R_{r_+,{\rm out}} \right] \Bigg|_{r_+ + \epsilon} = 0 \qquad (2.12)$$

determines the eigenvalue equation for integer n:

$$(\omega - \Omega_{\rm H} m) r_{*,{\rm H}} + \alpha_0(\omega) + \beta_0(\omega) = -\pi n . \qquad (2.13)$$

In these expressions, phase functions and tortoise coordinate on the horizon are defined:

$$\alpha_0(\omega) = \frac{\ell^2 r_+ (\omega - \Omega_{\rm H} m)}{2(r_+^2 - r_-^2)} \ln \frac{4 r_+^2}{r_+^2 - r_-^2} ,$$

$$\beta_0(\omega) = \arg \left(\frac{\Gamma(c-1)}{\Gamma(c-a)\Gamma(c-b)} \right) ,$$

$$r_{*,{\rm H}} \simeq \frac{\ell^2}{2(r_+^2 - r_-^2)} \left(r_+ \ln \frac{\epsilon}{2 r_+} - r_- \ln \frac{r_+ - r_-}{r_+ + r_-} \right) , \qquad (2.14)$$

where the small regularization parameter ϵ is introduced.

2.3. *Numerical analysis*

In this subsection, eigenvalue equations in Eqs.(2.11) and (2.13) are studied numerically. Parameter values are taken as $M = 1$, $\mu = 0$, $\ell = 1, z_\epsilon = (r^2 - r_+^2)/(r^2 - r_-^2) \,|_{r_+ + \epsilon} = 0.01$ throughout this subsection. Parameter value of the black hole rotation J will be indicated in each cases.

First we consider the case of Dirichlet B.C.. We show eigenvalues of ω for each m as the eigenvalue map in (m, ω)-plane without black hole rotation ($J = 0$) in Fig.1. From Fig.1, we know that eigenvalue points ω_n ($n = 0, 1, \ldots, m = 0, \pm 1, \pm 2, \ldots$) form convex curves with respect to horizontal line.

The rotation effect for Dirichlet B.C. is studied for the value $J = 0.2$ in Fig.2 and $J = 0.4$ in Fig.3. We indicate the zero mode line ($\omega - \Omega_{\rm H} m = 0$) in the figures. From these figures, we know that all eigenvalues are above the zero mode line and lie in the region $0 < \omega - \Omega_{\rm H} m$.

The Neumann B.C. cases show the similar tendency as the Dirichlet B.C. cases, in which all the eigenvalues lie above the zero mode line too.

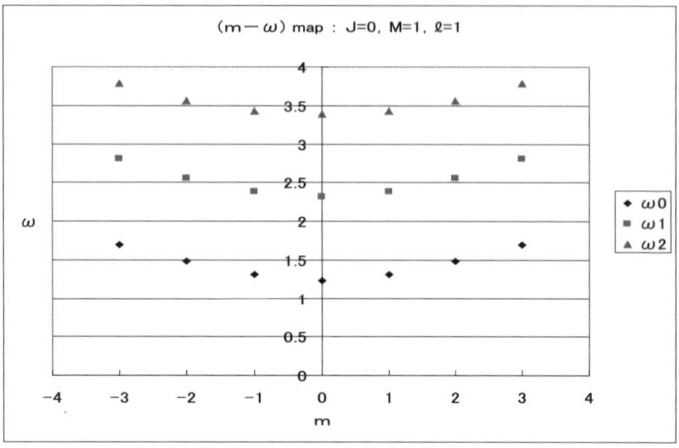

Fig. 1. (m-ω) map for Dirichlet B.C. without black hole rotation.

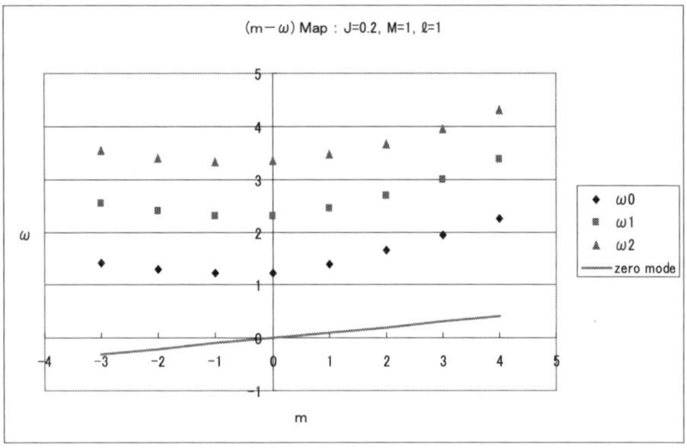

Fig. 2. The black hole rotation effect (J=0.2) for the Dirichlet B.C..

2.4. *Zero and super-radiant modes in BTZ spacetime*

In this subsection, some properties are shown for zero modes and super-radiant modes in BTZ spacetime. The zero mode states defining $0 = \omega - \Omega_{\rm H} m$ $(-\infty < m < \infty)$ play an important role to determine the physical modes defining to satisfy the correct boundary condition and normalization. We give some statements for zero modes and super-radiance modes in the following.

Statement 1. The radial function for zero modes with the correct boundary conditions does not exist.

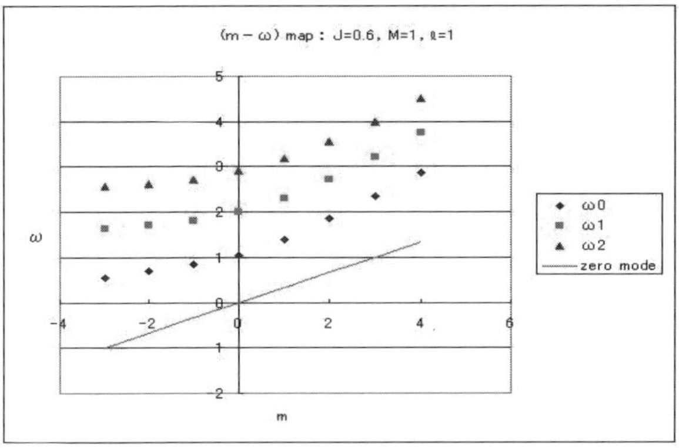

Fig. 3. The black hole rotation effect (J=0.4) for the Dirichlet B.C..

Proof. The radial function for $0 = \omega - \Omega_{\rm H} m$ satisfying the convergent boundary condition at infinity is expressed as

$$R_{\rm zero} = \left(\frac{r_+^2 - r_-^2}{r^2 - r_-^2}\right)^b \frac{1}{\Gamma(2b)} F\left(b - ic, b + ic, 2b \,\middle|\, \frac{r_+^2 - r_-^2}{r^2 - r_-^2}\right)$$

with parameters $b = 1 + \sqrt{1 + \mu/2}$, $c = \ell m/2r_+$. This solution diverges logarithmically and does not satisfy the boundary condition on the horizon, which means that zero modes do not exist as physical modes.

Statement 2. The physical region for normal modes is above the zero mode line, that is $0 < \omega - \Omega_{\rm H} m$ with $-\infty < m < \infty$.

Proof. For the case without black hole rotation $J = 0$, the allowed physical region for normal modes is in the positive frequency region, that is $0 < \omega$ with $-\infty < m < \infty$. Analyticity of wave functions with respect to the rotation parameter J is assumed. After switching on the rotation with $J \neq 0$, the allowed physical region for normal modes shifts from $0 < \omega$ to $0 < \omega - \Omega_{\rm H} m$ because any normal mode cannot cross the zero mode line. As a consequence, normal modes are divided into two regions by zero mode line, the physical region ($0 < \omega - \Omega_{\rm H} m$ with $-\infty < m < \infty$) and the unphysical region ($0 > \omega - \Omega_{\rm H} m$ with $-\infty < m < \infty$).

Statement 3. From statements 1 and 2, super-radiant instability modes $\omega - \Omega_{\rm H} m < 0, 0 < \omega$ shown to be unphysical and the super-radiant stability modes $0 < \omega - \Omega_{\rm H} m < 0, \omega < 0$ shown to be physical in BTZ spacetime.

3. Scalar fields in Kerr-AdS spacetime

3.1. *Normal modes in K-AdS spacetime*

In this subsection, we study normal modes, zero modes and super-radiant modes in (3+1)-dimensional Kerr-AdS spacetime according to our recent paper [7]. The line element for Kerr-AdS spacetime is given by Carter [9]:

$$ds^2 = -\frac{\Delta_r}{\rho^2}\left(dt - \frac{a\sin^2\theta}{\Xi}d\phi\right)^2 + \frac{\Delta_\theta \sin^2\theta}{\rho^2}\left(adt - \frac{r^2+a^2}{\Xi}d\phi\right)^2 + \frac{\rho^2}{\Delta_r}dr^2 + \frac{\rho^2}{\Delta_\theta}d\theta^2 , \tag{3.1}$$

where

$$\Delta_r = (r^2+a^2)(1+r^2\ell^{-2}) - 2Mr , \quad \Delta_\theta = 1 - a^2\ell^{-2}\cos^2\theta , \quad \rho^2 = r^2 + a^2\cos^2\theta , \quad \Xi = 1 - a^2\ell^{-2} , \tag{3.2}$$

with the cosmological parameter $\ell = \sqrt{-3/\Lambda}$ and the rotation parameter $a = J/M$. Writing scalar field in the form of separation variables:

$$\Phi = \mathrm{e}^{-i\omega t}\mathrm{e}^{im\phi}S(\theta)R(r) , \tag{3.3}$$

with the frequency ω and the azimuthal angular momentum m, angular and radial equations are obtained:

$$\left(\frac{\partial_\theta \sin\theta\Delta_\theta\partial_\theta}{\sin\theta} - \frac{(a\omega\sin\theta - \Xi m/\sin\theta)^2}{\Delta_\theta} - \bar{\mu}^2a^2\cos^2\theta + \lambda\right)S(\theta) = 0$$
$$\left(\partial_r\Delta_r\partial_r + \frac{((r^2+a^2)\omega - \Xi am)^2}{\Delta_r} - \bar{\mu}^2r^2 - \lambda\right)R(r) = 0 , \tag{3.4}$$

where λ and $\bar{\mu}$ denote the separation parameter and the effective mass of scalar field respectively. Boundary conditions for R(r) are imposed to be zero for infinity ($r \to \infty$) and the Dirichlet or Newmann B.C. on the horizon ($r = r_{\mathrm{H}}$). From field equations and boundary conditions orthonormal relations are obtained:

$$\int_{r_+}^{\infty} dr \int_0^{\pi} d\theta\sqrt{-g}\left(-g^{tt}(\omega+\omega') + 2g^{t\phi}m\right) \times S^*_{\omega,m,\lambda}(\theta)S_{\omega',m,\lambda'}(\theta)R^*_{\omega,m,\lambda}(r)R_{\omega',m,\lambda'}(r) = \delta_{\omega,\omega'}\delta_{\lambda,\lambda'} . \tag{3.5}$$

Note that the norm is positive for $0 < \omega - \Omega_{\mathrm{H}}m$ because the contribution to the integration is large near the horizon. Scalar field is expanded in the form:

$$\Phi(t,x) = \sum_\alpha \left(a_\alpha f_\alpha(t,x) + b^\dagger_\alpha f^*_\alpha(t,x)\right) , \tag{3.6}$$

using the full eigenfunction

$$f_\alpha := \frac{1}{\sqrt{2\pi}} e^{-i\omega t} e^{im\phi} S_\alpha(\theta) R_\alpha(r) \quad \text{where} \quad \alpha = (\omega, m, \lambda) \ , \tag{3.7}$$

with annihilation and creation operators a_α and $b_\alpha^\dagger$ in the quantized filed theory. Combining the energy and the angular Momentum of scalar field defined as in the standard method

$$\begin{aligned} E &= \int_\Sigma d^3x \sqrt{-g}(-g^{tt}\partial_t\Phi^\dagger\partial_t\Phi + g^{\phi\phi}\partial_\phi\Phi^\dagger\partial_\phi\Phi \\ &\quad + g^{rr}\partial_r\Phi^\dagger\partial_r\Phi + g^{\theta\theta}\partial_\theta\Phi^\dagger\partial_\theta\Phi) \\ L &= \int_\Sigma d^3x\sqrt{-g}(g^{tt}(\partial_t\Phi^\dagger\partial_\phi\Phi + \partial_\phi\Phi^\dagger\partial_t\Phi) + 2g^{t\phi}\partial_\phi\Phi^\dagger\partial_\phi\Phi) \ , \end{aligned} \tag{3.8}$$

the effective energy is expressed by the creation and the annihilation operators in a compact form:

$$E - \Omega_{\rm H} L = \sum_\alpha (\omega - \Omega_{\rm H} m)(a_\alpha^+ a_\alpha + b_\alpha b_\alpha^\dagger) \ , \tag{3.9}$$

where $\Omega_{\rm H} = a\Xi/(r_+^2 + a^2)$ denotes the angular velocity at r_+. Note that the effective energy is positive definite for $0 < \omega - \Omega_{\rm H} m$.

3.2. *Zero and super-radiance modes in K-AdS spacetime*

In this subsection, we study zero modes satisfying $\omega - \Omega_{\rm H} m = 0$ and super-radiance instability modes satisfying $\omega - \Omega_{\rm H} m < 0$ with $0 < \omega$, or super-radiance stable modes satisfying $0 < \omega - \Omega_{\rm H} m$ with $\omega < 0$ in K-AdS spacetime under the criterion of the boundary conditions and the normalization conditions.

(1) Radial wave function for zero modes near horizon is obtained from Eq.(3.4):

$$R_{\rm zero} \simeq d_1 + d_2 \ln(r - r_+) \ , \tag{3.10}$$

where d_1, d_2 denote integration constants. In Eq.(3.10), the first term satisfies the Neumann B.C. but both terms do not satisfy the Dirichlet B.C. on the horizon.

(2) Normalization condition for zero modes is obtained from Eq.(3.5) in near horizon approximation:

$$\int_{r_+}^\infty dr \int_0^\pi d\theta \sqrt{-g} \frac{-2g^{tt}\Xi am(r^2 - {r_+}^2)}{(r_+^2 + a^2)(r^2 + a^2)} |S_{\rm zero}(\theta) R_{\rm zero}(r)|^2 \simeq 1 \tag{3.11}$$

The extra suppression factor $r^2 - {r_+}^2$ appears in this normalization condition and the first term in Eq.(3.10) cannot satisfy the normalization condition. Note that the near horizon approximation is justified due to the enhanced factor g^{tt} in the integrand.

3.3. *Preliminary numerical analysis*

We have studied normal modes numerically as eigenvalue problem and obtained the preliminary result shown in Fig.4 [10]. Parameter values are taken as $M = 4, J = 4$ and $\Lambda = 0$ under the special relation $\mu = \omega$. Our preliminary calculation shows that all normal modes are above the zero mode line.

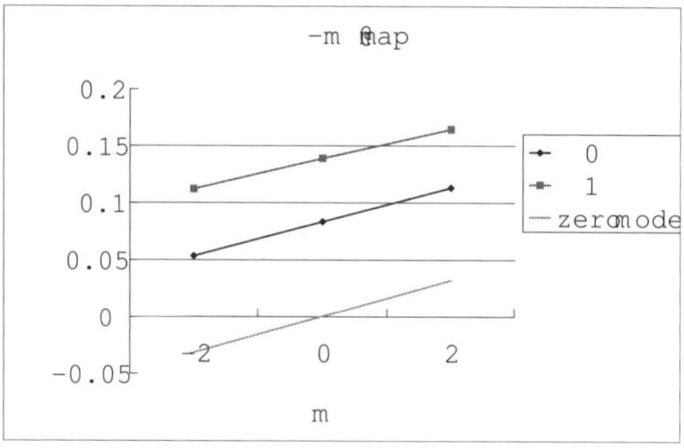

Fig. 4. (m-ω) map for Dirichlet B.C. with black hole rotation for Kerr spacetime.

4. Summary

We have studied normal modes, zero modes and super-radiance modes extensively in BTZ and Kerr-AdS spacetimes. Main results are summarized.

(i) For BTZ spacetime, we have studied normal modes in numerically as eigenvalue problem and obtained the result that any normal modes are in the region $0 < \omega - \Omega_{\rm H} m$. We have also studied the zero modes ($\omega - \Omega_{\rm H} m = 0$) analytically and obtained the result that zero modes cannot satisfy the boundary condition at infinity or on the horizon, which indicates the non-existence of zero modes. From the non-existence of zero modes we obtain the result that super-radiance instability modes ($\omega - \Omega_{\rm H} m < 0$ with $0 < \omega$) are unphysical but super-radiance stable modes ($0 < \omega - \Omega_{\rm H} m$ with $\omega < 0$) are physical, in assuming the analyticity with respect to rotation parameter J. Note that the numerical study and the analytical study are completely consistent.

(ii) For K-AdS spacetime, we have derived orthonormal relations for normal modes and obtain the expression for the effective energy $E - \Omega_{\rm H} L$ using the normal mode expansion. We have derived the non-existence of zero modes rigorously for the Dirichlet B.C. and approximately for the Neumann B.C. considering the boundary condition and the normalization condition. Non-existence of zero modes indicate that super-radiant modes of $\omega - \Omega_{\rm H} m < 0$ with $0 < \omega$ are unphysical but super-radiant modes of $\omega - \Omega_{\rm H} m < 0$ with $0 < \omega$ are physical. The result is consistent with that for (2+1)-dimensional BTZ spacetime and the co-rotating consideration.

(iii) One of important applications of our study is statistical mechanics of scalar fields around Kerr-AdS BH. As physical allowed normal modes satisfy $0 < \omega - \Omega_{\rm H} m$, the total effective energy $E - \Omega_{\rm H} L$ in Eq.(3.9) should be positive and the partition function for scalar field around Kerr-AdS background spacetime becomes well-defined:

$$Z = \mathrm{Tr} \exp\left(-\beta_{\rm H}(E - \Omega_{\rm H} L)\right)$$

where $\beta_{\rm H}$ is the inverse of Hawking temperature and the trace is taken with respect to physical modes.

References

1. G. 't Hooft, Nucl. Phys. **B256** (1985) 727.
2. See for example, S. Mukohyama, Phys. Rev. **D61** (2000) 124021; I. Ichinose and Y. Satoh, Nucl. Phys. **447** (1995) 340 and others.
3. W.H. Press and S.A. Teukolsky, Nature **238** (1972) 211.
4. S.W. Hawking and H.S. Reall, Phys. Rev. **D61** (1999) 024014.
5. V. Cardoso and O.J.C. Dais, Phys. Rev. **D70** (2004) 084011; V. Cardoso, Ó.J.C. Dias, J.P.S. Lemos and S. Yoshida, Phys.Rev. **D70** (2004) 044039.
6. M. Kenmoku, M. Kuwata and K. Shigemoto, Class. Quant. Grav. **25** (2008) 145016; M. Kuwata, M. Kenmoku and K. Shigemoto, Prog. Theor. Phys. **119** (2008) 939.
7. M. Kenmoku, arXiv:0809.2634.
8. M. Bañados, C. Teitelboim and J. Zanelli, Phys. Rev. Lett. **69** (1992) 1849.
9. B. Carter, Commun. Math. Phys. **10** (1968) 280.
10. M. Kenmoku and K. Shigemoto, "Numerical analysis of super-radiance modes in Kerr anti-de Sitter black holes", (2009), in preparation.

AN ANALYSIS FOR THE EFFECTIVE SPECTRUM INDICES FOR FSRQs

JIANG-HE YANG

Department of Physics and Electronics Science,
Hunan University of Arts and Science, Changde, 415000, P.R.China
Center for Astrophysics, Guangzhou University,
Guangzhou, 510405, P.R.China
E-mail: yjianghe@163.com

JUN-HUI FAN

Center for Astrophysics, Guangzhou University,
Guangzhou, 510405, P.R.China
E-mail: jhfan_cn@yahoo.com.cn

RU-SHU YANG, JIAN-JUN NIE, JUN CHENG, YUE-LIAN ZHANG

Department of Physics and Electronics Science,
Hunan University of Arts and Science, Changde, 415000, P.R.China

In this work, the radio, X-ray and γ-ray emissions are compiled for sample of 42 flat spectrum radio quasars-FSRQs, to calculate the effective spectrum indices, namely the radio-X-ray spectrum index $\alpha_{\rm RX}$, the radio -γ-ray spectrum index $\alpha_{\rm RG}$ and the X-ray to γ-ray spectrum index $\alpha_{\rm XG}$. Also the correlations among $\alpha_{\rm RX}$, $\alpha_{\rm RG}$ and $\alpha_{\rm XG}$ are investigated. It is interesting that $\overline{\alpha}_{\rm RX} \simeq \overline{\alpha}_{\rm XG} \simeq \overline{\alpha}_{\rm RG} \approx 0.82$ for 42 FSRQs, which can be used to estimate the γ-ray emissions from the X-ray and the radio emissions. Clear correlations are found between $\alpha_{\rm XG}$ and $\alpha_{\rm RX}$ and between $\alpha_{\rm RG}$ and $\alpha_{\rm XG}$ respectively. A weak correlation between $\alpha_{\rm RX}$ and $\alpha_{\rm RG}$ is found. The results show that the γ-ray emissions are associated with both radio and the X-ray emissions in blazars.

Keywords: Active galactic Nuclei (AGNs); BL Lacs; Quasars; γ-ray emissions.

1. Introduction

Blazars are extremely active galactic nuclei (AGNs) showing rapid and large variation, high and variable polarization, superluminal motion and high energetic γ-ray emissions.[1] Blazars consist of two different subclasses as BL Lacertae objects (BL Lacs) and flat spectrum radio quasars (FSRQs) by their emission lines. BL Lacs have no or very weak emission line while FSRQs have strong emission lines. Both BL Lacs and FSRQs are strong γ-ray emitters ($\sim$ TeV Blazar), and their emission mechanisms are investigated by many authors.[2–5] Some γ-ray blazars are also superluminal radio sources.

At present, the origin of the strongly variable γ-ray emissions is still not clear. Various models have been proposed for the γ-ray emissions.[6] However, none of these models has proved convincing. The various relations among the emissions of different wave bands might be used to distinguish the variety of emission mechanisms. Therefore, simultaneous multi-wavelength observations and the variability in various bands are important for investigating the γ-ray-radio correlation. However, it is difficult to achieve simultaneous multi-band observations, and we have to use data relating to different states to investigate the existence of correlation between γ-ray and lower energy band emissions.

In this paper, we will compile the relevant data for a sample of γ-ray loud blazars, calculate the effective spectrum index, and investigate their correlation amongst the spectrum index for FSRQs. In the 2nd section the data of the sample and calculation results are given, in the 3th section, we give some discussion and a brief conclusion.

2. Sample and results

2.1. *Sample*

In this paper, a sample including 42 FSRQs is compiled from the available literatures (see Table 1). The descriptions of the columns in Table 1 are as follows: Col.(1) name of the source, Col.(2) redshift, Col.(3) radio flux at 5 GHz in units of Jy, Col.(4) radio spectral index, Col.(5) reference for the radio flux, Col.(6) X-ray flux at 1 keV in μJy, Col.(7) X-ray spectral index, Col.(8) reference for the X-ray flux, Col.(9) the observed average γ-ray photon flux (> 100 MeV) in units of 10^{-8} photon $\cdot$ cm$^{-2}\cdot$ s^{-1}, Col.(10) γ-ray spectral index, Col.(11) reference for γ-ray data.

2.2. *Method of Analysis*

To calculate the spectrum index, we first convert γ photon flux to flux densities at 1 GeV according to $\frac{\mathrm{d}F}{\mathrm{d}E} = F_0 E^{-\alpha_{\mathrm{ph}}}$. Secondly the flux densities at 5 GHz, 1 keV and the γ-ray band are k-corrected according to $f_\nu = f_\nu^{\mathrm{ob.}}(1+z)^{\alpha_\nu - 1}$, where α_ν is the spectral index at frequency ν, and z is the redshift. f_ν is the true flux, and $f_\nu^{\mathrm{ob.}}$ is the observation flux.

Table 1. The sample of 42 FSRQs.

Name (1)	z (2)	f_5, α_R (3),(4)	Ref. (5)	f_X, α_X (6), (7)	Ref. (8)	f_γ, α_G (9), (10)	Ref. (11)
0204+1458	1.202	2.714, -0.40	[7]	0.06, 0.68	[8]	8.7, 1.23	[9]
0210-5055	1.003	3.198, -0.20	[7]	1.20, 1.66	[8]	85.5, 0.99	[9]
0239+2815	1.213	3.356, 0.30	[7]	0.19, ...	[10]	13.8, 1.53	[9]
0340-0201	0.852	3.014, 0.30	[7]	0.25, ...	[11]	15.1, 0.84	[9]
0422-0102	0.915	6.992, 1.10	[7]	0.28, 1.86	[8]	16.3, 1.44	[9]
0440-003	0.844	1.620, ...	[12]	0.11, ...	[11]	12.5, 1.37	[12]
0456-2338	1.009	1.863, 0.10	[7]	0.06, ...	[10]	8.1, 2.14	[9]
0458-4635	0.858	1.653, -0.60	[7]	0.16, ...	[10]	7.7, 1.75	[9]
0500-0159	2.286	3.317, 1.00	[7]	0.10, ...	[13]	11.2, 1.45	[9]
0512-6150	1.093	1.211, -0.80	[7]	0.28, ...	[10]	7.2, 1.40	[9]
0530+1323	2.070	2.978, 0.30	[7]	1.59, 1.54	[8]	93.5, 1.46	[9]
0530-3626	0.055	8.180, -0.70	[7]	2.12, 1.89	[8]	15.8, 1.63	[9]
0531-2940	3.104	1.231, 0.90	[7]	0.18, 1.36	[8]	6.9, 1.47	[9]
0808+4844	1.433	1.229, 0.30	[7]	0.17, 1.56	[8]	10.7, 1.15	[9]
0829+2413	2.046	0.670, 0.00	[7]	0.34, ...	[11]	24.9, 1.42	[9]
0845+7049	2.172	2.436, -0.40	[7]	1.27, 1.32	[8]	10.2, 1.62	[9]
0906+430	0.670	1.800, ...	[14]	0.11, 1.57	[8]	32.0, ...	[14]
0917+4427	2.180	1.003, 0.20	[7]	0.47, 1.39	[8]	13.8, 1.19	[9]
0952+5501	0.909	2.260, -0.20	[7]	0.10, 2.17	[8]	9.1, 1.12	[9]
1134-1530	1.187	4.209, -0.60	[7]	0.66, 3.06	[15]	9.9, 1.70	[9]
1200+2847	0.729	1.542, -0.10	[7]	0.44, 2.30	[8]	7.5, 0.98	[9]
1224+2118	0.435	1.261, -0.40	[7]	0.41, ...	[10]	13.9, 1.28	[9]
1229+0210	0.158	44.940, -0.10	[7]	20.42, 1.51	[8]	15.4, 1.58	[9]
1230-0247	1.045	1.020, -0.30	[7]	0.08, ...	[16]	6.9, 1.85	[9]
1255-0549	0.538	11.192, -0.10	[7]	1.50, 1.65	[8]	74.2, 0.96	[9]
1317+520	1.060	0.660, ...	[16]	0.06, ...	[16]	7.7, ...	[16]
1329+1708	2.084	0.708, 0.70	[7]	0.05, ...	[10]	4.4, 1.41	[9]
1512-0849	0.361	3.000, 0.00	[7]	0.83, 1.38	[8]	18.0, 1.47	[9]
1608+1055	1.227	1.688, 0.00	[7]	0.08, ...	[10]	25.0, 1.63	[9]
1614+3424	1.404	2.843, -0.10	[7]	0.24, 1.76	[8]	26.5, 1.42	[9]
1625-2955	0.815	1.920, 0.10	[7]	0.08, ...	[13]	47.4, 1.07	[9]
1635+3813	1.814	3.198, 0.40	[7]	0.42, 1.53	[11]	58.4, 1.15	[9]
1641+339	0.595	12.400, ...	[17]	8.30, ...	[18]	25.0, ...	[19]
1733-1313	0.902	6.991, 0.80	[7]	0.63, ...	[10]	36.1, 1.23	[9]
1738+5203	1.375	1.134, -0.40	[7]	0.16, ...	[10]	18.2, 1.42	[9]
1744-0310	1.054	2.369, 0.30	[7]	2.21, ...	[11]	11.7, 1.42	[9]
1830-210	1.000	7.920, ...	[12]	0.43, ...	[10]	26.6, 1.59	[12]
1935-4022	0.966	1.129, -0.30	[7]	0.38, ...	[20]	8.5, 1.86	[9]
2055-4716	1.489	2.520, ...	[10]	0.28, ...	[10]	9.6, 1.04	[9]
2232+1147	1.037	3.765, -0.50	[7]	0.73, 1.51	[8]	19.2, 1.45	[9]
2254+1601	0.859	15.859, 0.10	[7]	1.37, 1.62	[8]	53.7, 1.21	[9]
2359+2041	1.066	0.704, 0.20	[7]	0.28, ...	[10]	8.3, 1.09	[9]

Thirdly, we calculate the effective spectrum index values, α_{RX}, α_{RG} and α_{XG} by $\alpha_{12} = -\frac{\log(f_1/f_2)}{\log(\nu_1/\nu_2)}$, with 1 and 2 stand for radio(R), X-ray(X) and

γ-ray(G) band respectively.

2.3. *Results*

Using the data in Table 1, we can get $\alpha_{RX} = 0.821 \pm 0.010$, $\alpha_{RG} = 0.819 \pm 0.006$ and $\alpha_{XG} = 0.817 \pm 0.015$. In addition, we can investigate the linear correlation between them. When the linear-regression analysis is performed to the data, we get the results shown in Fig.1. The correlations between α_{RX}, α_{RG} and α_{XG} are as follows.

$\alpha_{XG} = -(0.37 \pm 0.09)\alpha_{RX} + (1.12 \pm 0.07)$, with the correlation coefficient $r = -0.55$, and the chance probability $p = 1.7 \times 10^{-4}$.

$\alpha_{RG} = (1.52 \pm 0.33)\alpha_{XG} - (0.42 \pm 0.26)$, with $r = 0.60$, and $p < 10^{-4}$.

$\alpha_{RX} = (0.20 \pm 0.09)\alpha_{RG} + (0.65 \pm 0.07)$, with $r = 0.35$, and $p = 0.025$.

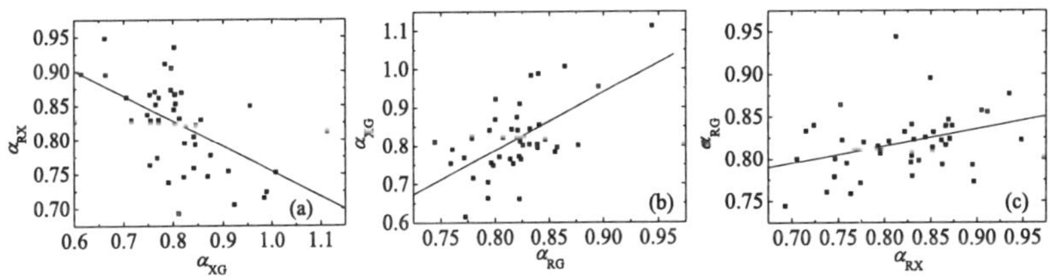

Fig. 1. The correlations between α_{RX}, α_{RG} and α_{XG}.

3. Discussion and Conclusions

Blazars consist of two subclasses with quite different emission line features (BL Lacs and FSRQs). Their relationship is still an open question. The γ-ray emission mechanisms are also not quite clear.

For the discussion of the γ-ray emissions, many authors have studied the correlation between the γ-rays and the low-energy waveband emissions, particularly between the γ-rays and the radio emissions.[21] We also discussed the correlation between the γ-ray and radio emissions for their maximum data, the γ-ray emission is more closely correlated with the high frequency(1.3 mm, 230 GHz) radio emission than with the lower frequency (5 GHz, 6 cm) radio emission.[13,22] Schachter & Elvis (1993)[23] reported that there is a correlation between the γ-ray and 5 GHz radio emissions. A good correlation between the radio and γ-ray luminosities for the γ-ray-loud blazars was found by Dondi & Ghisellini (1995).[24] Most correlation analysis was done for the γ-ray and other lower energy band, and the correlation analysis suggest that the γ-rays are more correlated with the radio than with other lower energy bands.

In the present paper, in order to understand the association for the blazars emissions from lower frequency to γ-ray, we studied the correlations between effective spectrum indices of radio to X-ray and to γ-ray.

For spectral index, we found that the spectral index are almost the same value of $\alpha = 0.82$, $\alpha_{\rm RX} = 0.821 \pm 0.010$, $\alpha_{\rm XG} = 0.817 \pm 0.015$, and $\alpha_{\rm RG} = 0.819 \pm 0.006$ for FSRQs, which can be used to estimate the γ-ray emissions from the X-ray and the radio emissions.

For the relations of $\alpha_{\rm RX}$, $\alpha_{\rm RG}$ and $\alpha_{\rm XG}$, there is also a clear correlation between $\alpha_{\rm XG}$ and $\alpha_{\rm RX}$ with correlation coefficient $r = -0.55$, and chance probability $p = 1.7 \times 10^{-4}$(Fig. 1(a)) and between $\alpha_{\rm RG}$ and $\alpha_{\rm XG}$, with $r = 0.60$, $P < 10^{-4}$(Fig. 1(b)). A weak correlation between $\alpha_{\rm RX}$ and $\alpha_{\rm RG}$ can also be found with $r = 0.35$ and $P = 0.025$ (Fig. 1(c)). Those correlations imply that the γ-ray emissions are associated with the radio and X-ray emissions.

In this paper, we compiled the data for 42 FSRQs and calculated the effective spectrum indices, then we investigated the correlation between the spectrum index. The correlations suggest that the γ-rays are associated with the radio and the X-ray bands in FSRQs. We also found that the effective spectrum index from radio to X-ray and γ ray is 0.82 for FSRQs, which can be used to estimate the γ-ray emissions in FSRQs.

Acknowledgments

This work is partially supported by the National Natural Science Foundation of China (10573005, 10633010), the 973 project(2007CB815405), and the Fund of the 11th Five-year Plan for Key Construction Academic Subject (Optics) of Hunan Province. We also think the Guangzhou City Education Bureau, which supports our research in astrophysics.

References

1. Urry C M, Padovani Paolo, *PASP*, **107**, 803 (1995).
2. Ghisellini G, Tavecchio F, Bodo G, et al., *MNRAS*, **393**, 16 (2009).
3. Dermer Charles D, Finke Justin D, Krug Hannah, et al., *ApJ*, **692**, 32 (2009).

4. Graff Philip B, Georganopoulos Markos, Perlman Eric S, et al., *ApJ*,**689**, 68 (2008).
5. Böttcher Markus, Dermer charles D, Finke Justin D, *ApJ*,**679**, 9 (2008).
6. Maraschi L, Ghisellini G, Celotti A, *ApJ*, **397**, L5 (1992).
7. Mattox J R, Hartman R C, Reimer O, *ApJSS*, **135**, 155 (2001).
8. Donato D, Ghisellini G, Tagliaferri G, et al., *A&A*, **375**, 739 (2001).
9. Hartman R C, Bertssch D L, Bloom A W, et al., *ApJS*, **123**, 79 (1999).
10. Cheng K S, Zhang X, Zhang L, *ApJ*, **537**, 80 (2000).
11. Fossati G, Maraschi L, Celotti A, et al., *MNRAS*, **299**, 433 (1998).
12. Zhang L, Cheng K S, Fan J H, *PASJ*, **47**, 265 (2001)
13. Andrea C, Giovanni F, Gabriele G, et al., *aspho-ph*/9612041, (1996).
14. Fan J H, Adam G, Xie G Z, et al., *A&A*, **338**: 27 (1998).
15. Siebert J, Brinkmann W, Drinkwater M J, et al., *MNRAS*, **301**, 261 (1998).
16. Comastri A, Fossti G, et al., *APJ*, **480**, 534 (1997).
17. Brown L M J, Gear W K, Smith M G, *APJ*, **340**, 150 (1989).
18. Urry C M, Sambruna R M, Worrall D M, et al., *APJ*, **463**, 424 (1996).
19. Von Montigny C, Bertsch D L, Chiang J, et al., *ApJ*, **440**, 525 (1995).
20. Brinkmann W, Maraschi L, Treves A, et al., *A&A*, **288**, 433 (1994).
21. Zhou Y Y, Lu Y J, Wang T G, et al., *ApJ*, **484**, L47 (1997).
22. Yang J H, Fan J H, *ChJAA*, **5**, 229 (2005).
23. Schachter J, Elvis M, *ApJS*, **92**, 623 (1993).
24. Dondi L, Ghisellini G, *MNRAS*, **273**, 583 (1995).

REFINEMENTS OF TRAPPED SURFACES

SEAN A. HAYWARD

Center for Astrophysics, Shanghai Normal University,
100 Guilin Road, Shanghai 200234, China
**E-mail: sean_a_hayward@yahoo.co.uk*

Various refinements of trapped surfaces are summarized, intended to apply near the outer horizon of a black hole, together with their relations. Assuming the null energy condition, minimal trapped implies outer trapped, which implies increasingly trapped. Variations of these three definitions form an interwoven hierarchy.

Keywords: black holes

1. Introduction

Trapped surfaces as originally defined by Penrose[1] play an important role in gravitational physics, both for black holes and in cosmology, e.g. in the singularity theorems.[1–3] Marginal surfaces, a limit of trapped surfaces, admit a local, dynamical theory of black holes.[4–9] One might therefore expect the boundary of the region of trapped surfaces to consist of marginal surfaces, i.e. to be a trapping horizon. However, this is not so: trapped surfaces can poke through a spherically symmetric trapping horizon.[10–12] On the other hand, this boundary does not have the special physical properties that trapping horizons have, such as a first law involving surface gravity[13] and a local Hawking temperature.[14]

There is thus a conflict between the mathematics and physics. The physics seems to be clear, so the mathematics must yield. That is, trapped surfaces as simply defined need to be refined in some way. Then one may conjecture that the boundary of a region of suitably refined trapped surfaces is a trapping horizon.

The following gives eight refinements, which are related: they form an interwoven hierarchy, with some direct relations and some relations which assume the Einstein equation, or more exactly just the null energy condition (NEC). This is essentially a summary of earlier papers,[15,16] to which one is referred for omitted proofs.

2. Minimal trapped surfaces

Consider spatial surfaces S embedded in a given space-time, and normal vectors η, with L_η denoting the Lie derivative along η, and $L(\eta) = L_\eta$. The *expansion 1-form* θ is defined by

$$\star\theta = L(\star 1) \tag{1}$$

where $\star$ is the Hodge operator induced on S by the space-time metric g, i.e. $\star 1$ is the area form and θ its logarithmic normal derivative. The *expansion vector* (a.k.a. mean-curvature vector) is

$$H = g^{-1}(\theta). \tag{2}$$

There is also a Hodge operator $*$ in the normal space, e.g. $*\theta$ is the *dual expansion 1-form.* This induces a duality operation on normal vectors by $g(\eta_*) = *g(\eta)$, or equivalently

$$g(\eta_*, \eta) = 0, \quad g(\eta_*, \eta_*) = -g(\eta, \eta). \tag{3}$$

In particular, there is the *dual expansion vector* H_*.

A surface is *trapped* if H is temporal, or equivalently if H_* is spatial. Assuming a time-orientable space-time, the surface is *future* (respectively *past*) trapped if H is future (respectively past) temporal. Any surface is extremal in the H_* direction:

$$H_* \cdot \theta = 0. \tag{4}$$

Then one may ask whether the surface is not merely extremal but minimal.

Definition 1. A (strictly) *minimal* trapped surface is a trapped surface for which, for some variation,

$$Q = H_*^a H_*^b \nabla_b \theta_a > 0 \tag{5}$$

where ∇ is the covariant derivative operator of g. The strict sign will turn out to be convenient. Rewriting as

$$Q = -H^a H_*^b \nabla_b (*\theta)_a \tag{6}$$

is convenient for calculations using differential forms.

3. Outer trapped surfaces

In spherical symmetry, outer trapped spheres can be defined by $\kappa > 0$, where κ is *surface gravity.*[13,15,17] The relevant object in general turns out to be the curvature

$$K = *d*\theta + \tfrac{1}{2}g^{-1}(\theta,\theta) \tag{7}$$

where $*d*$ is the normal codifferential or divergence. A previous definition of quasi-local surface gravity was[18]

$$\kappa = \frac{1}{16\pi R}\oint_S \star K \tag{8}$$

where $R = \sqrt{A/4\pi}$ is the area radius, i.e. area is

$$A = \oint_S \star 1 = 4\pi R^2. \tag{9}$$

This κ enters a quasi-local first law for trapping horizons involving the Hawking mass.[18]

Definition 2. An *outer* trapped surface is a trapped surface for which, for some variation,

$$K > 0. \tag{10}$$

Lemma 1. Assuming the Einstein equation with units $G = 1$,

$$2Q = -g^{-1}(\theta,\theta)K - 16\pi H \cdot \Psi \tag{11}$$

where

$$\Psi = (T + \Theta)\cdot H + w\theta \tag{12}$$

in terms of the energy tensor T,

$$w = -\tfrac{1}{2}\mathrm{tr}\,(T + \Theta) \tag{13}$$

is an energy density, where the trace is in the normal space, and Θ is an *effective energy tensor for gravitational radiation.*[9,19–28] The proofs of the lemmas are all calculations using the null-null components of the Einstein equation.[16]

Proposition 1. *NEC and minimal trapped implies outer trapped.*

Proof. NEC $\Rightarrow H \cdot \Psi \geq 0$. For a trapped surface, $g^{-1}(\theta,\theta) < 0$, then inspect signs in Lemma 1.

4. Increasingly trapped surfaces

Since $-g^{-1}(\theta,\theta)$ vanishes for marginal surfaces and is positive for trapped surfaces, it can be taken as a measure of how trapped a surface is. Then one may ask whether it is increasing to the future (respectively past) for a future (respectively past) trapped surface.

Definition 3. An *increasingly* trapped surface is a trapped surface for which, for some variation,

$$H \cdot dg^{-1}(\theta,\theta) < 0. \tag{14}$$

Lemma 2. Assuming the Einstein equation,

$$-H \cdot dg^{-1}(\theta,\theta) = 16\pi H \cdot \Psi - g^{-1}(\theta,\theta)(K - g^{-1}(\theta,\theta)). \tag{15}$$

Proposition 2. *NEC and outer trapped implies increasingly trapped.*

Proof. For a trapped surface, $g^{-1}(\theta,\theta) < 0$, NEC $\Rightarrow H\cdot\Psi \geq 0$ as before, then inspect signs in Lemma 2.

Definition 4. An *anyhow increasingly* trapped surface is a future (respectively past) trapped surface for which, for all variations along a future (respectively past) causal normal vector ζ,

$$\zeta \cdot dg^{-1}(\theta,\theta) < 0. \tag{16}$$

Definition 5. A *somehow increasingly* trapped surface is a future (respectively past) trapped surface for which, for some variation along a future (respectively past) causal normal vector ζ,

$$\zeta \cdot dg^{-1}(\theta,\theta) < 0. \tag{17}$$

Clearly anyhow increasingly trapped implies increasingly trapped, which implies somehow increasingly trapped.

5. Doubly outer trapped surfaces

Outer trapped implies anyhow increasingly trapped in spherical symmetry,[15] but this does not hold in general. Instead, a stricter version of outer trapped has this property, as follows. Introduce two more curvatures:

$$K_{(\pm)} = *d{*}\theta \mp *d\theta + \tfrac{1}{2}g^{-1}(\theta,\theta) \tag{18}$$

where $(\pm)$ indicates a label rather than an index. Then $2K = K_{(+)} + K_{(-)}$.

Definition 6. A *doubly outer* trapped surface is a trapped surface for which, for all variations,

$$K_{(+)} > 0, \quad K_{(-)} > 0. \tag{19}$$

Clearly doubly outer trapped implies outer trapped.

Lemma 3. Assuming the Einstein equation,

$$-\zeta\cdot dg^{-1}(\theta,\theta) = 16\pi\zeta\cdot\Psi + g^{-1}(\theta,\theta)\zeta\cdot\theta - \tfrac{1}{2}\zeta\cdot(\theta - *\theta)K_{(-)} - \tfrac{1}{2}\zeta\cdot(\theta + *\theta)K_{(+)}. \quad (20)$$

Proposition 3. *NEC and doubly outer trapped implies anyhow increasingly trapped.*

Proof. For a trapped surface, $g^{-1}(\theta,\theta) < 0$, while for ζ in the appropriate causal quadrant, $\zeta\cdot\theta < 0$ and $\zeta\cdot(\theta \pm *\theta) \leq 0$, as is best seen in null-null components,[16] then NEC $\Rightarrow \zeta\cdot\Psi \geq 0$, then inspect signs in Lemma 3. The proof also makes clear that outer trapped generally does not imply anyhow increasingly trapped.

6. Involute trapped surfaces

Minimal trapped generally does not imply doubly outer trapped. But if it is refined further, such a result can be obtained.

Definition 7. An *involute* trapped surface is a future (respectively past) trapped surface for which, for all variations along a future (respectively past) causal normal vector ζ,

$$Q = -\zeta^a H^b_* \nabla_b (*\theta)_a > 0. \quad (21)$$

Clearly involute trapped implies minimal trapped. Involute means curved or curled inwards, as of a leaf.

Lemma 4. Assuming the Einstein equation,

$$-2Q = 16\pi\zeta\cdot\Psi + \tfrac{1}{2}\zeta\cdot(\theta - *\theta)K_{(+)} + \tfrac{1}{2}\zeta\cdot(\theta + *\theta)K_{(-)}. \quad (22)$$

Proposition 4. *NEC and involute trapped implies doubly outer trapped.*

Proof. As before, $\zeta\cdot(\theta \pm *\theta) \leq 0$ and NEC $\Rightarrow \zeta\cdot\Psi \geq 0$. Considering the null normal vectors $\zeta = l_\pm$, both $K_{(\pm)}$ must be positive. The proof also makes clear that minimal trapped generally does not imply doubly outer trapped.

Definition 8. A *somehow involute* trapped surface is a future (respectively past) trapped surface for which, for some variation along a future (respectively past) causal normal vector ζ,

$$Q = -\zeta^a H^b_* \nabla_b (*\theta)_a > 0. \quad (23)$$

Clearly minimal trapped implies somehow involute trapped.

One might ask whether somehow involute trapped implies somehow increasingly trapped. The answer is negative, except for a special case.

Proposition 5. *NEC and somehow involute trapped implies somehow increasingly trapped if* $K_{(+)} = K_{(-)}$.

Proof. The involute condition gives

$$\zeta \cdot (\theta - *\theta)K_{(+)} + \zeta \cdot (\theta + *\theta)K_{(-)} < 0. \tag{24}$$

However, to get somehow increasingly trapped generally requires

$$\zeta \cdot (\theta - *\theta)K_{(-)} + \zeta \cdot (\theta + *\theta)K_{(+)} < 0. \tag{25}$$

Thus it works if $K_{(+)} = K_{(-)}$.

7. Summary

The hierarchy of trapped surfaces may be illustrated as follows:

$$\begin{array}{lllll}
\text{involute} \Rightarrow & \text{doubly outer} & \Rightarrow & \text{anyhow increasingly} & \\
\Downarrow \quad \text{NEC} & \Downarrow & \text{NEC} & \Downarrow & \\
\text{minimal} \Rightarrow & \text{outer} & \Rightarrow & \text{increasingly} & \\
\Downarrow \quad \text{NEC} & & \text{NEC} & \Downarrow & \\
\text{somehow involute} & & \Rightarrow & \text{somehow increasingly} & \\
 & & \text{NEC}, K_{(+)} = K_{(-)} & &
\end{array} \tag{26}$$

where the vertical implications are straightforward, while the horizontal implications require NEC, and in the last case, the symmetry where the curvatures $K_{(\pm)}$ are equal. Otherwise, the threads are respectively geometrical warp and physical weft.

Acknowledgments

Research supported by the National Natural Science Foundation of China under grants 10375081, 10473007 and 10771140, by Shanghai Municipal Education Commission under grant 06DZ111, and by Shanghai Normal University under grant PL609.

References

1. R. Penrose, Phys. Rev. Lett. **14**, 57 (1965).
2. S. W. Hawking, Proc. R. Soc. London **A300**, 187 (1967).
3. S. W. Hawking & R. Penrose, Proc. R. Soc. London **A314**, 529 (1970).
4. S. A. Hayward, Phys. Rev. **D49**, 6467 (1994).
5. A. Ashtekar & B. Krishnan, Living Rev. Relativity **7**, 10 (2004).
6. I. Booth, Can. J. Phys. **83**, 1073 (2005).

7. B. Krishnan, Class. Quantum Grav. **25**, 114005 (2008).
8. E. Gourgoulhon & J. L. Jaramillo, New Astron. Rev. **51**, 791 (2008).
9. S. A. Hayward, Adv. Sci. Lett. **2**, 205 (2009).
10. E. Schnetter & B. Krishnan, Phys. Rev. **D73**, 021502 (2006).
11. I. Ben-Dov, Phys. Rev. **D75**, 064007 (2007).
12. I. Bengtsson & J. M. M. Senovilla, Phys. Rev. **D79**, 024027 (2009).
13. S. A. Hayward, Class. Quantum Grav. **15**, 3147 (1998).
14. S. A. Hayward, R. Di Criscienzo, L. Vanzo, M. Nadalini & S. Zerbini Class. Quantum Grav. **26**, 062001 (2009) .
15. S. A. Hayward, Involute, minimal, outer and increasingly trapped spheres, arXiv:0905.3950.
16. S. A. Hayward, Involute, minimal, outer and increasingly trapped surfaces, arXiv:0906.2528.
17. S. A. Hayward, Phys. Rev. Lett. **81**, 4557 (1998).
18. S. Mukohyama & S. A. Hayward, Class. Quantum Grav. **17**, 2153 (2000).
19. S. A. Hayward, Class. Quantum Grav. **17**, 1749 (2000).
20. S. A. Hayward, Class. Quantum Grav. **18**, 5561 (2001).
21. S. A. Hayward, Phys. Lett. **A294**, 179 (2002).
22. S. A. Hayward, Phys. Rev. Lett. **93**, 251101 (2004).
23. S. A. Hayward, Phys. Rev. **D70**, 104027 (2004).
24. S. A. Hayward, Phys. Rev. **D74**, 104013 (2006).
25. S. A. Hayward, Class. Quantum Grav. **23**, L15 (2006).
26. S. A. Hayward, Class. Quantum Grav. **24**, 923 (2007).
27. H. Bray, S. A. Hayward, M. Mars & W. Simon, Comm. Math. Phys. **272**, 119 (2007).
28. S. A. Hayward, Phys. Rev. **D78**, 044027 (2008).

ANALYTICAL SPECTRA OF RGW AND ITS INDUCED CMB ANISOTROPIES AND POLARIZATION

YANG ZHANG*

Key Laboratory for Researches in Galaxies and Cosmology, CAS
Center for Astrophysics
Department of Astronomy
University of Science and Technology of China
Hefei, Anhui 230026, China
**E-mail: yzh@ustc.edu.cn*

We present the results from a series of analytical studies on relic gravitational waves (RGW) and the anisotropies and polarization of cosmic background radiation (CMB). The analytical spectrum $h(\nu)$ of RGW shows the influences of the dark energy, neutrino free-streaming (NFS), quantum chromodynamical (QCD) phase transition, e^+e^- annihilation, and inflation. Various possible detections of, and constraints on RGW are examined. The resulting $h(\nu)$ is then used to analytically calculate the spectra C_l^{XX} of CMB anisotropies and polarizations. The influences of the inflation index, NFS, and baryon on C_l^{XX} are demonstrated. We also extend analytical calculation of C_l^{XX} to the case with reionization. The explicit dependence of C_l^{XX} on the optical depth is obtained, whose degeneracies with the amplitude and index of RGW are shown, and the consequential implications in extracting RGW signal from observed C_l^{XX} are explored.

Keywords: relic gravitational waves; cosmic microwave radiation background.

1. Introduction

As a major prediction of General Relativity, the existence of gravitational waves has not yet been directly detected. On other hand, inflationary models predict, among other things, a stochastic background of RGW generated during inflation.[1–3] Therefore, direct detections of RGW will play a double role in relativity and in cosmology. For ongoing or planning gravitational wave detections,[4–13] the spectrum of RGW is one of the major targets. RGW is basically determined by the behavior of expansion of the Universe. First of all, inflation will determine the RGW spectrum, including the amplitude, the spectral index, and the running spectral index[14] as the initial condition. Besides, the existence of cosmic dark energy will modify the overall amplitude of RGW.[15] Several other physical processes in the expanding universe will also modify RGW, such as neutrino free-streaming,[16,17] QCD phase transition, and e^+e^- annihilation.[18,19] These will have impacts on the outcome of RGW. A compilation of these modifications upon RGW will be presented here, and examinations on possible detections by the current major detectors will be given.

The observations on CMB[20–22] are in good agreement with the cosmological picture of a spatially flat universe with nearly scale-invariant primordial adiabatic perturbations predicted by inflationary models. Both density perturbations and RGW will contribute to the CMB anisotropies and polarizations.[2,23–25] As a special feature, RGW can give rise to magnetic type of polarizations C_l^{BB} of CMB,[26,27] and this could provide a distinguished channel to directly detect RGW of very long wavelength comparable to the Hubble radius $\sim 1/H_0$. The spectra of CMB anisotropies and polarizations generated by RGW have been computed.[28,29] Analytical studies have been made.[30–32] Compared with numeric results, these approximate, analytic spectra have relatively large errors and are valid only in a limited range of $l \leq 300$. The temperature spectrum C_l^{TT} and the cross spectrum C_l^{TE} were not given. Our recently work[33] has improved these results, obtained all the four spectra C_l^{XX} valid over an extended range $l \leq 600$. Moreover, the RGW spectrum modified by NFS has been used to calculate C_l^{XX}, so that the NFS effect upon CMB is also demonstrated.

Reionization is a cosmological process around the redshift $z = (20 \sim 6)$, and is essential in shaping the profiles of CMB spectra on large scales, only secondary to the decoupling. On large angular scales, C_l^{XX} exhibit reionization bumps as distinguished signatures. The reionization parameters are entangled with other parameters, thus biasing our interpretation of CMB.[34] For three simple reionization models, we have obtained the analytical, reionized C_l^{XX},[35] which consist of two parts, from the decoupling and reionization, respectively. The optical depth κ_r of reionization appears explicitly in the expressions of C_l^{XX}. This result is quite useful in investigating the impact of reionization and in analyzing the degeneracy of κ_r with the amplitude and index of RGW.[35]

First I shall present briefly the result of the analytical spectrum of RGW. Discussions on possible detections and constraints are given. Next, using the RGW spectrum, I shall sketch the analytical calculation of C_l^{XX}. The dependence of C_l^{XX} upon inflation, NFS, and baryons are demonstrated. Finally I shall extend the calculation to the reionized C_l^{XX} for three models.

The reionization bumps in C_l^{EE} and C_l^{BB} are examined. The degeneracy of κ_r with the amplitude and index of RGW are explored. These results have substantially updated what were reviewed in Ref.36.

2. Analytical Solution of RGW in Expanding Universe

A spatially flat Robertson-Walker spacetime has a metric

$$ds^2 = a^2(\tau)[d\tau^2 - (\delta_{ij} + h_{ij})dx^i dx^j], \tag{1}$$

where h_{ij} is transverse-traceless $\partial_i h^{ij} = 0$, $\delta^{ij} h_{ij} = 0$, representing RGW, τ is the conformal time. The scalar factor $a(\tau)$ is given for the following various stages. The inflationary stage

$$a(\tau) = l_0 \mid \tau \mid^{1+\beta}, \quad -\infty < \tau \leq \tau_1, \tag{2}$$

where $1 + \beta < 0$, and $\tau_1 < 0$. The special case of $\beta = -2$ is the de Sitter expansion of inflation. The reheating stage

$$a(\tau) = a_z(\tau - \tau_p)^{1+\beta_s}, \quad \tau_1 \leq \tau \leq \tau_s, \tag{3}$$

allowing a general reheating epoch.[1,17] The radiation-dominated stage

$$a(\tau) = a_e(\tau - \tau_e), \quad \tau_s \leq \tau \leq \tau_2. \tag{4}$$

The matter-dominated stage

$$a(\tau) = a_m(\tau - \tau_m)^2, \quad \tau_2 \leq \tau \leq \tau_E, \tag{5}$$

where τ_E is the time when the dark energy density ρ_Λ is equal to the matter energy density ρ_m. The redshift z_E at the time τ_E is given by $1 + z_E = (\frac{\Omega_\Lambda}{\Omega_m})^{1/3}$. The accelerating stage (up to the present time τ_H)

$$a(\tau) = l_H|\tau - \tau_a|^{-\gamma}, \quad \tau_E \leq \tau \leq \tau_H, \tag{6}$$

where γ is a parameter, and $\gamma = 1.0$ for $\Omega_\Lambda = 1$ and $\Omega_m = 0$. By numerically solving the Friedman equation, the expression of (6) gives a good fitting with $\gamma = 1.05$ for $\Omega_\Lambda = 0.7$, $\gamma = 1.048$ for $\Omega_\Lambda = 0.75$, respectively.[15,17]

There are ten constants in $a(\tau)$, except β and β_s that are imposed upon as the model parameters. By the continuity of $a(\tau)$ and $a(\tau)'$ at the four given joining points τ_1, τ_s, τ_2, and τ_E, one can fix only eight constants, the other two can be fixed by the overall normalization of a and by the observed Hubble constant. Specifically, we put $a(\tau_H) = l_H$, i.e.

$$|\tau_H - \tau_a| = 1, \tag{7}$$

and the constant l_H is fixed by the following calculation

$$\frac{1}{H_0} \equiv \left(\frac{a^2}{a'}\right)_{\tau_H} = \frac{l_H}{\gamma}. \tag{8}$$

The time instants τ_1, τ_2, τ_s, and τ_E are specified by $a(\tau_H)/a(\tau_E) = 1.33$, $a(\tau_E)/a(\tau_2) = 3454$, $a(\tau_2)/a(\tau_s) = 10^{24}$, and $a(\tau_s)/a(\tau_1) = 300$.[15] The physical wavelength λ is related to the comoving wave number k by $\lambda \equiv \frac{2\pi a(\tau)}{k}$. The wave number corresponding to the present Hubble radius is $k_H = 2\pi a(\tau_H)/l_H = 2\pi$.

The gravitational wave wave equation for a fixed wave vector $\mathbf{k}$ and a fixed polarization state + or ×, reduces to

$$h_k^{''} + 2\frac{a'}{a}h_k^{'} + k^2 h_k = 0. \tag{9}$$

For the scale factor $a(\tau) \propto \tau^\alpha$, the solution is a linear combination of Bessel functions

$$h_k(\tau) = \tau^{\frac{1}{2}-\alpha}\left[C_1 J_{\alpha-\frac{1}{2}}(k\tau) + C_2 N_{\alpha-\frac{1}{2}}(k\tau)\right], \tag{10}$$

where the coefficients C_1 and C_2 for each stage are determined by the continuity of h_k and of h_k' at the joining points τ_1, τ_s, τ_2 and τ_E. From the mode $h_k(\tau)$ one defines the spectrum $h(k,\tau)$ of RGW

$$h(k,\tau) = \frac{\sqrt{2}}{\pi}k^{3/2}|h_k(\tau)|, \tag{11}$$

and $h(\nu,\tau_H)/\sqrt{2}$ is just the characteristic amplitude, denoted by $h_c(f)$ in Ref.40, and the spectral energy density

$$\Omega_g(k) = \frac{\pi^2}{3}h^2(k,\tau_H)\left(\frac{k}{k_H}\right)^2. \tag{12}$$

The analytical $h_k(\tau)$ is completely fixed, once the initial condition during the inflation is given. For a given wave number k, the corresponding wave crossed over the horizon at a time τ_i, i.e. when the wave length $\lambda_i = 2\pi a(\tau_i)/k$ was equal to the Hubble radius $1/H(\tau_i)$. The initial condition is taken to be of a form[14]

$$h(k,\tau_i) = \Delta_{\Re}(k_0) r^{\frac{1}{2}} (\frac{k}{k_0})^{2+\beta+\frac{1}{4}\alpha_t \ln{(k/k_0)}}. \tag{13}$$

In this expression, several quantities are explained in the following. k_0 is a pivot wavenumber, corresponding to a physical wavenumber $k_0^p = k_0/a(\tau_H) = 0.002$ Mpc^{-1} by WMAP5.[37] $\Delta_{\Re}(k_0)$ is the amplitude of the curvature spectrum at $k = k_0$, and $\Delta^2_{\Re}(k_0) = (2.41 \pm 0.11) \times 10^{-9}$ by

WMAP5,[38] and $\Delta^2_{\Re}(k_0) = (2.445 \pm 0.096) \times 10^{-9}$ by WMAP5+BAO+SN Mean.[37] r is the tensor/scalar ratio,

$$r \equiv \frac{\Delta^2_h(k_0)}{\Delta^2_{\Re}(k_0)}, \tag{14}$$

with $\Delta^2_h(k_0) \equiv h^2(k_0, \tau_i)$ represents the amplitude normalization of RGW. α_t is the running index, reflecting an extra bending from the simple power-law spectrum. When $\alpha_t = 0$, Eq.(13) reduces to the power-law form.

The index β is most influential in determining the spectrum. Taking $r = 0.37$, $\alpha_t = 0$, and $\gamma = 1.05$, Fig. 1 shows the analytical spectra $h(k, \tau_H)$ and $\Omega_g(\nu)$ for three inflationary models with $\beta = -1.8, -1.9$, and -2.0, and $\beta_s = 0.598, -0.552$, and -0.689, respectively. A larger β yields a higher spectrum, increasing with ν.

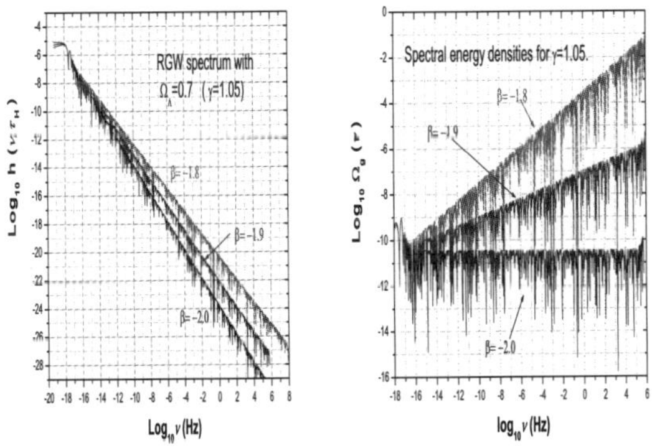

Fig. 1. Left: For $r = 0.37$ and $\alpha_t = 0$, the analytic $h(\nu, \tau_H)$ for $\beta = -1.8, -1.9, -2.0$, respectively. Right: the spectral energy density $\Omega_g(\nu)$.

The influence of dark energy Ω_Λ upon RGW is through γ. The amplitude in the model $\gamma = 1.06$ is about $\sim 50\%$ greater than that in the model $\gamma = 1.05$. This provides a new possible channel to detect the dark energy.

The running index α_t also changes the shape of $h(\nu, \tau_H)$ substantially. Fig. 2 demonstrates this for the model of $r = 0.22$ and $\beta = -2.015$. A greater α_t tilts the spectrum increasingly with frequency ν. The modification is quite drastic in high-frequency range. For instance, in going from $\alpha_t = -0.01$ to $\alpha_t = 0.01$, the amplitude of $h(\nu, \tau_H)$ gets enhanced by 3 orders of magnitudes at $\nu \simeq 10^{-2}$ Hz falling the range for LISA,[8,9] 5 orders at $\nu \simeq 10^2$ Hz for LIGO,[4] 6 orders at $\nu \simeq 10^4$ Hz for MAGO,[11] and 9 orders at $\nu \simeq 10^9$Hz for the waveguide[12] and the Gauss beam.[39]

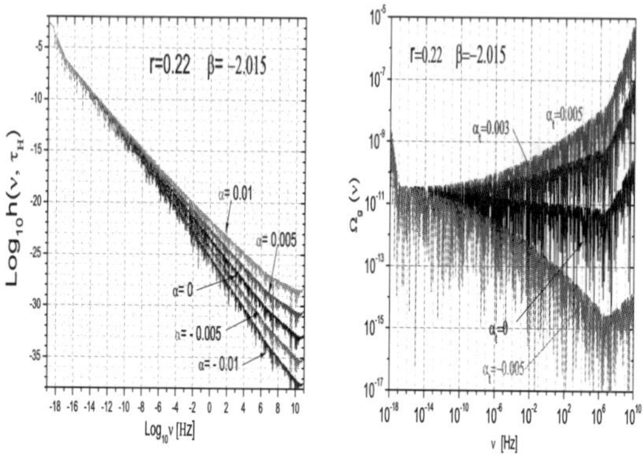

Fig. 2. Left: $h(\nu, \tau_H)$ for various values of the running index α_t. Right: $\Omega_g(\nu)$.

Let us examine possible detections and constraints by the major detectors. Left panel in Figure 3 gives the comparison of the sensitivity curves of LIGO[4] and Advanced LIGO[5] with the theoretical spectra of RGW for various parameters $(r,\ \beta,\ \alpha_t)$, in the frequency range $(10^1, 10^4)$ Hz. The amplitude per root Hz, $h(\nu, \tau_H)/\sqrt{\nu}$, has been used, in order to compare to the strain sensitivity $\tilde{h}_f(\nu)$[40] of the detectors. It is seen that, for $r = 0.55$ and $\beta = -1.956$, the LIGO I SRD[4] has already put a constraint on the running index: $\alpha_t \leq 0.013$. Advanced LIGO will be able to detect the RGW from models with $r = 0.55$ and $\beta > -1.956$ and $\alpha_t > 0$. Right panel in Fig.3 compares the theoretical rms spectrum in the band $\Delta\nu$,[1,40]

$$h(\nu, \tau_H, \Delta\nu) = h(\nu, \tau_H)\sqrt{\frac{\Delta\nu}{\nu}}. \tag{15}$$

with the LISA sensitivity curve[8,9] in a range $(10^{-7}, 10^0)$ Hz, where one year observation time is assumed, corresponding to a frequency bin $\Delta\nu \simeq 3\times10^{-8}$Hz around each frequency. It is seen that LISA will be quite effective in detecting RGW around a range of $(10^{-6}, 10^{-1})$ Hz, broader than LIGO. (r, β, α_t) have a degeneracy, since a larger value of each of them tends to enhance the amplitude of $h(\nu, \tau_H)$. Detectors operating over a broader frequency range, such as LISA, will have a better chance to break the degeneracy.

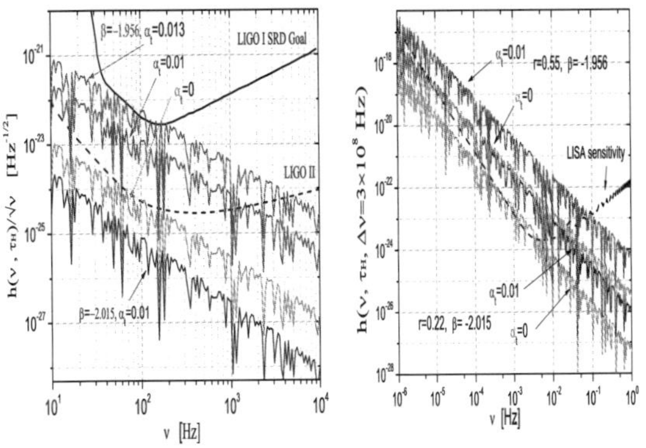

Fig. 3. Left: Comparisons of the spectra with the sensitivity of the LIGO I SRD [4] and of Adv LIGO.[5] Right: Comparison of the rms spectrum with the LISA sensitivity.[9]

For lack of direct detection of RGW, cosmological considerations can be effective in constraining. Often used is the present energy density parameter

$$\Omega_{gw} = \int_{\nu_{low}}^{\nu_{upper}} \Omega_g(\nu)\frac{d\nu}{\nu}, \tag{16}$$

where the cutoffs of frequencies can be taken to be $\nu_{low} \simeq 2\times10^{-18}$ Hz and $\nu_{upper} \simeq 10^{10}$ Hz.[17] The energy density of RGW should not be too large to significantly affect the outcome of the Big Bang nucleogenesis process (BBN). Measured abundances of light-element constrain $h_0^2\Omega_{gw} < 7.8\times10^{-6}$.[40,41] Just recently, LIGO has given a constraint on the energy density $\Omega_g(\nu) < 6.9\times10^{-6}$ at 95% confidence around 100Hz.[42] Left panel in Fig.4 shows the α_t-dependence of Ω_{gw} for various values of parameters. Adopting the BBN bound, the constraint on α_t is found to be $\alpha_t \leq 0.0014$ for $r = 0.55$ and $\beta = -1.956$, $\alpha_t \leq 0.0056$ for $r = 0.22$ and $\beta = -2.015$, and $\alpha_t \leq 0.0077$ for $r = 0.001$ and $\beta = -2.0$, respectively. BBN puts a constraint $\alpha_t < 0.008$ for any cosmological model with reasonable parameters.

Observations of millisecond pulsars can serve as a gravitational wave detector.[43] By analyzing the uncertainty ϵ in the arrival timing of pulses for a duration T of observation, the pulsar will be sensitive to gravitational waves $h(\nu,\tau_H) \sim \epsilon/T$ with $\nu \sim 1/T$. For PSR B1855+09, a bound has been:[44] $\Omega_g(\nu_*)h_0^2 < 4.8\times10^{-9}(\nu/\nu_*)^2$ for $\nu > \nu_*$, where $\nu_* = 4.4\times10^{-9}$ Hz. Applying this bound to compare with the calculated $\Omega_g(\nu)$, we obtain the constraint on RGW, shown in the right panel of Fig.4. For the parameters $r = 0.55$ and $\beta = -1.956$, the pulsar detector puts a constraint $\alpha_t < 0.01$,

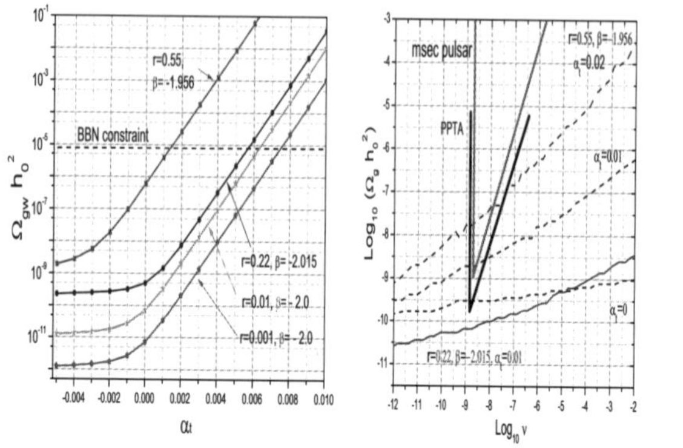

Fig. 4. Left: The α_t-dependence of Ω_{gw} for various r and β, and the constraint from BBN. Right: The constraint from millisecond pulsars[44] and PPTA.[45]

which is less stringent than the BBN constraint. As an extension of this technique, Parkes Pulsar Timing Array (PPTA)[13,45] consisting of a sample of 20 millisecond pulsars will have a chance to detect RGW of $r = 0.55$, $\beta = -1.956$, and $\alpha_t = 0$.

Let us demonstrate the modifications of RGW by neutrino free-streaming, QCD transition and e^+e^- annihilation.

At a temperature $T \sim 2$ MeV, cosmic neutrinos decoupled from electrons and photons, and started free-streaming in space. This will give rise to an inhomogeneous term π_k as a source of Eq.(9),

$$h_k'' + 2\frac{a'}{a}h_k' + k^2 h_k = 16\pi G a^2 \pi_k, \tag{17}$$

causing a damping effect on RGW.[16,17] We have taken into account of the effects of the accelerating Universe, used a perturbation method to solve Eq.(17) by iteration, valid for an arbitrary wavelength. Left panel in Fig.5 shows the damping effect of NFS on $\Omega_g(\nu)$ for the model $\beta = -2.02$ up to the first order approximation to Eq.(17). The effect is pronounced in the range $(10^{-16} \sim 10^{-10})$ Hz, where the amplitude of $\Omega_g(\nu)$ drops visibly by a factor of $\sim 36\%$, leaving observable imprints on CMB spectra.[17]

According to QCD, around $T \sim 190$ MeV in the early universe, a phase transition occurs, during which quarks are combined into hadrons, causing a change in the expansion behavior of the universe. The e^+e^- annihilation around $T \sim 0.5$ Mev also has a similar effect. Right panel in Fig.5 shows that the transition causes a reduction of the amplitude of RGW by $\sim 20\%$

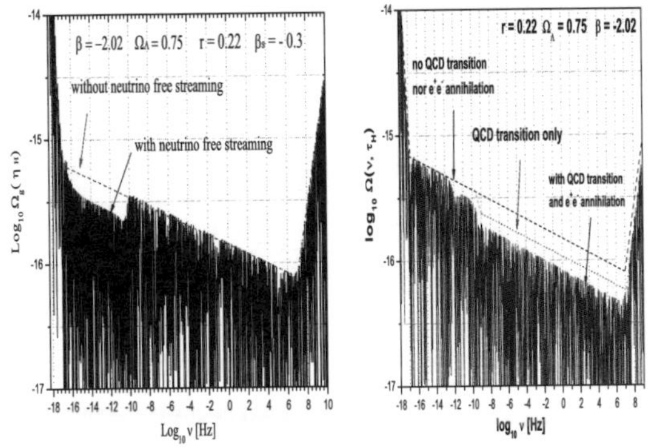

Fig. 5. Left: NFS reduces $\Omega_g(\nu)$ in the range $10^{-17} \sim 10^{-10}$Hz. Right: $\Omega_g(\nu)$ is reduced by QCD transition and e^+e^- annihilation.

in the range $> 10^{-9}$ Hz, and the annihilation causes a reduction $\sim 10\%$ in the range $> 10^{-12}$ Hz.[19] The reductions are in high frequencies, and will not be detected by current observations of CMB on large angular scales.

3. Analytical Spectra of CMB Anisotropies and Polarization

The Boltzmann equation of the CMB photon has the formal solution:[32,33]

$$\xi(\tau,\mu) = \int_0^\tau h'(\tau')e^{-\kappa(\tau,\tau')}e^{ik\mu(\tau'-\tau)}d\tau', \tag{18}$$

$$\beta(\tau,\mu) = \int_0^\tau G(\tau')q(\tau')e^{-\kappa(\tau,\tau')}e^{ik\mu(\tau'-\tau)}d\tau', \tag{19}$$

where β_k is the linear polarization, $\alpha_k \equiv \xi_k - \beta_k$ is the temperature anisotropy, $\mu = \cos\theta$, q is the differential optical depth, and $G_k(\tau) = \frac{3}{16}\int_{-1}^{1} d\mu'[(1+\mu'^2)^2\beta_k - \frac{1}{2}(1-\mu'^2)^2\xi_k]$. $\kappa(\tau',\tau) \equiv \int_\tau^{\tau'} q d\tau = \kappa(\tau) - \kappa(\tau')$ with $\kappa(\tau) \equiv \kappa(\tau_H,\tau)$ being the optical depth, such that $q(\tau) = -d\kappa(\tau_H,\tau)/d\tau$. ξ and β are usually expressed in terms of their Legendre components

$$\xi_l(\tau_H) = i^l \int_0^{\tau_H} e^{-\kappa(\tau)}h'(\tau)j_l(k(\tau-\tau_H))d\tau, \tag{20}$$

$$\beta_l(\tau_H) = i^l \int_0^{\tau_H} G(\tau)V(\tau)\, j_l(k(\tau-\tau_H))d\tau, \tag{21}$$

both being evaluated at the present time τ_H, where

$$V(\tau) = q(\tau)e^{-\kappa(\tau)} \tag{22}$$

is the visibility function for the decoupling, peaked around the decoupling time τ_d with a width $\Delta\tau_d$. It can be fitted by two pieces of half gaussian function[33]

$$V(\tau) = \begin{cases} V(\tau_d)\exp\left(-\frac{(\tau-\tau_d)^2}{2\Delta\tau_{d1}^2}\right), & (\tau \le \tau_d), \\ V(\tau_d)\exp\left(-\frac{(\tau-\tau_d)^2}{2\Delta\tau_{d2}^2}\right), & (\tau > \tau_d), \end{cases} \tag{23}$$

where $\tau_d \simeq 0.0707$ corresponding to the redshift $z_d \simeq 1100$, $\Delta\tau_{d1} = 0.00639$, $\Delta\tau_{d2} = 0.0117$, and $(\Delta\tau_{d1} + \Delta\tau_{d2})/2 = \Delta\tau_d$ is the thickness of the decoupling. The coefficient $V(\tau_d)$ in Eq.(23) will be fixed by

$$\int_0^{\tau_H} V(\tau)d\tau = 1. \tag{24}$$

Substituting Eq.(23) into Eq.(21), after some treatment of the time integration, yields an approximate solution valid up to the second order of a small $1/q^2$ in the tight coupling limit

$$\beta_l(\tau_H) = \frac{1}{17}\ln\frac{20}{3} i^l \Delta\tau_d D(k) h'(\tau_d) j_l(y_d), \tag{25}$$

where $y_d \equiv k(\tau_d - \tau_H)$, $h'(\tau_d)$ is the time derivative of RGW at τ_d, and

$$D(k) = \frac{1}{2}[e^{-c(k\Delta\tau_{d1})^b} + e^{-c(k\Delta\tau_{d2})^b}] \tag{26}$$

is a fitting formula, with $c \simeq 0.6$ and $b \simeq 0.85$. The integrand in (20) contains a factor $e^{-\kappa(\tau)}$, which can be treated approximately as

$$e^{-\kappa(\tau)} \simeq \begin{cases} 0 & (\tau < \tau_d), \\ 1 & (\tau_d < \tau < \tau_H). \end{cases} \tag{27}$$

Substituting Eq.(27) into Eq.(20) yields

$$\xi_l(\tau_H) = -i^l h(\tau_d) j_l(y_d). \tag{28}$$

From Eqs.(25) and (28) follows the temperature anisotropies[33]

$$\alpha_l(\tau_H) = -i^l j_l(y_d)\left[h(\tau_d) + \frac{1}{17}\ln\frac{20}{3}\Delta\tau_d D(k) h'(\tau_d)\right]. \tag{29}$$

From α_l and β_l, one calculates C_l^{XX}[27]

$$C_l^{TT} = \frac{1}{8\pi}\frac{(l+2)!}{(l-2)!}\int k^2 dk P_{Tl}^2(y_d)\left|h(\tau_d) + \frac{1}{17}\ln\frac{20}{3}\Delta\tau_d D(k) h'(\tau_d)\right|^2 \tag{30}$$

$$C_l^{EE} = \frac{1}{16\pi}\left(\frac{1}{17}\ln\frac{20}{3}\right)^2 \int k^2 dk P_{El}^2(y_d)\Delta\tau_d^2 D^2(k)\left|h'(\tau_d)\right|^2, \tag{31}$$

$$C_l^{BB} = \frac{1}{16\pi}\left(\frac{1}{17}\ln\frac{20}{3}\right)^2 \int k^2 dk P_{Bl}^2(y_d)\Delta\tau_d^2 D^2(k)\left|h'(\tau_d)\right|^2, \quad (32)$$

$$\begin{aligned} C_l^{TE} = & \frac{1}{136\sqrt{2}\pi}\ln\frac{20}{3}\sqrt{\frac{(l+2)!}{(l-2)!}}\int k^2 dk P_{Tl}(y_d)P_{El}(y_d) \\ & \frac{1}{2}\left\{\left[-h(\tau_d)-\frac{1}{17}\ln\frac{20}{3}\Delta\tau_d D(k)h'(\tau_d)\right]h'^*(\tau_d)\right. \\ & \left.+\left[-h^*(\tau_d)-\frac{1}{17}\ln\frac{20}{3}\Delta\tau_d D(k)h'^*(\tau_d)\right]h'(\tau_d)\right\}\Delta\tau_d D(k), (33) \end{aligned}$$

where the projection factors P_{Tl}, P_{El}, P_{Bl} are defined in Ref 33.

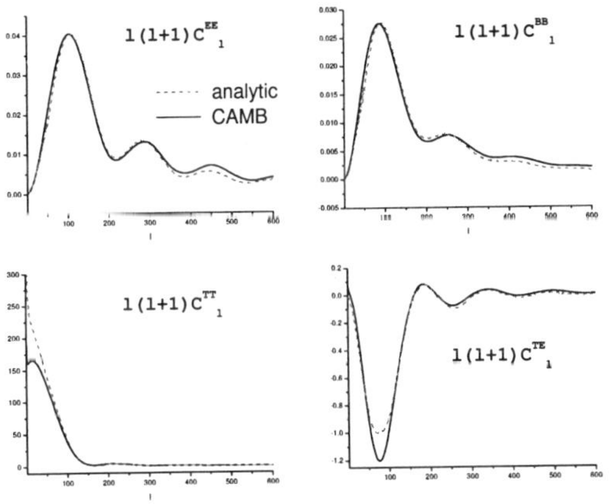

Fig. 6. The analytic C_l^{XX} generated by RGW are compared with the numeric ones.[29]

Fig. 6 shows the result, for the model with $r = 0.37$, $\beta = -2.02$, $\Omega_\Lambda = 0.75$, the baryon density $\Omega_b = 0.045$, the neutrino species $N_\nu = 3$. In the range $l \leq 600$ covering the first three peaks, the analytic C_l^{EE} and C_l^{BB} agree very well with the numerical ones from CAMB,[29] and the error is only $\sim 2\%$. The spectra C_l^{TT} and C_l^{TE} are also obtained. Comparing with Refs.31,32, the new result not only extends the range of validity from $l \lesssim 300$ to $l \lesssim 600$, but also improves accuracy substantially.

Fig. 7 shows the effect of NFS on C_l^{XX}. The third peak of C_l^{TT} is reduced by $\sim 25\%$ and the fourth peak by $\sim 35\%$. Similar modifications also occur in the spectra C_l^{EE}, C_l^{BB}, and C_l^{TE}. Besides, NFS causes a slight shift $\Delta l \sim 4$ of the peak locations of C_l^{XX} to larger l. These effects reveal important information of cosmic neutrinos in the Universe.[46]

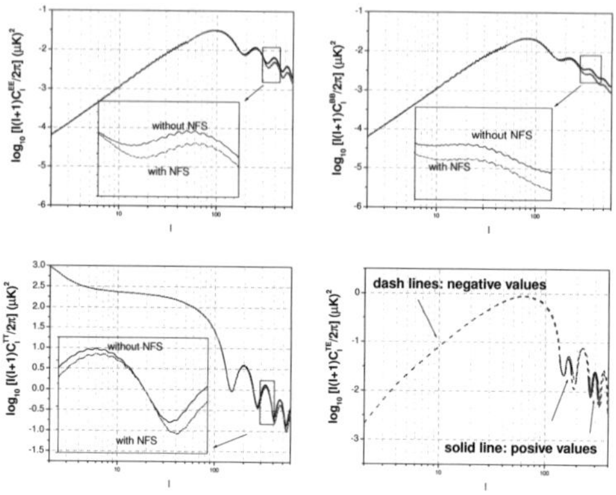

Fig. 7. NFS reduces the amplitudes by $\sim 35\%$ and shifts the peaks.

The CMB spectra depend sensitively on the index β. Fig. 8 shows that a larger index β yields higher amplitudes of C_l^{EE} and C_l^{BB}.

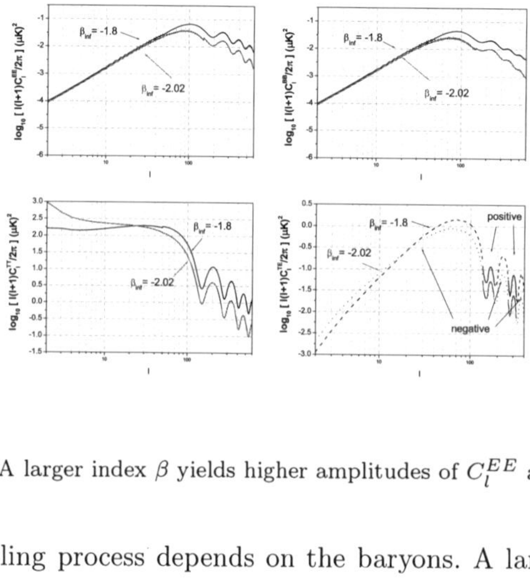

Fig. 8. A larger index β yields higher amplitudes of C_l^{EE} and C_l^{BB}.

The decoupling process depends on the baryons. A larger Ω_b yields a larger decoupling time τ_d and and a smaller decoupling width $\Delta\tau_d$. Fig. 9 shows that a larger Ω_b yields lower C_l^{EE} and C_l^{BB}, agreeing with the previous calculations.[32]

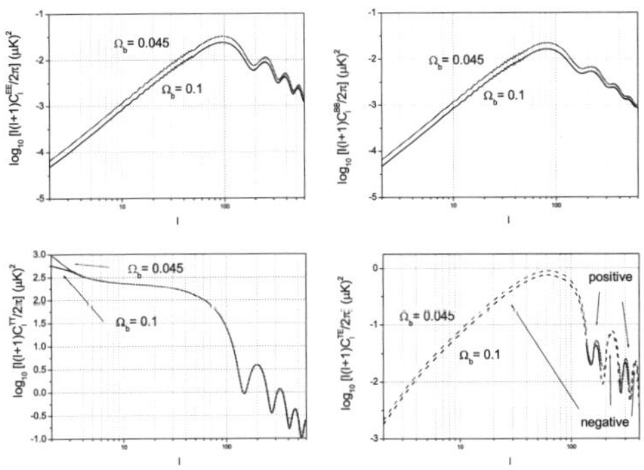

Fig. 9. A larger baryon fraction Ω_b yields lower amplitudes.

4. Analytic Spectra C_l^{XX} with Re-ionization

When the reionization is included, $V(\tau)$ will have two peaks, one for decoupling and another for reionization, as shown in the Panel (d) in Fig.10. The time integration of Eq.(24) is split into two parts

$$\int_0^{\tau_{split}} V_d(\tau)d\tau + \int_{\tau_{split}}^{\tau_H} V_r(\tau)d\tau = 1, \tag{34}$$

where $V_d(\tau)$ is for the decoupling, $V_r(\tau)$ for the reionization, and τ_{split} is some point between decoupling and reionization with $V(\tau_{split}) \simeq 0$. Eq.(34) has a physical interpretation: if more CMB photons are last scattered around reionization, less will be last scattered around decoupling. We study three homogeneous reionization models with a respective ionization fraction $X_e(\tau)$.[35]

The first is the sudden reionization model with

$$X_e(\tau) = \begin{cases} 0, & \text{for } \tau < \tau_r, \\ 1, & \text{for } \tau \geq \tau_r, \end{cases} \tag{35}$$

where $\tau_r = 0.915$ corresponds to a redshift $z_r = 11$. There are accumulating evidences that reionization is an extended process, from $z \simeq 6$ up to $z \sim 11$,

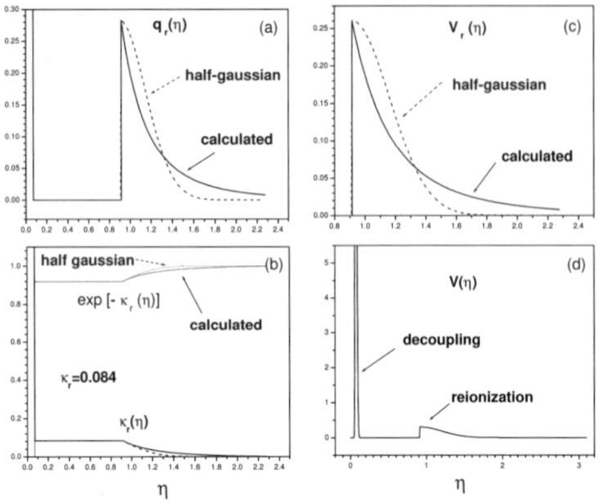

Fig. 10. The sudden reionization model and its fitting. Panel (d) shows $V(\tau)$, including both reionization and decoupling.

even up to $z \sim 20$.[20–22] One extended model is the τ-linear reionization

$$X_e(\tau) = \begin{cases} 0, & \text{for } \tau < \tau_{r1} \\ \frac{\tau - \tau_{r1}}{\tau_{r2} - \tau_{r1}}, & \text{for } \tau_{r1} < \tau < \tau_{r2}, \\ 1, & \text{for } \tau > \tau_{r2}. \end{cases} \tag{36}$$

where $\tau_{r1} = 0.685$ and $\tau_{r2} = 1.207$, corresponding to $z_{r1} = 20$ and $z_{r2} = 6$, respectively. Another extended model is the z-linear model:[47]

$$X_e(z) = \begin{cases} 0, & \text{for } z > z_{r1} \\ 1 - \frac{z - z_{r2}}{z_{r1} - z_{r2}}, & \text{for } z_{r1} > z > z_{r2}, \\ 1, & \text{for } z \le z_{r2}. \end{cases} \tag{37}$$

Fig. 11 shows $X_e(\tau)$ for three models, from which follow the differential optical depth $q_r(\tau)$,[47] the optical depth $\kappa_r(\tau) = \int_\tau^{\tau_H} q_r(\tau')d\tau'$, and the visibility function $V_r(\tau) = q_r(\tau)e^{-\kappa_r(\tau)}$. The value $\kappa_r \equiv \kappa_r(\tau_b)$ of the optical depth is from τ_H back up to some time before the reionization. WMAP5 + SN + BAO yields $\kappa_r = 0.084 \pm 0.016$,[21,22] which will be used in our computation. For the two extended models, $V_r(\tau)$ can be also fitted by two pieces of half gaussian functions, and by a half piece of Gaussian function for the sudden model.

By applying the same technique as before, one obtains[35]

$$\beta_l(\tau_H) = -\frac{1}{17} \ln \frac{20}{3} i^l \left[a_1(\kappa_r) D_d(k) \Delta\tau_d h'(\tau_d) j_l(y_d) + a_2(\kappa_r) D_r(k) \Delta\tau_r h'(\tau_r) j_l(y \right. \tag{38}$$

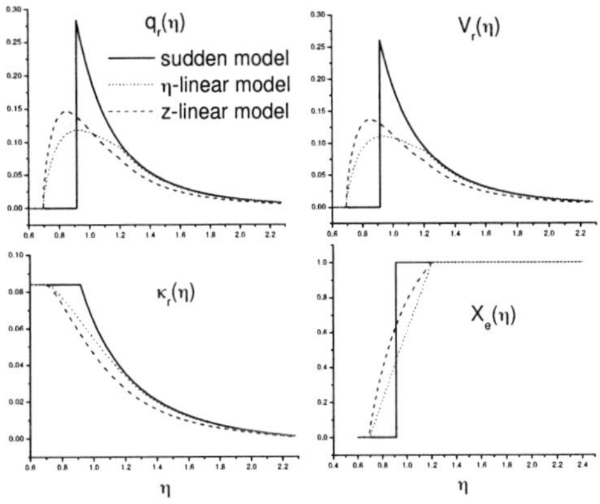

Fig. 11. The three models of reionization with a fixed optical depth $\kappa_r = 0.084$.

where $y_r \equiv k(\tau_r - \tau_H)$, $D_r(k) = \frac{1.4}{2}[e^{-c(k\Delta\tau_{r1})^b} + e^{-c(k\Delta\tau_{r2})^b}]$ for the extended models, $D_r(k) = \frac{1.4}{2}e^{-c(k\Delta\tau_r)^b}$ for the sudden model, the κ_r-dependence coefficients

$$a_1(\kappa_r) = (\frac{10}{17}\ln\frac{20}{3})^{-1}\int_1^\infty \frac{dx}{x(\frac{17}{10}+\frac{3}{10}x)} e^{-(\frac{17}{10}+\frac{3}{10}x)\kappa_r}, \tag{39}$$

$$a_2(\kappa_r) = (\frac{10}{17}\ln\frac{20}{3})^{-1}\int_1^\infty \frac{dx}{x(\frac{17}{10}+\frac{3}{10}x)} \left[1 - e^{-(\frac{17}{10}+\frac{3}{10}x)\kappa_r}\right]. \tag{40}$$

$a_1(\kappa_r)$ is the probability that a polarized photon we perceive was last scattered during decoupling, and $a_2(\kappa_r)$ is that during the time interval from the beginning of reionization up to the present time τ_H. Fig. 12 shows that, $a_1(\kappa_r)$ is a decreasing function of κ_r, and $a_2(\kappa_r)$ is an increasing one. That is, if more CMB photons are scattered during the reionization, the optical depth κ_r acquires a larger value, giving rise to a higher $a_2(\kappa_r)$ and a lower $a_1(\kappa_r)$. The $a_1(\kappa_r)$ part will give rise to the primary peaks, the $a_2(\kappa_r)$ part will yield the reionization bumps of C_l^{EE} and C_l^{BB}.

The factor $e^{-\kappa(\tau)}$ in Eq.(20) can be approximated by

$$e^{-\kappa(\tau)} \simeq \begin{cases} 0 & (\tau < \tau_d); \\ e^{-\kappa_r} & (\tau_d < \tau < \tau_r); \\ 1 & (\tau_r < \tau < \tau_H). \end{cases} \tag{41}$$

By integration, one obtains

$$\xi_l(\tau_H) = -i^l\left[e^{-\kappa_r}h(\tau_d)j_l(y_d) + (1-e^{-\kappa_r})h(\tau_r)j_l(y_r)\right], \tag{42}$$

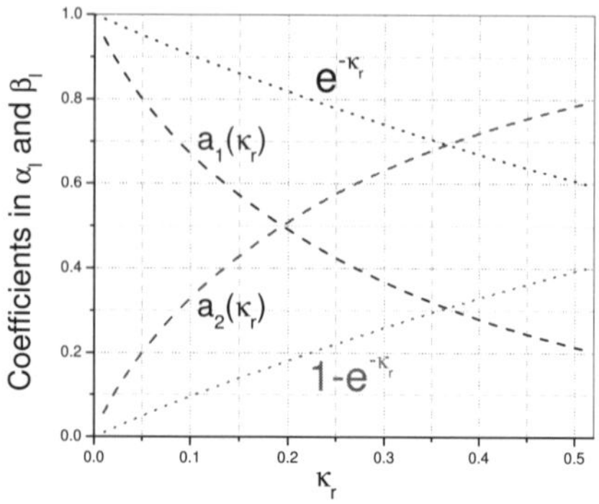

Fig. 12. The coefficients $a_1(\kappa_r)$ and $a_2(\kappa_r)$ for β_l. Also plotted are the coefficients $e^{-\kappa_r}$ and $(1-e^{-\kappa_r})$ for α_l.

and the mode of CMB temperature anisotropies[35]

$$\begin{aligned}\alpha_l(\tau_H) &= -i^l j_l(y_d)\left[e^{-\kappa_r}h(\tau_d) - \frac{1}{17}\ln\frac{20}{3}a_1(\kappa_r)D_d(k)\Delta\tau_d h'(\tau_d)\right] \\ &\quad -i^l j_l(y_r)\left[(1-e^{-\kappa_r})h(\tau_r) - \frac{1}{17}\ln\frac{20}{3}a_2(\kappa_r)D_r(k)\Delta\tau_{dr}h'(\tau_r)\right] \quad (43)\end{aligned}$$

where the first and the second term are generated during the decoupling and the reionization, respectively.

Fig. 13 shows the reionized C_l^{XX} for the three models of reionization. On large scales $l \leq 600$ our analytical C_l^{EE} and C_l^{BB} agree with the numerical ones. C_l^{EE} and C_l^{BB} in the sudden model have lower bumps than those in the extended models.

5. Effects of Reionization on C_l^{XX}

The main effects of reionization on C_l^{XX} are summarized below. 1.The most prominent consequence of reionization is the bump at $l \sim 5$ for C_l^{EE} and C_l^{BB}. Its location is a reflection of the horizon scale at reionization. Fig.14 shows that the bumps of C_l^{EE} and C_l^{BB} are generated by $h'(\tau_r)$, and the primary peaks are due to $h'(\tau_d)$.

2. The bumps depend on the reionization history. Fig. 15 shows that the bump location depends on the the reionization time τ_r, and the height of bump depends on the width $\Delta\tau_r$.

3. C_l^{XX} are also sensitive to κ_r. In particular, κ_r is strongly degenerated with the amplitude A of RGW, and this makes probing the reionization

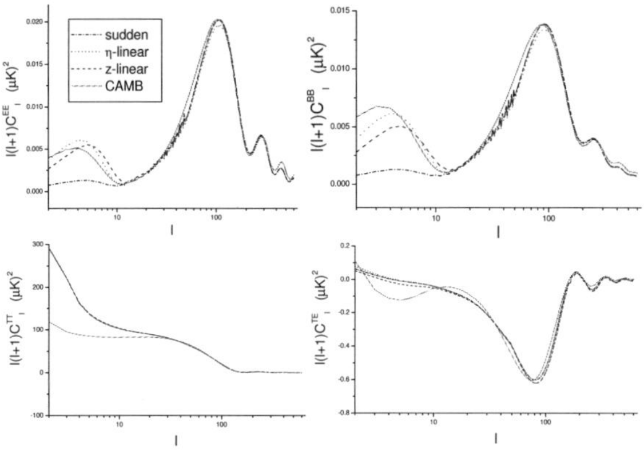

Fig. 13. C_l^{XX} in three reionization models. $\kappa_r = 0.084$ and $r = 0.37$ are taken. The numerical result is from CAMB with the same set of parameters.[29]

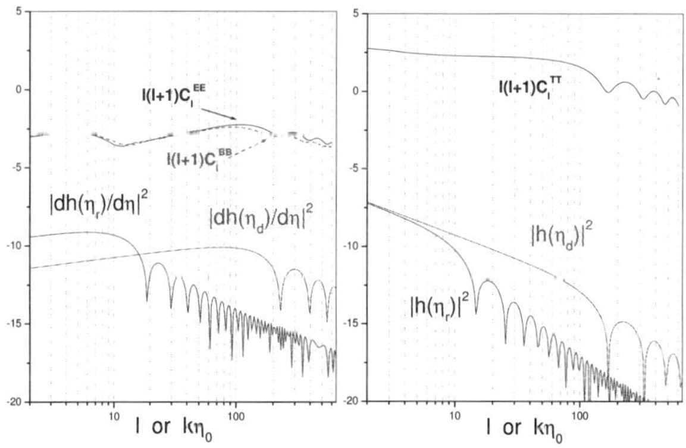

Fig. 14. The profiles of C_l^{XX} with that of RGW . $h'(\tau_r)$ is responsible for the bumps $l \sim 5$, while $h'(\tau_d)$ is responsible the primary peaks for $l \geq 100$.

difficult.[48,49] We have demonstrated in Fig. 16 that the $\kappa_r - A$ degeneracy can be broken. For C_l^{EE} and C_l^{BB}, the ratio

$$\frac{\text{primary peak amplitude}}{\text{bump amplitude}} \propto \left(\frac{a_1(\kappa_r)}{a_2(\kappa_r)}\right)^2 \tag{44}$$

only depends on κ_r, and helps to infer the value of κ_r.

4. A larger index β brings about the same kind of effect on C_l^{XX} as a smaller κ_r does, leading to certain bias in determining κ_r.[49,50]

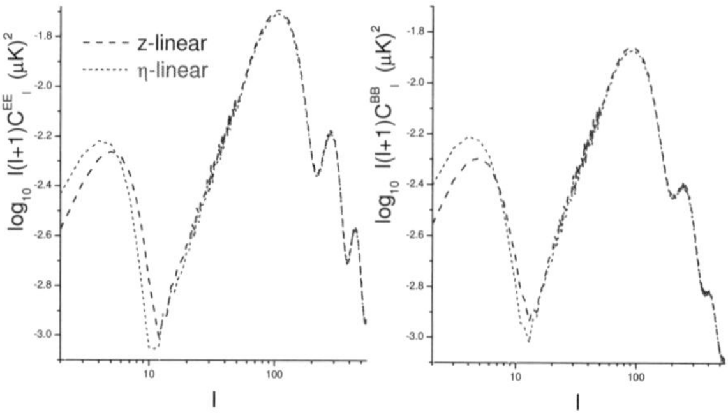

Fig. 15. The two models with different τ_r and $\Delta\tau_r$ yield different bumps at $l \sim 5$.

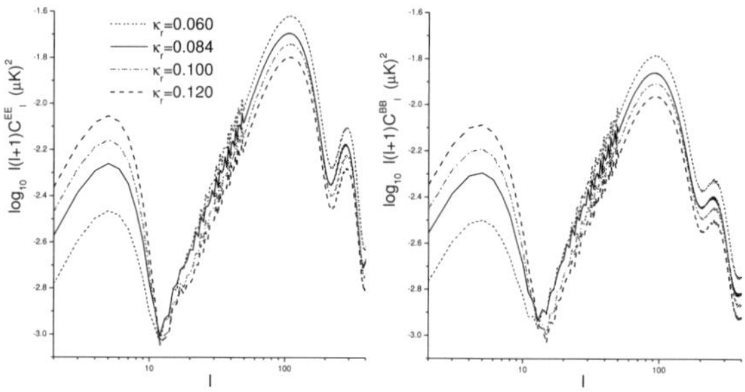

Fig. 16. The $\kappa_r - A$ degeneracy. The plot is made for the z-linear model.

5. The observed C_l^{TE} by WMAP5[22] is negative in a range $l \sim (50, 220)$. In the zero-multipole method,[51] one examines the impact of r upon the zero multipole l_0 around ~ 50, where C_l^{TE} first crosses the value 0 and turns negative. However, besides a change of l_0 by an amount of $\Delta l \sim 4$ due to NFS,[33] the reionization also shifts $\Delta l \sim 20$, increasing with the optical depth κ_r. This kind of shifting has to be incorporated into analysis before one can make an extraction of RGW via C_l^{TE}.

Acknowledgments

Y. Zhang would like to thank the ICGA9 organizers. The research is supported by CNSF No.10773009, SRFDP, and CAS.

References

1. L. P. Grishchuk, Sov.Phys.JETP **40**, 409 (1975); Class.Quant.Grav.**14**, 1445 (1997); in *Lecture Notes in Physics*, Vol.562, p.167, Springer-Verlag, (2001).
2. A. A. Starobinsky, JEPT Lett. **30** 682 (1979); Sov.Astron.Lett.**11**, 133 (1985);
3. V. Rubakov, M. Sazhin and A. Veryaskin, Phys. Lett. B **115**, 189 (1982) ; R. Fabbri and M. Pollock, Phys. Lett. B **125**, 445 (1983) ; L. Abbott and M. Wise, Nuc. Phys. B **237**, 226 (1984); B. Allen, Phys. Rev. D **37**, 2078 (1988).
4. http://www.ligo.caltech.edu/
5. http://www.ligo.caltech.edu/advLIGO
6. A. Freise, *et al.*, Class.Quant.Grav. **22**, S869 (2005).
7. B. Willke, *et al.*, Class.Quant.Grav. **19**, 1377 (2002); http://geo600.aei.mpg.de/
8. http://lisa.nasa.gov/ http://www.lisa.caltech.edu/
9. http://www.srl.caltech.edu/~shane/sensitivity/MakeCurve.html
10. P. Astone, *et al.*, Class.Quant.Grav.**25**, 114028 (2008); P. Astone, *et al.*, Class.Quant.Grav.**25**, 184012 (2008).
11. R. Ballatini, *et al.*, arXiv:gr-qc/0502054, INFN Technical Note INFN/TC-05/05, (2005).
12. A.M. Cruise, Class.Quant.Grav. **17**, 2525 (2000); A.M. Cruise and R.M.J. Ingley, Class.Quant.Grav.**23**, 6185 (2006); M.L. Tong and Y. Zhang, Chin. J. Astron. Astrophys. **8**, 314 (2008).
13. G. Hobbs, Class.Quant.Grav.**25**: 114032, (2008); J.Phys.Conf.Ser.**122**: 012003, (2008); F. A. Jenet, *et al.*, Astrophys.J.**653**, 1571, (2006); R.N. Manchester, AIP Conf. Series. Proc. 983: 584, (2008).
14. M.L. Tong and Y. Zhang, Phys.Rev. D**80**, 084022 (2009).
15. Y. Zhang *et al*, Class. Quant. Grav. **22** 1383 (2005); Class. Quant. Grav. **23** 3783 (2006); Y. Zhang and W. Zhao, Chin. Phys. Lett. **22** 1817 (2005).
16. S. Weinberg, Phys. Rev. D **69**, 023503 (2004); Y. Watanabe and E. Komatsu, Phys. Rev. D **73**, 123515 (2006).
17. H. X. Miao and Y. Zhang, Phys. Rev. D **75**, 104009 (2007).
18. D. J. Schwarz, Mod. Phys. Lett. A**13**, 2771 (1998).
19. S. Wang, Y. Zhang, T. Y. Xia and H. X. Miao, Phys. Rev. D**77**, 104016 (2008).
20. D. N. Spergel, et al., Astrophys.J.Suppl., **148**, 175 (2003); D. N. Spergel, et al., Astrophys.J.Suppl., **170**, 377 (2007).
21. G. Hinshaw, et. al., Astrophys. J. Suppl. **180**, 225 (2009).
22. E. Komatsu, et. al., Astrophys. J. Suppl. **180**, 330 (2009).
23. M. M. Basko and A. G. Polnarev, Mon. Not. R. Astron. Soc. **191**, 207 (1980); A. Polnarev, Sov. Astron. **29**, 6 (1985).
24. J.R. Bond and G. Efstathiou, Astrophys.J. **285**, L45 (1984); R. Crittenden, R. L. Davis and P. J. Steinhardt, Astrophys. J. **417**, L13 (1993); D. Coulson, R. Crittenden and N. Turok, Phys. Rev. Lett. **73**, 2390 (1994).
25. W. Zhao and Y. Zhang, Phys. Rev. D **74**, 043503 (2006).
26. M. Zaldarriaga and U. Seljak, Phys.Rev. D **55**, 1830 (1997).
27. M. Kamionkowski, A. Kosowsky and A. Stebbins, Phys. Rev. D **55**, 7368

(1997).
28. U. Seljak and M. Zaldarriaga, ApJ. **469**, 437 (1996).
29. A. Lewis, A. Challinor and A. Lasenby, ApJ, **538**, 473 (2000). http://lambda.gsfc.nasa.gov/toolbox/tb_camb_form.cfm
30. Y. Zhang, H. Hao and W. Zhao, *ChAA* **29** (2005) 250.
31. J. Pritchard and M. Kamionkowski, Annals. Phys. **318**, 2 (2005).
32. W.Zhao and Y. Zhang, Phys.Rev. D **74**, 083006 (2006).
33. T.Y Xia and Y. Zhang, Phys.Rev. D **78** 123005 (2008).
34. M. Zaldarriaga, Phys.Rev. D**55**, 1822 (1997); M. Zaldarriaga, D. Spergel, U. Seljak, Astrophys. J. **488**, 1 (1997); A. Venkatesan, Astrophys.J. **572**, 15 (2002); K.L. Ng and K.W. Ng, Astrophys.J. **456**, 413 (1996).
35. T.Y. Xia and Y. Zhang, Phys.Rev. D **80** 083002 (2009).
36. Y. Zhang, et al., Int.JMPD Vol **17**, 1105 (2008).
37. E. Komatsu, *et al*, Astrophys. J. Suppl. **180**, 330 (2009).
38. J. Dunkley, *et al*, Astrophys. J. Suppl. **180**, 306 (2009).
39. F.Y. Li, *et al.*, Phys.Rev.D **67**, 104008 (2003); Eur.Phys.J.C **56**,407 (2008); M.L. Tong, Y. Zhang, and F.Y. Li, Phys. Rev. D **78**, 024041 (2008).
40. M. Maggiore, Phys.Rept.**331**, 283 (2000).
41. R.H. Cyburt, B.D. Fields, K.A. Olive, and E. Skillman, Astropart.Phys.**23**, 313 (2005).
42. LIGO Scientific Collaboration & Virgo Collaboration, Nature Letters Vol **460**, 990 (2009).
43. M.V. Sazhin, Sov.Straon. **22**, 36 (1978); S. Detweiler, Astrophys.J.**234**, 1100 (1979); R.W. Hellings and G.S. Downs, Astrophys.J.**265**, L39 (1983).
44. V.M. Kaspi, J.H. Taylor, and M.F. Ryba, ApJ. **428**, 712 (1994); S. Thorsett and R. Dewey, Phys.Rev.D**53**, 3468 (1996).
45. R.N. Manchester, Ch.J.A.A. Suppl. 2, Vol.**6**, 139 (2006).
46. W. Zhao, Y. Zhang, and T.Y. Xia, Phys.Lett.B **677**, 235 (2009).
47. M. Mortonson and W. Hu, Astrophys.J. **657**, 1 (2007).
48. L.M. Griffiths and A. R. Liddle, Mon.Not.R.Astron.Soc. **324**, 769 (2001); M. Kaplinghat, et al., Astrophys. J. **583**, 24 (2003).
49. L.P.L. Colombo and E. Pierpaoli, New Astron. **14**, 269 (2009).
50. G. Jungman , M. Kaminokowski, A, Kosowsky, D.N. Spergel, Phys.Rev. D **54**, 1332 (1996); L.P.L. Colombo and S.A. Bonometto, New Astron. **8**, 313 (2003).
51. A.G. Polnarev, N.J. Miller, and B.G. Keating, Mon. Not. R. Astron. Soc., **386**, 1053 (2008).

EVOLUTION OF LARGE-SCALE MAGNETIC FIELDS AND STATE TRANSITIONS IN BLACK HOLE X-RAY BINARIES

DING-XIONG WANG, CHANG-YIN HUANG and JIU-ZHOU WANG

School of Physics, Huazhong University of Science and Technology, Wuhan, 430074, China

The state transitions of black hole (BH) X-ray binaries are discussed based on the evolution of large-scale magnetic fields, in which the combination of three energy mechanisms are involved: (1) the Blandford-Znajek (BZ) process related to the open field lines connecting a rotating BH with remote astrophysical loads, (2) the magnetic coupling (MC) process related to the closed field lines connecting the BH with its surrounding accretion disk, and (3) the Blandford-Payne (BP) process related to the open field lines connecting the disk with remote astrophysical loads. It turns out that each spectral state of the BH binaries corresponds to each configuration of magnetic field in BH magnetosphere, and the main characteristics of low/hard (LH) state, hard intermediate (HIM) state and steep power law (SPL) state are roughly fitted based on the evolution of large-scale magnetic fields associated with disk accretion.

1. Introduction

State transitions in black hole (BH) X-ray binaries involve a number of unresolved issues in astrophysics, displaying complex variations not only in the luminosities and energy spectra, but also in presence/absence of jets and quasi-periodic oscillations (QPOs). How to analyze and classify states in BH X-ray binaries from observations in multi-wavelength band is of foremost importance, and have been addressed by a number of authors.

Recently, Fender, Belloni and Gallo[1] proposed a unified semi-quantitative model for the disk-jet coupling in BH binaries, in which the states with jet and those with no jet are divided by a 'jet line' in an X-ray hardness-intensity diagram (HID). However, a detailed argument for producing jets in BH binaries has not been given.

In this paper, we intend to interpret the state transition of BH binaries based on the evolution of the magnetic field configurations related to the BZ, MC and BP processes. The conservation of magnetic flux is invoked to determine the transition from low/hard (LH) state to hard intermediate (HIM) state, and to steep power law (SPL) state. It turns out that the main characteristics of several BH binaries in LH, HIM and SPL states, such as the presence/absence of jets, X-ray luminosities and jet powers can be roughly fitted based on our model.

2. Magnetic Field Configurations and Basic Equations

First, we present a brief description of the magnetic field configurations related to LH and SPL states as shown in Figures 1a and 1b, respectively. The concerned equations, and some assumptions involved are given as follows.

(1) The accretion disk is perfectly conducting, and the magnetic field lines are frozen in the disk, which is thin and relativistic, lying in the equatorial plane of a rotating BH.

(2) The magnetic field is assumed to be constant on the horizon and to vary as a power law with disk radius, and the magnetic flux connecting the BH with its surrounding disk takes precedence over that connecting the BH with the remote load.

(3) The magnetic field configurations corresponding to LH and SPL states are given in Figures 1a and 1b, respectively. In Figure 1a the closed field lines connecting a rotating BH with the inner disk inside the radius r_{tr} (i.e., the MC region) correspond to the MC process, and the open field lines threading the BH horizon and those anchored in the disk outside the radius r_{tr} (i.e., the BP region) correspond to the BZ and BP processes, respectively. Corona is introduced above the MC region, which consists of tenuous hot plasma as shown in Figure 1.

(4) The magnetic field is assumed to be constant on the BH horizon, and it varies as a power law with disk radius as follows,

$$B_{\rm MC} \propto r^{-n}, \qquad r_{ms} < r < r_{tr}, \tag{1}$$

where $B_{\rm MC}$ is the magnetic field in the MC region, r_{tr} is the boundary radius between the MC and BP regions, r_{ms} is the radius of the innermost stable circular orbit (ISCO), and n is the power-law index indicating the degree of concentration of the magnetic field in the MC region. Following Blandford[2] the open magnetic field $B_{\rm BP}$ varies as a power law with disk radius as follows,

$$B_{\rm BP} \propto r^{-5/4}, \qquad r_{tr} < r < r_{out}, \tag{2}$$

where r_{out} is the outer radius of the BP region.

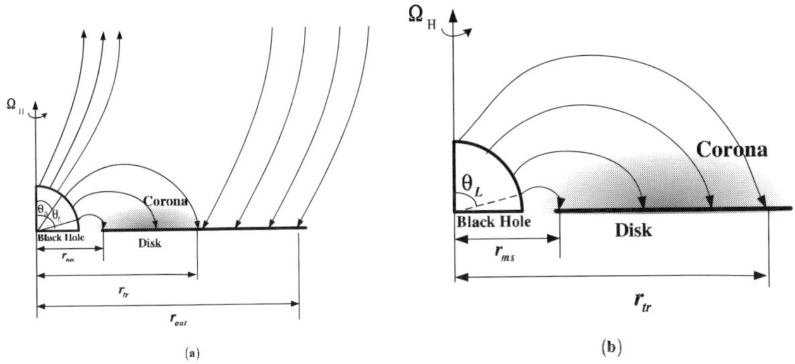

Fig. 1. Schematic drawings of the magnetic field configurations corresponding to (a) LH state and (b) SPL state.

The BZ power and torque, and the MC power and torque are given as follows,[3]

$$\widetilde{P}_{\rm BZ} \equiv P_{\rm BZ}/P_0 = 2a_*^2 \int_0^{\theta_S} \frac{k(1-k)\sin^3\theta d\theta}{2-(1-q)\sin^2\theta}, \tag{3}$$

$$\widetilde{T}_{\rm BZ} \equiv T_{\rm BZ}/T_0 = 4a_*(1+q) \int_0^{\theta_S} \frac{(1-k)\sin^3\theta d\theta}{2-(1-q)\sin^2\theta}, \tag{4}$$

where $k \equiv \Omega_F/\Omega_H$ is the ratio of the angular velocity of the open field line on the horizon to that of the horizon itself, and $k = 0.5$ is taken for the optimal BZ power in this paper. The quantities P_0 and T_0 are defined as

$$\begin{cases} P_0 \equiv B_H^2 M^2 \approx B_4^2 m_H^2 \times 6.59 \times 10^{28} erg \cdot s^{-1}, \\ T_0 \equiv B_H^2 M^3 \approx B_4^2 m_H^3 \times 3.26 \times 10^{23} g \cdot cm^2 \cdot s^{-2}, \end{cases} \tag{5}$$

where $m_H \equiv M/M_\odot$ and $B_4 \equiv B_H/(10^4 gauss)$ are defined as the dimensionless BH mass and magnetic field on the horizon, respectively.

It is assumed that the accretion rate of the MC region, $\dot{M}_{\rm MC}$, is independent of the disk radius, being equal to the accretion rate of the BP region at r_{tr} . Considering the balance between the pressure of the magnetic field and the ram pressure of the innermost parts of an accretion flow, we have the relation between $\dot{M}_{\rm MC}$ and B_H as follows,

$$B_H^2/(8\pi) = P_{ram} \sim \rho c^2 \sim \dot{M}_{\rm MC}/(4\pi r_H^2). \tag{6}$$

Equation (6) can be rewritten as

$$\dot{m}_{\rm MC} \equiv \dot{M}_{\rm MC}/(B_H^2 M^2) = \alpha_m (1+q)^2, \tag{7}$$

where α_m is a parameter to adjust the accretion rate, and

$$B_H^2 M^2 \equiv (B_4^2 m_H^2 \times 7.32 \times 10^7) g \cdot s^{-1}. \tag{8}$$

Assuming that the accretion rate in the BP region varies with disk radius in a power-law, $\dot{M}_{\rm BP} \propto r^S$, we obtain its expression as follows,

$$\dot{m}_{\rm BP} \equiv \dot{M}_{\rm BP}/(B_H^2 M^2) = \alpha_m (1+q)^2 (r/r_{tr})^S, \qquad \text{for} \qquad r_{tr} < r < r_{out}. \tag{9}$$

Thus the outflow rate from a ring with width $r - r + dr$ can be expressed as

$$d\dot{m}_{outflow} = \alpha_m (1+q)^2 S (r/r_{tr})^{S-1} d(r/r_{tr}), \tag{10}$$

and the total rate of outflow matter driven from the BP region is

$$\dot{m}_{outflow} = \alpha_m (1+q)^2 (\zeta_{out}^S - 1), \tag{11}$$

where we have $\dot{m}_{outflow} \equiv \dot{M}_{outflow}/(B_H^2 M^2)$ and $\zeta_{out} \equiv r_{out}/r_{tr}$.

The BP power and torque are given as follows,[4]

$$\widetilde{P}_{\rm BP} \equiv P_{\rm BP}/P_0 = (\lambda \alpha_m/2) \frac{(1+q)^2}{\xi_{tr} \chi_{ms}^2} \frac{S(\zeta_{out}^{S-1} - 1)}{S - 1}, \tag{12}$$

$$\widetilde{T}_{\rm BP} \equiv T_{\rm BP}/T_0 = (\lambda \alpha_m/2)(1+q)^2 \chi_{ms} \xi_{tr}^{1/2} \frac{S(\zeta_{out}^{S+1/2} - 1)}{S + 1/2}, \tag{13}$$

where we have $\xi_{tr} \equiv r_{tr}/r_{ms}$, and λ is defined as the ratio of the specific angular momentum of outflow to that of the matter at the midplane of the disk.

3. Transfer of Energy and Angular Momentum From MC Region To BP Region

As shown in Figure 1a the magnetic field configuration for LH state contains the BZ, MC and BP processes. Energy and angular momentum are transferred from a rotating BH to remote astrophysical loads in the Poynting flux regime by the BZ process, and they are extracted from an accretion disk to remote astrophysical loads in the hydromagnetic regime by the BP process. Thus the jet power in LH state can be regarded as the sum of the BZ and BP powers, i.e.,

$$L_J = L_{\rm BZ} + L_{\rm BP}. \tag{14}$$

The MC process plays an important role in fitting LH state due to the following features: (1) disk accretion is suppressed by tremendous angular momentum transferred from a fast-spinning BH to the MC region, (2) the jet power is strengthened due to the transfer of energy and angular momentum from the MC region to the BP region. The transfer of energy and angular momentum across the radius r_{tr} is illustrated in Figure 2.

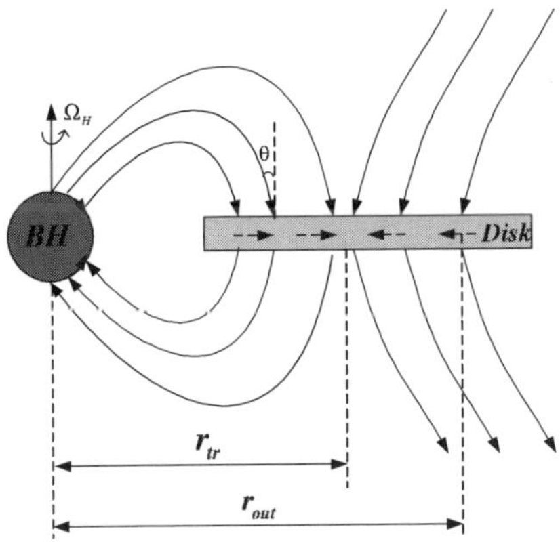

Fig. 2. Schematic drawing for interpreting the transfer of energy and angular momentum from the MC region to the BP region. The dashed arrowheads indicate the current in the disk, by which the direction of the magnetic torque exerted in the MC and BP regions can be determined.

Inspecting the direction of the poloidal magnetic field and the direction of the current in the MC and BP regions, we find that the magnetic torques exerted insides and outside the radius r_{tr} are always opposite. This result is independent of the direction of the large-scale magnetic field in Figure 2, because the magnetic torque $G_{\rm MC}^{EM}$ always accelerates the disk inside r_{tr} due to transfer of angular momentum from a fast-spinning BH to the MC region, while the magnetic torque $G_{\rm BP}^{EM}$ always decelerates the disk outside r_{tr} due to removal of angular momentum from the disk to remote astrophysical loads. According to the theory of accretion theory the rate of transferring energy across r_{tr} is given as follows,[5]

$$(G_{\rm BP}\Omega)_{r_{tr}+\varepsilon} - (G_{\rm MC}\Omega)_{r_{tr}-\varepsilon}, \tag{15}$$

where $G_{\rm MC}$ and $G_{\rm BP}$ are respectively the total torques exerted inside and outside r_{tr}, and they read

$$\begin{cases} G_{\rm MC} = G_{\rm MC}^{vis} - G_{\rm MC}^{EM}, \\ G_{\rm BP} = G_{\rm BP}^{vis} + G_{\rm BP}^{EM}. \end{cases} \tag{16}$$

In Eq. (16) $G_{\rm MC}^{vis}$ and $G_{\rm MC}^{EM}$ are respectively the viscous and magnetic torques exerted inside r_{tr}, while $G_{\rm BP}^{vis}$ and $G_{\rm BP}^{EM}$ are respectively the viscous and magnetic torques exerted outside r_{tr}. The opposite signs before $G_{\rm MC}^{EM}$ and $G_{\rm BP}^{EM}$ imply that their directions are opposite.

Incorporating Eqs. (15) and (16), and ignoring the difference between the angular velocity inside and that outside r_{tr}, we have the rate of transferring energy across the radius r_{tr} at the presence of the jet as follows,

$$(G_{\rm BP}^{vis} - G_{\rm MC}^{vis} + G_{\rm BP}^{EM} + G_{\rm MC}^{EM})\Omega_{tr}, \tag{17}$$

where Ω_{tr} is the angular velocity at r_{tr}.

Similarly, the rate of transferring energy across r_{tr} at the absence of the jet can be derived as follows,

$$(G_{\rm BP}^{vis} - G_{\rm MC}^{vis} + G_{\rm MC}^{EM})\Omega_{tr}. \tag{18}$$

The rates given by Eqs. (17) and (18) correspond to LH and SPL states, which are represented by the magnetic field configurations in Figures 1a and 1b, respectively. Thus the extra rate of transferring energy can be written as $G_{\rm BP}^{EM}\Omega_{tr}$, which is equal to the difference between Eqs. (17) and (18). Since $G_{\rm BP}^{EM}\Omega_{tr}$ is the extra rate of energy transferred from the MC region to the BP region, we infer that the luminosity of the MC region could be suppressed significantly in LH state.

Assuming that $(G_{\rm BP}^{EM})_{tr}$ and $(G_{\rm BP}^{EM})_{out}$ are respectively the magnetic torques exerted at r_{tr} and r_{out}, we have

$$(G_{\rm BP}^{EM})_{tr}\Omega_{tr} \approx (G_{\rm BP}^{EM})_{tr}\Omega_{tr} - (G_{\rm BP}^{EM})_{out}\Omega_{out} = T_{\rm BP}. \tag{19}$$

Since the magnetic field and the angular velocity in the BP region are proportional to $r^{-5/4}$ and $r^{-3/2}$, we neglect the term $(G_{\rm BP}^{EM})_{out}\Omega_{out}$ in Eq. (19) for a large outer radius r_{out} of the BP region. Based on the conservation of angular momentum the accretion rate $\dot{M}_{out}$ at r_{out} is related to $\dot{M}_{MC}$ by

$$\dot{M}_{out}(r_{out}^2\Omega_{out}) - \dot{M}_{\rm MC}(r_{tr}^2\Omega_{tr}) = T_{\rm BP} - \delta T_{\rm MC}, \tag{20}$$

where $\delta T_{\rm MC}$ is the angular momentum transferred from the MC region to the BP region across r_{tr}, and δ is a fraction parameter to be determined in fittings.

Substituting Eqs. (7) and (9) into Eq. (20), we have

$$\alpha_m(1+q)^2\chi_{ms}\zeta_{out}^S(\xi_{out}^{1/2} - \xi_{tr}^{1/2}) = \tilde{T}_{\rm BP} - \delta\tilde{T}_{\rm MC}. \tag{21}$$

It is noted that the value of the power-law index S in Eq. (13) can be determined by Eq. (21), provided that the parameters a_*, n, α_m and ζ_{out}

are given. In this paper we take a_* and n as two key parameters to fit the states of BH binaries, and $\alpha_m = 0.1$ and $\zeta_{out} = 100$ are assumed in calculations.

Assuming that X-ray luminosity in LH state is produced by disk accretion and by the MC process, we have

$$L_X = 4\pi \int_{r_{ms}}^{r_{tr}} [F_{\rm DA}(r) + (1-\delta)F_{\rm MC}(r)] r dr, \tag{22}$$

where the factor $(1-\delta)$ is given due to $\delta T_{\rm MC}$ transferred out of the MC region, and $F_{\rm DA}$ and $F_{\rm MC}$ are respectively the radiation flux arising from disk accretion and from the MC process, and they read

$$F_{\rm DA} = \dot{M}_{\rm MC} f_{\rm DA}/(4\pi r), \tag{23}$$

$$F_{\rm MC} = -\frac{d\Omega}{rdr}(E^\dagger - \Omega L^\dagger)^{-2} \int_{r_{ms}}^{r} (E^\dagger - \Omega L^\dagger) H_L r dr. \tag{24}$$

The function $f_{\rm DA}$ in Eq. (23) is the contribution to the radiation due to disk accretion, being derived by Page and Thorne,[6] and H_L in Eq. (24) is the flux of angular momentum transferred from the BH to the MC region, being related to the MC torque given by Ref. 3,

$$\partial T_{\rm MC}/\partial r = 4\pi r H_L. \tag{25}$$

The quantities $E^\dagger$ and $L^\dagger$ in Eq. (24) are respectively the specific energy and angular momentum of accreting matter.[7]

4. Evolution of Large-Scale Magnetic Fields and State Transitions

As argue in Ref. 4, the BZ, MC and BP processes can coexist, provided that a_* and n are greater than some critical values. By using Eqs. (14) and (22) we can fit the X-ray luminosities and jet powers of LH state of several BH binaries as listed in Table 1, in which the observational data are taken from Ref. 1. It is found from Table 1 that L_X is of a few percent of Eddington luminosity, which is about one order of magnitude less than L_J. These results are associated with the presence of a quasi-steady radio jet powered by the BZ and BP processes. In addition, a hard power-law component could be produced due to Comptonization of soft photons in corona above the MC region. Thus the main features of LH state of these BH binaries can be fitted.

One of the most remarkable results obtained in Ref. 1 is the discovery that the states with jet and those with no jet are divided by a 'jet line'

Table 1. Fitting LH states of the BH binaries.

Sources	m_H	L_X	L_J	η_L	$\tilde{r}_{tr}$	n	δ	a_*	B_4
A J1748-288	7	0.0880	1.9	1.137	3.887	6.93	0.937	0.85	8×10^4
B 1915+105	14	0.0753	0.6	1.865	2.214	5.00	0.891	0.99	3×10^4
C J1655-40	7	0.0815	1.0	1.239	5.338	7.16	0.892	0.70	7×10^4
D J1550-564	9	0.0450	0.3	4.864	4.376	7.05	0.692	0.80	3×10^4
E GX 339-4	7	0.0660	0.3	8.153	3.124	6.10	0.593	0.93	3×10^4

in HID. It turns out that the 'jet line' can be understood based on the criterion of the kink instability. By using the criterion of kink instability given in Ref. 8 we have a critical line (CL) between LH and SPL states as shown in Figure 3. The shaded region above the CL represents LH states with jets driven by the BZ and BP processes, and the region below the CL represents SPL states with no jet.

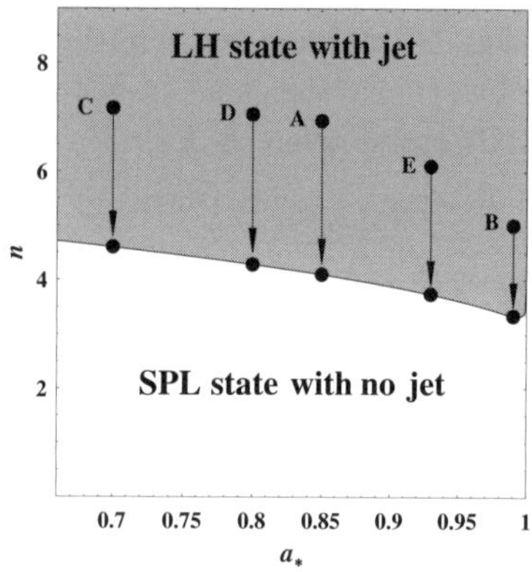

Fig. 3. The parameter space for interpreting the transitions from LH to HIM and SPL states. The dots above the CL represent LH states, and those at the CL represent HIM states. The arrowheads represent the direction of state transitions with the decreasing parameter n. The symbols A, B, C, D and E represent the five sources given in Tables 1-3.

The LH states evolve to the HIM states with the decreasing n, attaining the values indicated by the black dots on the CL as shown in Figure 3. As the parameter n continues to decrease, the black dots fall below the CL until $r_{tr} = r_{out}$, the HIM states evolve to the SPL states.

As shown in Tables 2 and 3, although the values of n corresponding to the SPL states are slightly less than those corresponding to the HIM states

Table 2. Fitting HIM states of the BH binaries.

Sources	m_H	L_X	L_J	η_L	$\tilde{r}_{tr}$	n	δ	a_*	B_4
A J1748-288	7	0.5411	0.3078	0.399	11.887	4.10	0.937	0.85	8×10^4
B 1915+105	14	0.2040	0.1693	1.074	4.172	3.34	0.891	0.99	3×10^4
C J1655-40	7	0.4342	0.1786	0.412	18.257	4.60	0.892	0.70	7×10^4
D J1550-564	9	0.1932	0.0451	1.591	15.077	4.28	0.692	0.80	3×10^4
E GX 339-4	7	0.2238	0.0551	3.488	7.722	3.74	0.593	0.93	3×10^4

Table 3. Fitting SPL states of the BH binaries.

Sources	m_H	L_X	L_J	η_L	$\tilde{r}_{tr}$	n	δ	a_*	B_4
A J1748-288	7	0.8690	0	0.3542	388.7	$4.10 - \varepsilon$	0	0.85	4×10^4
B 1915+105	14	0.6287	0	4.715	221.4	$3.34 - \varepsilon$	0	0.99	2×10^4
C J1655-40	7	0.9317	0	2.280	533.8	$4.60 - \varepsilon$	0	0.70	5×10^4
D J1550-564	9	0.5538	0	3.076	437.6	$4.28 - \varepsilon$	0	0.80	3×10^4
E GX 339-4	7	0.6046	0	4.399	312.4	$3.74 - \varepsilon$	0	0.93	3×10^4

(the difference of ε between the HIM and SPL states varies from $\sim 10^{-6}$ to 10^{-4}), the values of $\tilde{r}_{tr}$ of the SPL states are much greater than those of the HIM states.

5. Summary

In this paper, it is shown that the evolution of the large-scale magnetic fields plays a very important role in the state transition of the BH binaries. Although it is widely believed that the jets are produced and collimated by the large-scale magnetic fields, the origin of the large-scale magnetic fields in BH binaries remains unclear. Very recently, Zhao, Wang and Gan discussed the possible origins of large-scale magnetic fields based on a continuous distribution of toroidal electric current flowing in the inner region of the disk around a Kerr black hole (BH) in the framework of general relativity.[9] It turns out that four types of configuration of the magnetic connection (MC) can be generated, i.e., MC of the BH with the remote astrophysical load (MCHL), MC of the BH with the disk (MCHD), MC of the plunging region with the disk (MCPD) and MC of the inner and outer disk regions (MCDD). The BZ process can be regarded as one type of MC, i.e., MCHL. In addition, a scenario for fitting the quasi-periodic oscillations in BH binaries based on MCDD associated with the magnetic reconnection was proposed. We shall combine the origin of the large-scale magnetic fields with the state transitions in BH binaries in our future work.

Acknowledgements

This work is supported by the National Natural Science Foundation of China under grant 10873005, the Research Fund for the Doctoral Program of Higher Education under grant 200804870050 and National Basic Research Program of China under grant 2009CB824800.

References

1. R. P. Fender, T. Belloni and E. Gallo, *MNRAS* **355**, 1105 (2004).
2. R. D. Blandford, *MNRAS* **176**, 465 (1976).
3. D.-X. Wang, R.-Y. Ma, W.-H. Lei and G.-Z. Yao, *ApJ* **595**, 109 (2003).
4. D.-X. Wang, Y.-C. Ye, Y. Li and Z.-J. Ge, *MNRAS* **385**, 841 (2008).
5. J. Frank, A. R. King and D. L. Raine, *Accretion Power in Astrophysics*, 2nd edn. (Cambridge Univ. Press, Cambridge, 1992).
6. D. N. Page and K. S. Thorne, *ApJ* **191**, 499 (1974).
7. I. D. Novikov and K. S. Thorne, in *Black Holes*, ed. C. Dewitt (Gordon and Breach, New York, 1973).
8. D.-X. Wang, R.-Y. Ma, W.-H. Lei, and G.-Z. Yao, *ApJ* **601**, 1031 (2004).
9. C.-X. Zhao, D.-X. Wang and Z.-M. Gan, (2009 MNRAS in press).

PULSAR MASS AND RADIUS ESTIMATION BY THE KHZ QPO

C.M. ZHANG*, Y.Y. PAN and Y.H. ZHAO

National Astronomical Observatories, Chinese Academy of Sciences, Beijing 100012, China

**E-mail: zhangcm@bao.ac.cn*

RXTE satellite discovered many kHz quasi-periodic oscillations (QPOs) in X-ray neuron star systems, and these oscillations are often taken as the accreting matter circling around the central compact objects. As the models for kHz QPOs are constructed, the mass and/or radius of these neutron stars can be inferred, by which the comparisons with the star equations of states (EOSs) can be performed.

Keywords: pulsar; kHz QPOs; Binary system

1. Introduction

RXTE has observed the kilo-Hertz quasi-periodic oscillations (kHz QPOs) of about $\simeq$200 Hz - 1300 Hz in the X-ray flux in more than thirty accreting neutron stars in low-mass X-ray binaries (LMXBs), and the twin peak kHz QPOs, upper and lower frequencies (ν_2 and ν_1), are often detected.[1] These kHz QPO frequencies generally follow the tight correlations between themselves and with the other timing features of the X-ray emissions.[2–4]

The idea of estimating neutron star (NS) mass and radius is proposed for kHz QPOs by Miller et al. (c.f.[5]). Here, we estimate the NS mass and radius for the special QPO sources by the Alfven wave models,[6] where the upper kHz QPO frequency is ascribed as the Keplerian frequency of the accreting matter in the orbital motion at radius r,

$$\nu_2 = \nu_k = 1850 A X^{3/2} (\mathrm{Hz}) \tag{1}$$

with the parameters X=R/r and $A = (m/R_6^3)^{1/2}$, where $R_6 = R/10^6(cm)$ and $m = \frac{\mathrm{M}}{\mathrm{M}_\odot}$ are the stellar radius R and mass M in the units of 10^6 (cm) and solar masses, respectively. The quantity A^2 is the averaged mass density of star, and expressed as

$$\langle\rho\rangle = 3\mathrm{M}/(4\pi\mathrm{R}^3) \simeq 2.4 \times 10^{14}(\mathrm{g/cm^3})(\mathrm{A}/0.7)^2 \,. \qquad (2)$$

The lower kHz QPO is identified as the Alfvén wave oscillation frequency at the same radius r, expressed as

$$\nu_1 = \nu_2 X \sqrt{1 - \sqrt{1 - X}} \,. \qquad (3)$$

In Equations (1) and (3), there are two free parameters, mass density parameter A and position parameter $X = R/r$, If the twin kHz QPOs are known, then both parameters can be determined. The averages of them have been obtained for the typical Atoll and Z sources[19] to be $\langle A\rangle = 0.7$ and $\langle X\rangle = 0.8$,[6] which can infer the nuclear matter density $\sim 10^{14}(g/cm^3)$ and the emission position of the kHz QPO is so close to the stellar surface, $r \sim 12(R/10km)km$. We can also express the NS radius by the mass density parameter A and mass parameter m from Eq.(2),

$$R_6 = 1.27 m^{1/3}(A/0.7)^{-2/3}\,(10\ \mathrm{km}) \,. \qquad (4)$$

To know the matter compositions inside the compact star, or the equation of state (EOS),[7–9] the stellar mass and radius should be known, from which we can know if the star is composed of the neutrons or quarks (c.f.[10–14] The NS mass can be measured in the binaries,[3] and radius may be known by estimations or by measuring M-R relations,[14,15] or by exploiting the spectral of X-rays and the hydrogen atmosphere model.[16,17]

2. Results and Summaries

The first known accreting powered millisecond pulsar SAX J1808.4-3658, is a peculiar one, and it has the low kHz QPO frequencies ($\nu_1 = 499$, $\nu_2 = 694 Hz$), implying the frequency separation about 195 Hz, which is just the half of its spin frequency 401 Hz.[22] Using the Alfven wave model, we find that its stellar mass density parameter A=0.4 is less than the typical value of other kHz QPO sources with A$\sim$0.7, which has been shown in Fig.1. In Fig.2, we find that the low value of the parameter A means that the star EOS is not possible for a quark star. The maximum mass for the neutron or quark star without rotation is 3.2 $M_{\odot}$,[18] over which the black hole will be implied. If the mass of star in SAX J1808.4-3658 is 1.0 (1.4) $M_{\odot}$, then its radius is about 18 (22) km.

For another accretion powered millisecond pulsar XTE 1807-294 , its kHz QPOs are very similar to those of SAX J1808.4-3658 , as shown in

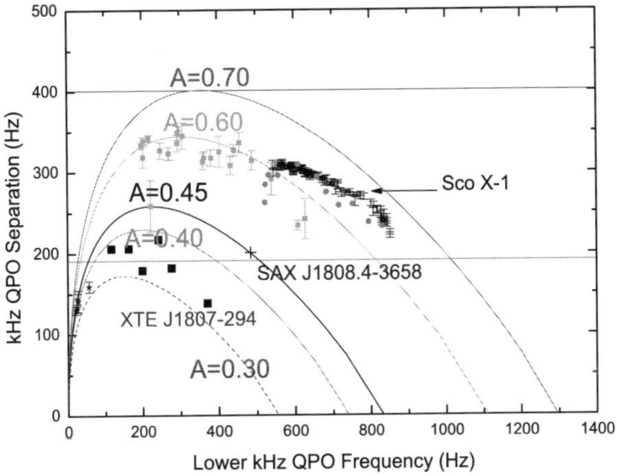

Fig. 1. Diagram of the kHz QPO separation vs. lower kHz QPO frequency. The parabola curves represent for the different mass density parameters from the Alfven wave model for kHz QPOs. For the typical Atoll and Z sources,[19] the parameter A = 0.7 is achieved. However, for two special cases SAX J1808.4-3658 and XTE 1807-294 ,[20] a small parameter of about 0.4 is obtained.

Fig.1, and its mass density parameter is very low, about 0.3, which means that either its mass is very low, e.g. $0.7M_{\odot}$for R=15km, or its size is big, e.g. 27km for the mass of $1.4M_{\odot}$.

However, why the two millisecond pulsars have the low masses is still a puzzle. From the X-ray observations, both SAX J1808.4-3658 and XTE 1807-294 [20] are low luminosity sources; There have not yet been found any other sources with the kHz QPOs between SAX J1808.4-3658 and the other Atoll/Z sources. The future detections are needed to confirm what makes the low kHz QPOs of SAX J1808.4-3658 . In Fig.1, we notice that the kHz QPOs of Z source Sco X-1 is different from that of SAX J1808.4-3658 , the former with the higher kHz QPO separations.

Acknowledgements

This research was supported by NSFC(No.10773017) and National Basic Research Program of China (2009CB824800).

References

1. van der Klis, M.: 2000, ARA&A 38, 717
2. Belloni T., Méndez M., Homan J., 2007, MNRAS , 376, 1133

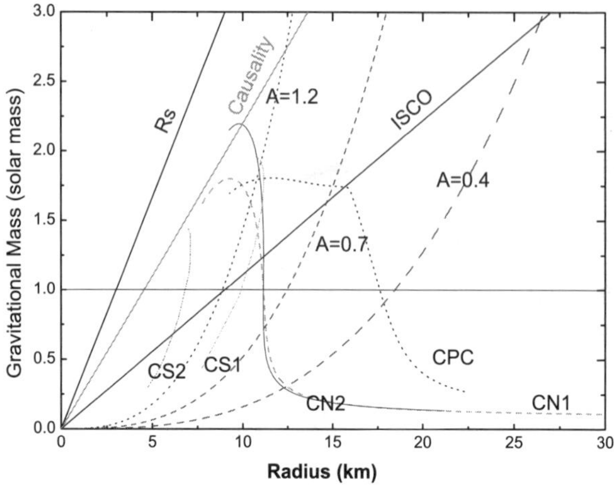

Fig. 2. The mass versus radius diagram. The parabola curves represent for the different mass density parameters. The five theoretical curves of equations of states are the same meanings of plots in the reference.[21] The straight lines labeled by R_s (ISCO) represents the radius equal to one (three) Schwarzschild radius. The parabola curves represent A = 1.2, 0.7 and 0.4, respectively.

3. Zhang, C. M., Yin, H. X., Zhao, Y. H., *et al.*: 2006a, MNRAS 366, 1373.
4. Stella, L., Vietri, M.: 1999, Phys. Rev. Lett. 82, 17
5. Miller, M. C., Lamb, F. K., Psaltis, D.: 1998, ApJ 508, 791
6. Zhang, C.M.: 2004, A&A 423, 401 (astro-ph/0402028)
7. Lattimer, J. M., Prakash, M.: 2004, Science 304, 536
8. Lattimer, J. M., & Prakash, M., 2007, PhR., 442, 109L.
9. Haensel, P., Potekhin, A.Y., & Yakovlev, D.G. 2007, *NEUTRON STARS, Equation of state and structure*, (Springer, Berlin)
10. Li, X.D., Bombaci, I., Dey, M., *et al.*: 1999, Phys. Rev. Lett. 3776, 83
11. Xu, R. X. 2005, MNRAS, 356, 359
12. D. P. Menezes, C. Providencia, 2004, Phys. Rew. C, 69, 045801.
13. D. P. Menezes et al., 2009, Phys. Rev. C, 79, 035807.
14. Melrose, D.B., Fok, R., Menezes, D.P.: 2006, MNRAS, 371, 204
15. Özel, F. 2006, Nature, 441, 1115
16. Webb, N.A., Barret, D. 2008, ApJ in press, arXiv:0708.3816
17. Cottam, J., Paerels, F., & Méndez, M. 2002, Nature , 420, 51
18. Rhoades, C. E., Jr., & Ruffini, R. 1974, Phys. Rev. Lett., 32, 324
19. Hasinger, G., van der Klis, M.: 1989, A&A 225, 79
20. Zhang, F., Qu, J. L., Zhang, C. M., *et al.*: 2006b, ApJ 646, 1116
21. Miller, M. C.: 2002, Nature 420, 31
22. Wijnands, R., van der Klis, 1998, Nature, 394, 344

THE CENTRAL BLACK HOLE MASSES FOR γ-RAY LOUD BLAZARS

JIANG-HE YANG

Department of Physics and Electronics Science,
Hunan University of Arts and Science, Changde, 415000, P.R.China
Center for Astrophysics, Guangzhou University,
Guangzhou, 510405, P.R.China
E-mail: yjianghe@163.com

JUN-HUI FAN

Center for Astrophysics, Guangzhou University,
Guangzhou, 510405, P.R.China
E-mail: jhfan_cn@yahoo.com.cn

The Large Area Telescope (LAT) on the Fermi Gamma-ray Space Telescope (formerly GLAST) provides an increase of sensitivity and has detected rapid variation in the γ-ray region. In this work, the variation timescales detected from the γ-ray loud blazars by LAT and EGRET are used to estimate the central black hole masses for 14 blazars. The obtained masses are in a range of $(2.07 \sim 15.54) \times 10^7 M_\odot$, the Doppler factors are in the range of $\delta = 1.12 \sim 4.63$, which are comparable with those obtained by other authors.

Keywords: active galactic nuclei (AGNs); doubling time scale; central black hole mass.

1. Introduction

Many blazars show variation in the γ-ray band on timescales from days to months,[1] but large flux variation on short timescales of < 1 day are also detected. Some authors claimed that the γ-rays are produced at a region of[2] $\sim 100 R_g$ and hundreds of Schwarzshild radii[3] from the central black hole, it is also found to be 205 R_g.[4] The distance is an important parameter, if it is known, then the central black hole masses can be estimated.

The emission mechanisms of AGNs are investigated by many authors.[5–8] The central black hole masses are very important for the emission and the evolution in blazers.[9–14] There are many methods of mass determination.[9,14–21]The power for all types of AGNs is almost universally ascribed to accretion onto supermassive black holes with masses $M = (10^6 \sim 10^9)M_\odot$.[22]

The Large Area Telescope (LAT) on the Fermi Gamma-ray Space Telescope (formerly GLAST, launched on June 11, 2008) provides higher sensitivity than EGRET. The Fermi Gamma Ray Space Telescope has provided new data about γ-ray activity of AGNs.[23] In the first three months of operation, the Fermi Gamma Ray Space Telescope revealed more than one hundred blazars with some sources displaying large flares.[23]

In this paper, we will use the observation data (γ-ray emissions, short time scales) to derive the central black hole masses. In Sec. 2, we will introduce a method to estimate the central black hole masses and the Doppler factor. In Sec. 3, we show the LAT on Fermi observations of the γ-ray loud blazars with available GeV variable timescale, and present results. In Sec. 4, some discussion and a brief summary are given.

2. Method

Here, we describe our method of estimating the basic parameters, the mass of the central black hole and the Doppler factor, of some blazars with variation timescale of hours to days in the γ-ray band. If we take the variation timescale as the measurements of the size, R, of the emission region, then the R in the jet obeys the inequality,

$$R \leqslant \mathrm{c}\Delta T_{\mathrm{D}} \frac{\delta}{1+z} (\mathrm{cm}), \tag{1}$$

where c is the speed of light in units of $\mathrm{cm} \cdot \mathrm{s}^{-1}$, δ the Doppler factor, z the redshift, ΔT_{D} the doubling time scale in units of second, and $\Delta T_{\mathrm{D}} = (F_{\mathrm{initial}}/\Delta F) \cdot \Delta T$. F is the γ-ray flux (> 100 MeV). Some authors argued that the γ-rays are from a region of some hundreds of Schwarzschild radii ($R_{\mathrm{g}} \equiv \frac{GM}{\mathrm{c}^2} = 1.48 \times 10^5 \frac{M}{M_\odot} (\mathrm{cm})$) from the center.[2–4] Considering an accretion disk surrounding a supermassive black hole, when $R < 200R_g$, the electrons in the accretion flow become ultrarelativistic, and $200R_g$ is an important critical point. So, we scale the size, R, of the emission region to 200 gravitational radii. Namely, $R < 200R_{\mathrm{g}}$. So, by relations (1) and $R = 200R_{\mathrm{g}}$, we have

$$\frac{M}{M_\odot} \leqslant 1 \times 10^3 \frac{\delta}{1+z} \Delta T_{\mathrm{D}}. \tag{2}$$

For an object with a mass of M, the Eddington limit gives[24] $L_{\mathrm{Edd.}} \approx 1.26 \times 10^{38} (\frac{M}{M_\odot}) (\mathrm{erg} \cdot \mathrm{s}^{-1})$. So, we have that the intrinsic luminosity, $L^{\mathrm{in.}}$, of a source with a mass of M should satisfy $L_{\rm in} \leqslant L_{\rm Edd.}$. In a relativistic beaming frame, the observed luminosity is associated with the intrinsic luminosity, $L^{\rm ob.} = \delta^{(4+\alpha_\gamma)} L^{\rm in.}$. Here, α_γ is the energy spectral index, and $\alpha_\gamma = \alpha_{\rm ph} - 1$, $\alpha_{\rm ph}$ is the γ-ray photon spectrum index.

From above relations, we have

$$L^{\rm ob.} \leqslant 1.26 \times 10^{41} \frac{\delta^{(5+\alpha_\gamma)}}{1+z} \Delta T_{\rm D} ({\rm erg \cdot s^{-1}}). \qquad (3)$$

The γ-ray flux is not always dominant over the flux in lower energy bands.[1] For PKS 0528+134 and 3C 279, their γ-ray luminosity, L_γ, is 0.80 $L_{\rm bol.}(L^{\rm ob.})$ and 0.5 $L_{\rm bol.}$, respectively.[2,25] In addition, we consider the flare states of the selected objects. Therefore, we can take the γ-ray luminosity to stand for half of the bolometric luminosity approximately, i.e. $L_\gamma \simeq 0.5 L_{\rm bol.}$. From relations (3) we have $L_\gamma \leqslant 6.3 \times 10^{40} \frac{\delta^{(5+\alpha_\gamma)}}{1+z} \Delta T_{\rm D} ({\rm erg \cdot s^{-1}})$, which gives

$$\delta \geqslant \left(\frac{L_\gamma (1+z)}{6.3 \times 10^{40} \Delta T_{\rm D}}\right)^{\frac{1}{5+\alpha_\gamma}} = \left(\frac{L_\gamma (1+z)}{6.3 \times 10^{40} \Delta T_{\rm D}}\right)^{\frac{1}{4+\alpha_{\rm ph}}}. \qquad (4)$$

From the relations (2) and (4), we can obtain the central black hole mass,

$$\frac{M}{M_\odot} \leq 1 \times 10^3 \frac{\Delta T_{\rm D}}{1+z} \cdot \left[\frac{L_\gamma (1+z)}{6.3 \times 10^{40} \Delta T_{\rm D}}\right]^{\frac{1}{4+\alpha_{\rm ph}}}. \qquad (5)$$

The γ-ray emission luminosities at 1 GeV are calculated by equation $L_\gamma^{48} = 4\pi d_{\rm L}^2 \nu F_\nu$. Here, $d_{\rm L}$ is the luminosity distance computed by[26] $d_{\rm L} = (1+z) \cdot \frac{c}{H_0} \cdot \int_1^{1+z} \frac{1}{\sqrt{\Omega_{\rm M} x^3 + 1 - \Omega_{\rm M}}} {\rm d}x$, values $H_0 = 73 {\rm km \cdot s^{-1} \cdot Mpc^{-1}}$ and $\Omega_{\rm M} = 0.27$ are adopted.

3. Data and Results

3.1. *Data*

In this section, we consider only those γ-ray loud blazars that are detected by Fermi and have short timescales of variation. 14 blazars with short timescales of variation are selected from previous literature and down to date data observed by LAT. They are listed in Table 1.

3.2. *Results*

When the available data are applied to the relations, we can estimate the central black hole masses and the Doppler factors. The results and the available data are listed in Table 1. Col.(1): name of the sources; Col.(2):

redshift; Col.(3): the γ-ray photon flux (> 100 MeV) and their uncertainty in units of 10^{-8}photon$\cdot$cm$^{-2}\cdot$s^{-1}; Col.(4): the γ-ray photon spectrum index and their uncertainty obtained by reference [23]; Col.(5): the doubling time scale (hours); Col.(6): the reference for ΔT_D; Col.(7): the Doppler factor calculated by Eq. (4); Col.(8): the central black hole mass ($10^7 M_\odot$) obtained by Eq. (5).

Table 1. The observations and results for 14 gamma-ray loud blazars.

Name (1)	z (2)	F (3)	α_{ph} (4)	ΔT_D (5)	Ref. (6)	δ (7)	M_7 (8)
0219+428	0.444	49.6 ± 4.8	1.97 ± 0.04	30	[22]	1.70	6.35
0235+164	0.94	104.8 ± 7.1	2.05 ± 0.02	72	[27]	2.33	15.54
0528+134	2.06	39.5 ± 6.7	2.54 ± 0.09	24	[22]	3.13	4.42
0537-441	0.894	49.7 ± 5.6	2.19 ± 0.04	16	[2]	2.47	3.76
0650+453	0.933	56.5	2.32 ± 0.06	66	[23]	1.98	12.27
0735+178	0.424	7.5 ± 2.4	2.10 ± 0.14	29	[22]	1.19	4.33
1156+295	0.729	16.0 ± 3.8	2.47 ± 0.13	11	[22]	1.84	2.12
1226+023	0.158	137 ± 13	2.71 ± 0.05	24	[22]	1.12	4.19
1253-055	0.537	46.3 ± 6.8	2.35 ± 0.05	12	[28]	1.92	2.70
1502+106	1.83	180 ± 30	2.17 ± 0.02	12	[29]	4.63	3.53
1510-089	0.361	400	2.41 ± 0.05	8	[30]	2.36	2.50
1520+31	1.487	190 ± 70	2.39 ± 0.06	6.7	[31]	4.28	2.07
1633+382	1.814	49.8 ± 6.0	2.44 ± 0.07	16	[32]	3.31	3.39
2230+114	1.04	24.6 ± 6.2	2.61 ± 0.12	48	[33]	1.81	7.65

4. Discussion and Conclusion

In this paper, we have estimated the central black hole mass for 14 blazars. The masses of central black holes are in a range of $(2.07 \sim 15.54) \times 10^7 M_\odot$ as listed in Table 1. The maximum mass is $15.54 \times 10^7 M_\odot$ for 0235+164 while the minimum mass is $2.07 \times 10^7 M_\odot$ for 1520+31. The mass range is quite consistent with the arguments by Dermer & Gehrels(1995),[22] who indicated that the values of black hole masses are in the range $(10^6 \sim 10^8) M_\odot$. Our results are also comparable with others.

For 0528+134, our result $4.42 \times 10^7 M_\odot$ is consistent with those by Cheng et al. (1999),[18] who obtained $5.09 \times 10^7 M_\odot$, by Fan et al. (1999),[17] who obtained $6.69 \times 10^7 M_\odot$ and by Dermer & Gehrels(1995),[22] who gave $3.9 \times 10^7 M_\odot$.

For 1226+023 (3C273), our result is $4.19 \times 10^7 M_\odot$ while Dermer & Gehrels (1995)[22] gave $9.3 \times 10^7 M_\odot$, and Woo et al. (2002)[34] give $1.66 \times 10^7 M_\odot$. A mass of $493 \times 10^7 M_\odot$ was obtained by Cao et al. (2002)[9,35]

For 1253-055 (3C279), Our result is $2.70 \times 10^7 M_{\odot}$, the mass is $4.0 \times 10^7 M_{\odot}$ by Cheng et al. (1999)[18] and $1.9 \times 10^7 M_{\odot}$ by Dermer & Gehrels(1995).[22] A mass of $126 \times 10^7 M_{\odot}$ was obtained by Cao et al. (2002).[9,35]

For 1633+382, our result $3.39 \times 10^7 M_{\odot}$ is consistent with $3.81 \times 10^7 M_{\odot}$ by Cheng et al. (1999)[18] and $1.1 \times 10^7 M_{\odot}$ by Dermer & Gehrels(1995).[22]

From our calculations, the timescales are in a range of $(6.7 \sim 72)$ hours and the γ-ray luminosities are in a range of $(0.0051 \sim 6.13) \times 10^{48} \mathrm{erg} \cdot \mathrm{s}^{-1}$. So, the mass obtained from our method does not depend on the flux so sensitively, but our method depends on the timescales. This can be expected from relation (5).

The rapid variation, high luminosity, superluminal motion γ-ray in blazars are believed from the beaming effect. The present work shows that the Doppler factors are in the range of $\delta = 1.12$ to $\delta = 4.63$. Our Doppler factor results are consistent with other authors' for some blazars.[28,36,37]

In this paper, the mass of central black hole, M, Doppler factor, δ, are determined for 14 γ-ray loud blazars with available short γ-ray timescales. The masses of central black holes are in range of $(2.07 \sim 15.54) \times 10^7 M_{\odot}$. The mass obtained from our method depends on the timescales. The Doppler factors are in the range of $\delta = 1.12 \sim 4.63$. The mass and the Doppler factor results are consistent with other authors' results. The LAT will give more data for blazars, we can use them to probe the intrinsic nature of blazars in detail in our future work (Yang et al. in preparation).

Acknowledgments

This work is partially supported by the National Natural Science Foundation of China (10573005, 10633010), the 973 project(2007CB815405), and the Fund of the 11th Five-year Plan for Key Construction Academic Subject (Optics) of Hunan Province. We also think the Guangzhou City Education Bureau, which supports our research in astrophysics.

References

1. Mukherjee R, Bertsch D L, Bloom S D, et al., *ApJ*, **490**, 116 (1997).
2. Hartman R C, *ASPC*, **110**, 333 (1996).
3. Ghisellini Gabriele, Madau Piero, *MNRAS*, **280**, 67 (1996).
4. Xie G Z, Bai J M, Zhang X, et al., *A&A*, **334**, 29 (1998).
5. Ghisellini G., Maraschi L, Tavecchio F, *MNRAS*, **396**, 105 (2009).
6. Dermer Charles D, Finke Justin D, Krug Hannah, et al., *ApJ*, **692**, 32 (2009).
7. Graff Philip B, Georganopoulos Markos, Perlman Eric S, et al., *ApJ*, **689**, 68 (2008).

8. Böttcher Markus, Dermer Charles D, Finke Justin D, *ApJ*, **679**, 9 (2008).
9. Cao Xinwu, Jiang D R, *MNRAS*, **331**, 111 (2002).
10. Fan J H, Li J, Zhou J L, et al., *Published by World Scientific Publishing Co., Pte. Ltd., Singapore*, **137** (2007).
11. Fan Jun-Hui, Huang Yong, Yuan Yu-Hai, et al., *RAA*, **9**, 751 (2009).
12. Wang Jian-Min, Chen Yan-Mei, Ho Luis C, et al., *ApJ*, **642**, 111 (2006).
13. Wang Jian-Min, Chen Yan-Mei, Zhang Fan, *ApJ*, **647**, 17 (2006).
14. Wu Xue-Bing, Liu F K, Zhang T Z., *A&A*, **389**, 742 (2002).
15. Kaspi Shai, Smith Paul S, Netzer Hagai, et al., *ApJ*, **533**, 631 (2000).
16. Genzel R, Eckart A, Ott T, et al., *MNRAS*, **291**, 219 (1997).
17. Fan J H, Xie G Z, Bacon R., *A&AS*, **136**, 13 (1999).
18. Cheng K S, Fan J H, Zhang L, *A&A*, **352**, 32 (1999).
19. Fan J H, *A&A*, **436**, 799 (2005).
20. Vestergaard M., *ApJ*, **571**, 733 (2002).
21. Barth Aaron J, Ho Luis C, Sargent Wallace L W., *ApJ*, **566**, 13 (2002).
22. Dermer Charles D, Gehrels Neil, *ApJ*, **447**, 103 (1995).
23. Abdo A A, Ackermann M, Ajello M, et al., *arXiv*0902.1559 (2009).
24. Frank J, King A R, Rain D J, *Cambridge Uni.*, (1985).
25. Sambruna Rita M, Urry C Megan, Maraschi L, et al., *ApJ*, **474**, 639 (1997).
26. Pedro R Capelo, Priyamvada Natarajan., *NJPh*, **9**, 445 (2007).
27. Madejski Greg, Takahashi Tadayuki, Tashiro Makoto, et al., *ApJ*, **459**, 156 (1996).
28. Kniffen D A, Bertsch D L, Fichtel C E, et al., *ApJ*, **411**, 133 (1993).
29. Horan D, Hays E, 22 Jan 2009; 22:59 UT: **ATel** # 1905
30. Vercellone S, D'Ammando G, Pucella F., et. al. 19 Mar 2009; 12:07 UT: **ATel** # 1976
31. Cutini S, Hays E. 21 Apr 2009; 22:45 UT: **ATel** # 2026
32. Mattox J R, Bertsch D L, Chiang J, et al., *ApJ*, **410**, 609 (1993).
33. Pica Andrew J, Smith Alex G, Webb James R, et al., *AJ*, **96**, 1215 (1988).
34. Woo Jong-Hak, Urry C Megan, *ApJ*, **579**, 530 (2002).
35. Cao Xinwu, *ApJ*, **570**, 13 (2002).
36. Wehrle A E, Pian E, Urry C M, et al., *ApJ*, **497**, 178 (1998).
37. Henri G, Pelletier G, Roland J., *ApJ*, **404**, 41 (1993).

HAWKING RADIATION AND THERMALIZATION PHENOMENA IN OPEN QUANTUM SYSTEMS

HONGOWEI YU* and JIA-LIN ZHANG†

Department of Physics and Institute of Physics,
Hunan Normal University, Changsha, Hunan 410081, China
**Email: hwyu@hunnu.edu.cn,* †*Email: zhjl5a7s8d@gmail.com*

We analyze, in the framework of open quantum systems, the reduced dynamics of a static detector (a two-level atom) outside a two-dimensional black hole in weak interaction with a bath of massless quantum scalar fields in the Hartle-Hawking vacuum. We find that the detector outside the black hole is asymptotically driven to a thermal state at the temperature T, which reduces to the Hawking temperature in the spatial asymptotic region, regardless of its initial state. Our discussion therefore shows that the Hawking effect can be understood as a manifestation of thermalization phenomena in the framework of open quantum systems.

Keywords: Hawking radiation; Theory of open quantum systems; Thermalization

Ever since Hawking's discovery that black holes are not, after all, completely black, but, quantum mechanically, emit radiation with a thermal spectrum [1], black holes have been considered as a "Rosetta stone" to relate gravity, quantum theory and thermodynamics. As a result, Hawking radiation, as one of the most striking effects that arise from the combination of quantum theory and general relativity, has attracted widespread interest in physics community and it has been understood and derived in different physical contexts, including, but not limited to, quantum field theory in curved spacetime [1,2], the Euclidean quantum gravity [3,4], the approach based upon string theory [5,6], the proposal which ties its existence to the cancellation of gravitational anomalies at the horizon [7], and our recent study which reveals an interesting relationship between the existence of Hawking radiation and the spontaneous excitation of atoms using the DDC formalism [8] that separates the contributions of vacuum fluctuations and radiation reaction to the rate of change of the mean atomic energy [9].

In the current paper, we shall try to understand the Hawking radiation by analyzing the time behavior of a static detector (modelled by a two-level atom) outside a two-dimensional Schwarzschild black hole immersed in a bath of massless scalar fields in the Hartle-Hawking vacuum [10] in the framework of the theory of open quantum systems which has been fruitfully applied in atomic physics, quantum optics, quantum information and so on. A static detector in vacuum outside a black hole can be treated as open quantum system because it is in constant interaction with vacuum fluctuations of quantum fields. As for any open system, the full dynamics of the detector can be obtained from the complete time evolution of the total system (detector plus fluctuation vacuum fields) by tracing over the field degrees of freedom, which are in fact never observed. We shall show that the detector outside the black hole is asymptotically driven to a thermal state at the temperature T, which reduces to the Hawking temperature in the spatial asymptotic region, regardless of its initial state. The open quantum system approach in our paper therefore demonstrates that the Hawking radiation can be understood as a manifestation of thermalization phenomena in the framework of open quantum systems.

We shall consider the evolution in the proper time of a static detector (two-level atom) interacting with vacuum massless scalar fields outside a two-dimensional spherically symmetric black hole, of which the metric is given by

$$ds^2 = \left(1 - \frac{2M}{r}\right) du\, dv = \frac{2M}{r} e^{-r/2M} d\bar{u}\, d\bar{v} \ , \tag{1}$$

where

$$u = t - r^*, \quad v = t + r^*, \quad r^* = r + 2M \ln[(r/2M) - 1],$$

$$\bar{u} = -e^{-\kappa u}/\kappa, \quad \bar{v} = e^{-\kappa v}/\kappa \ . \tag{2}$$

Here $\kappa = 1/4M$ is the surface gravity of the black hole. In a curved space-time, however, one first has to specify how the vacuum state of the quantum fields is defined, which is related to specification of time and non-occupation of positive frequency modes. Here we shall choose the Hartle-Hawking vacuum defined by taking the incoming modes to be positive frequency with respect to $\bar{v}$, the canonical affine parameter on the future horizon, and outgoing modes to be positive frequency with respect to $\bar{u}$. We assume the combined system (detector + external vacuum fields) to be initially prepared in a factorized state, with the detector held static in the exterior region of the black hole and the fields in the Hartle-Hawking vacuum, and the detector to be fully described in terms of a two-dimensional Hilbert space, so that its states can be represented by a 2×2 density matrix ρ, which is Hermitian $\rho^{+} = \rho$, and normalized $\mathrm{Tr}(\rho) = 1$ with $\det(\rho) \geq 0$. Without loss of generality, we take the total Hamiltonian for the complete system to have the form

$$H = H_s + H_\phi + \lambda\, H' \ . \tag{3}$$

Here H_s is the Hamiltonian of the atom, which is taken, for simplicity, to be

$$H_s = \frac{\omega_0}{2}\sigma_3 \ , \tag{4}$$

where σ_3 is the Pauli matrix and ω_0 the energy level spacing. H_ϕ is the standard Hamiltonian of massless, free scalar fields, details of which need not be specified here and H' is the interaction Hamiltonian of the detector with the external scalar fields and is assumed to be given by

$$H' = \sigma_3 \Phi(x) \ . \tag{5}$$

In order to achieve a rigorous, mathematically sound derivation of the reduced dynamics of the detector, we shall assume that the interaction between the detector and the scalar fields are weak so that the finite-time evolution describing the dynamics of the detector takes the form of a one-parameter semigroup of completely positive maps [11,12]. It should be pointed out that the coupling constant λ in (3) should be small, and this is required by our assumption that the interaction of the atom with the scalar fields is weak.

Initially, the complete system is described by the total density $\rho_{tot} = \rho(0) \otimes |0\rangle\langle 0|$, where $\rho(0)$ is the initial reduced density matrix of the detector, and $|0\rangle$ is the Hartle-Hawking vacuum for field $\Phi(x)$. In the frame of the detector, the evolution in the proper time τ of the total density matrix, ρ_{tot}, of the complete system satisfies

$$\frac{\partial \rho_{tot}(\tau)}{\partial \tau} = -i L_H[\rho_{tot}(\tau)] \ , \tag{6}$$

where the symbol L_H represents the Liouville operator associated with H

$$L_H[S] \equiv [H, S] \ . \tag{7}$$

The dynamics of the atom can be obtained by tracing over the field degrees of freedom, i. e., by applying the trace projection to the total density matrix $\rho(\tau) = \mathrm{Tr}_\Phi[\rho_{tot}(\tau)]$.

In the limit of weak-coupling which we assume in the present paper, the reduced density is found to obey an equation in the Kossakowski-Lindblad form[13,14]

$$\frac{\partial\rho(\tau)}{\partial\tau} = -i\big[H_{\rm eff},\, \rho(\tau)\big] + \mathcal{L}[\rho(\tau)]\ , \tag{8}$$

where

$$\mathcal{L}[\rho] = \frac{1}{2}\sum_{i,j=1}^{3} a_{ij}\big[2\,\sigma_j\rho\,\sigma_i - \sigma_i\sigma_j\,\rho - \rho\,\sigma_i\sigma_j\big]\ . \tag{9}$$

The matrix a_{ij} and the effective Hamiltonian $H_{\rm eff}$ are determined by the Fourier transform, $\mathcal{G}(\lambda)$, and Hilbert transform, $\mathcal{K}(\lambda)$, of the field vacuum correlation functions (Wightman functions)

$$G^+(x-y) = \langle 0|\Phi(x)\Phi(y)|0\rangle\ , \tag{10}$$

which are defined as

$$\mathcal{G}(\lambda) = \int_{-\infty}^{\infty} d\tau\, e^{i\lambda\tau}\, G^+\big(x(\tau)\big)\ , \tag{11}$$

$$\mathcal{K}(\lambda) = \frac{P}{\pi i}\int_{-\infty}^{\infty} d\omega\ \frac{\mathcal{G}(\omega)}{\omega-\lambda}\ . \tag{12}$$

Then the coefficients of the Kossakowski matrix a_{ij} can be written explicitly as

$$a_{ij} = A\delta_{ij} - iB\epsilon_{ijk}\delta_{k3} + C\delta_{i3}\delta_{j3} \tag{13}$$

with

$$A = \frac{1}{2}[\mathcal{G}(\omega_0)+\mathcal{G}(-\omega_0)]\ ,\quad B = \frac{1}{2}[\mathcal{G}(\omega_0)-\mathcal{G}(-\omega_0)]\ ,\quad C = \mathcal{G}(0)-A\ . \tag{14}$$

Meanwhile, the effective Hamiltonian, $H_{\rm eff}$, contains a correction term, the so-called Lamb shift, and one can show that it can be obtained by replacing ω_0 in H_s with a renormalized energy level spacing Ω as follows[15]

$$H_{\rm eff} = \frac{\Omega}{2}\sigma_3 = \{\omega_0 + i[\mathcal{K}(-\omega_0) - \mathcal{K}(\omega_0)]\}\sigma_3\ , \tag{15}$$

where a suitable substraction is assumed in the definition of $\mathcal{K}(-\omega_0)-\mathcal{K}(\omega_0)$ to remove the logarithmic divergence which would otherwise be present.

For a two-dimensional black hole, one can show that the Wightman function for the scalar fields in the Hartle-Hawking vacuum is given by[16]

$$D_H{}^+(\Delta\tau) = -\frac{1}{4\pi}\ln[(\Delta\overline{u}-i\varepsilon)(\Delta\overline{v}-i\varepsilon)] = -\frac{1}{4\pi}\ln\frac{4e^{2\kappa r_*}\sinh^2\big[\kappa_r\Delta\tau/2-i\varepsilon\big]}{\kappa^2}\ , \tag{16}$$

where

$$\Delta\tau = \Delta t\sqrt{g_{00}} = \Delta t\sqrt{1-2M/r}, \quad \kappa_r = \kappa/\sqrt{1-2M/r}\,. \tag{17}$$

Its Fourier transform can be calculated as follows

$$\begin{aligned}\mathcal{G}_H(\lambda) &= \int_{-\infty}^{\infty} D_H{}^{+}(\tau)e^{i\lambda\tau}d\tau \\ &= -\ln\frac{2e^{r*\kappa}}{\kappa}\delta(\lambda) + \frac{\ln 2}{2}\delta(\lambda) + \frac{1+\coth(\pi\lambda/\kappa_r)}{2\lambda}\,.\end{aligned} \tag{18}$$

This leads to

$$A_H = 1/2[\mathcal{G}_H(\omega) + \mathcal{G}_H(-\omega)] = \frac{\coth(\pi\omega/\kappa_r)}{2\omega}\,, \tag{19}$$

$$B = 1/2[\mathcal{G}_H(\omega) - \mathcal{G}_H(-\omega)] = \frac{1}{2\omega}\,.$$

In order to find out how the reduced density evolves with proper time, we express it in terms of the Pauli matrices,

$$\rho(\tau) = \frac{1}{2}\left(1 + \sum_{i=1}^{3}\rho_i(\tau)\sigma_i\right)\,. \tag{20}$$

Substituting Eq. (20) into Eq. (9), it is easy to show that the Bloch vector $|\rho(\tau)\rangle$ of components $\{\rho_1(\tau), \rho_2(\tau), \rho_3(\tau)\}$ satisfies

$$\frac{\partial}{\partial\tau}|\rho(\tau)\rangle = -2\mathcal{H}|\rho(\tau)\rangle + |\eta\rangle\,, \tag{21}$$

where $|\eta\rangle$ denotes a constant vector $\{0, 0, -4B\}$. The exact form of the matrix $\mathcal{H}$ reads

$$\mathcal{H} = \begin{pmatrix} 2A+C & \Omega/2 & 0 \\ -\Omega/2 & 2A+C & 0 \\ 0 & 0 & 2A \end{pmatrix}\,. \tag{22}$$

This matrix is nonsingular and its three eigenvalues are $\lambda_1 = 2A$, $\lambda_\pm = (2A+C) \pm i\Omega/2$. Since the real parts of these eigenvalues are positive, so at later times, $|\rho(\tau)\rangle$ will reach an equilibrium state $|\rho(\infty)\rangle$ [17], which can be found by formally solving the Eq. (21)

$$|\rho(\tau)\rangle = e^{-2\mathcal{H}\tau}|\rho(0)\rangle + (1 - e^{-2\mathcal{H}\tau})|\rho_\infty\rangle\,, \tag{23}$$

with

$$|\rho_\infty\rangle = \frac{1}{2}\mathcal{H}^{-1}|\eta\rangle = -\frac{B}{A}\begin{pmatrix}0\\0\\1\end{pmatrix}\,. \tag{24}$$

If we let $\beta = 1/T = 2\,\text{arctanh}(B/A)/\omega_0$, we can easily show that Eq. (24) can be rewritten in a purely thermal form

$$\rho_\infty = \frac{e^{-\beta H_s}}{\text{Tr}[e^{-\beta H_s}]} \,. \tag{25}$$

Making use of Eq. (19), we find the temperature of the thermal state is

$$T = \frac{\omega}{2\text{arctanh}(B/A)} = \frac{\kappa}{2\pi}\frac{1}{\sqrt{1-\frac{2M}{r}}} = (g_{00})^{-1/2}\,T_H \,, \tag{26}$$

where $T_H = \kappa/2\pi$ is just the Hawking temperature of the black hole. For an detector at spatial infinity ($r \to \infty$), the temperature, T, approaches T_H. However, as the detector approaches the horizon ($r \to 2M$), the temperature T diverges. This can be understood as a result of the fact that the detector must be in acceleration relative to the local free-falling frame to maintain at a fixed distance from the black hole, and this acceleration, which blows up at the horizon, gives rise to additional thermal effect.[18] Thus, regardless of its initial state, a static two-level detector outside the black hole is asymptotically driven to a thermal state at the temperature T. Therefore, there must exist a bath of thermal radiation outside a black hole. Our open system approach thus reveals that the existence of Hawking radiation is simply a manifestation of thermalization phenomena in the framework of open system dynamics.

Further aspects of the Hawking radiation in terms of the thermalization phenomena can be studied by examining the behavior of the finite time solution (23). For this purpose, let us note that

$$e^{-2\mathcal{H}\tau} = \frac{4}{\Omega^2+4C^2}\left\{e^{-4A\tau}\Lambda_1 + 2e^{-2(2A+C)\tau}\left[\Lambda_2\cos(\Omega\tau) + \Lambda_3\frac{\sin(\Omega\tau)}{\Omega}\right]\right\}, \tag{27}$$

where

$$\begin{aligned}
\Lambda_1 &= \left[(2A+C)^2 + \frac{\Omega^2}{4}\right]I - 2(2A+C)\mathcal{H} + \mathcal{H}^2 \,, \\
\Lambda_2 &= -2A(A+C)I + (2A+C)\mathcal{H} - \frac{1}{2}\mathcal{H}^2 \,, \\
\Lambda_3 &= 2A\left[\frac{\Omega^2}{4} - C(2A+C)\right]I + \left[C(4A+C) - \frac{\Omega^2}{4}\right]\mathcal{H} - C\mathcal{H}^2 \,.
\end{aligned} \tag{28}$$

Eq. (27) reveals that a static detector outside a black hole is subjected to the effects of decoherence and dissipation characterized by the exponentially decaying factors involving the real parts of the eigenvalues of $\mathcal{H}$ and oscillating terms associated with the imaginary part. Therefore, the

Hawking radiation as a manifestation of thermalization phenomena in the framework of open quantum systems actually involves phenomena of decoherence and dissipation. In this regard, the open-quantum-system approach to the derivation of the Hawking effect seems to shed new light on the issue as compared to other traditional treatments.

In summary, we have analyzed, using the well-known techniques in the study of open quantum systems, the time evolution of a static detector (a two-level atom) outside a two-dimensional black hole in weak interaction with a bath of massless quantum scalar fields in the Hartle-Hawking vacuum. The detector has been shown to be asymptotically driven to a thermal state at the temperature T, which reduces to the Hawking temperature in the spatial asymptotic region, regardless of its initial state. Our open-system-approach to the issue therefore shows that the Hawking radiation can be understood as a manifestation of thermalization phenomena in the framework of open quantum systems, which actually involves the effects of decoherence and disspation. It is worthwhile to note that the general techniques developed in the theory of open quantum systems may also be applicable to studying other phenomena in curved space-times, such as particle creation, and may thus provide new insights in the physical understanding of these phenomena.

Acknowledgments

One of us (HY) would like to thank C.P. Sun, for interesting discussions at ITP-CAS. This work was supported in part by the National Natural Science Foundation of China under Grants No. 10775050, 10847125 and 10935013, and the SRFDP under Grant No. 20070542002.

References

1. S. Hawking, Nature (London) **248**, 30 (1974).
2. S. Hawking, Commun. Math. Phys. **43**, 199 (1975).
3. G. Gibbons and S. Hawking, Phys. Rev. D **15**, 2752 (1977).
4. M. Parikh and F. Wilczek, Phys. Rev. Lett. **85**, 5042 (2000).
5. A. Strominger and C. Vafa, Phys. Lett. **B 379**, 99(1996).
6. A. Peet, hep-th/0008241.
7. S. P. Robinson and F. Wilczek, Phys. Rev. Lett. **95**, 011303 (2005); S. Iso, H. Umetsu, and F. Wilczek, Phys. Rev. Lett. **96**, 151302 (2006); Phys. Rev. **D74**, 044017 (2006).
8. J. Dalibard, J. Dupont-Roc and C. Cohen-Tannoudji, J. Phys. (France) **43**, 1617(1982);J. Dalibard, J. Dupont-Roc and C. Cohen-Tannoudji, J. Phys. (France) **45**, 637(1984).

9. H. Yu and W. Zhou, Phys. Rev. D **76**, 027503 (2007); *ibid* **76**, 044023 (2007).
10. J. Hartle and S. Hawking, Phys. Rev. **D13**, 2188 (1976).
11. E.B. Davies, *Quantum Theory of Open Systems* (Academic Press, New York, 1976).
12. H.-P. Breuer and F. Petruccione, *The Theory of Open Quantum Systems* (Oxford University Press, Oxford, 2002).
13. F. Benatti, R. Floreanini and M. Piani, Phys. Rev. Lett. **91**, 070402 (2003).
14. V. Gorini, A. Kossakowski, and E. C. G. Surdarshan, J. Math. Phys. **17**, 821 (1976); G. Lindblad, Commun. Math. Phys. **48**, 119 (1976).
15. F. Benatti and R. Floreanini , Phys. Rev. A **70**, 012112 (2004).
16. N. D. Birrell and P. C. W. Davies, *Quantum Field Theory in Curved Space* (Cambridge Univ. Press, Cambridge, 1982).
17. K. Lendi, J. Phys. A20, 15(1987).
18. W. G. Unruh, Phys. Rev. D **14**, 870 (1976).

REPULSIVE CASIMIR FORCE, REALIZABLE OR NOT?

XIANG-HUA ZHAI

Shanghai United Center for Astrophysics(SUCA), Shanghai Normal University,
100 Guilin Road,Shanghai 200234, China
**E-mail: zhaixh@shnu.edu.cn*

Casimir energies and forces have been calculated in various configurations and boundary conditions. The calculations indicated that the Casimir energy might change its sign depending not only on the boundary conditions but also on geometry and topology of the configuration. With the development of nanotechnology, it is known that repulsive Casimir force is very important in nanodevices. In this paper, we review some research results on repulsive Casimir force, and discuss whether it could be realizable theoretically and experimentally.

Keywords: Casimir force; repulsion; piston; nanodevices.

1. Introduction

Casimir predicted in 1948 that an attractive force should act between two plane-parallel uncharged perfectly conducting plates in vacuum.[1] In the development of several decades, considerable progress has been made both in experiment and theory. Especially in recent 10 years, the effect has been paid more attention because of the development of precise measurements.[2] At the same time, Casimir energies and forces have been calculated theoretically in various configurations and for real media.[3] But there are still some interesting topics and unsolved problems related to this effect such as the puzzle of the thermal Casimir force, Casimir effect as a test for new physics and the application of the Casimir effect in cosmology.

In recent years, with the development of the nanotechnology, the influence of the Casimir force in nanomechanical device fabrication becomes more and more important.[4] The Casimir force dominates over other forces at distances of a few nanometers but the attraction Casimir force would create undesirable effects on nanodevices known as stiction, which would degrade or deteriorate the functionality of the devices. So there is an impelling need to search for repulsive Casimir force. But does the repulsive

Casimir force really exist?

About this topic, theoretically, a lot of work has been done but it is still a controversial one. The earliest work of repulsive Casimir force was given by Boyer for a conducting spherical shell[5]. And then it was claimed that the Casimir energy inside rectangular cavities can be either positive or negative depending on the ratio of the sides [6]. Furthermore, the repulsive Casimir force could also exist for appropriate boundary conditions. In ideal case, the Casimir force between a perfectly conducting plate and an infinitely permeable one (associated to Dirichlet-Neumann boundary conditions for scalar field) is repulsive [7]. For real boundary materials, the repulsive Casimir force could appear for materials with nontrivial magnetic permeability (metamaterials)[8]. Another possibility to get repulsive force is the use of two solid material boundaries immersed in a liquid whose permittivities satisfy certain condition. This situation was firstly studied by Lifshitz [9] and this is a measurable case. It was verified recently and it will be mentioned in the experimental section of this paper.

In this paper, I will mainly discuss about the repulsive Casimir force in rectangular boxes and its modified model-piston for the theory of repulsive Casimir force. For the experimental development, I will talk about the recent paper published in Nature on the measurement of repulsive Casimir force[10] which shows an exciting prospect in the application to nanoscale devices.

2. Theory of repulsive Casimir force -rectangular boxes and pistons

Firstly, let us review the result of the Casimir force in 2-dimensional box. For a massless scalar field with Dirichlet boundary conditions confined in the box $0 \leq x \leq a, 0 \leq y \leq b$, the vacuum energy (in units $\hbar = c = 1$) is given by

$$E_0(a,b) = \frac{1}{2}\sum_{j,k=1}^{\infty} \omega_{j,k} \tag{1}$$

where $\omega_{j,k} = \sqrt{(\frac{j\pi}{a})^2 + (\frac{k\pi}{b})^2}$ are eigenfrequencies and j, k are integers.

Using zeta function regularization, the summation is obtained as

$$E_{0,AR}(a,b) = -\frac{ab}{32\pi}Z_2(a,b:3) + \frac{\pi}{48}(\frac{1}{a} + \frac{1}{b}) \tag{2}$$

Where $Z_p(a_1,\cdots,a_p;s)$ is Epstein zeta function which is defined as $Z_p(a_1,\cdots,a_p;s) \equiv \sum_{n_1,\cdots,n_p=-\infty}^{\infty}{}' \left[(n_1a_1)^2+\cdots+(n_pa_p)^2\right]^{-\frac{s}{2}}$ and the prime means that the term $n_1 = n_2 = \cdots = n_p = 0$ has to be excluded.

The Casimir tension $T = -\partial E_{0,AR}/\partial A$ depends on the ratio b/a where $A = ab$ is the area of the box. The calculation tells us that when $1 \leq b/a \leq 2.74$ the force is repulsive and when $b/a > 2.74$ the force is attractive.

But the security of the conclusion needs to suspect because the calculations ignore the divergent term associated with the boundaries and the nontrivial contribution from the outside region of the box[11]. Recently, a modification of the rectangle–"Casimir piston" was introduced to avoid the above problems[12].The Casimir force on the piston is a well-defined finite force because the position of the piston is independent of the divergent terms in the internal vacuum energy and the external region. For a scalar field obeying Dirichlet boundary conditions on all surfaces, when the separation between the piston and one end of the cavity approaches infinity, the force on the piston is towards another end (the closed end), that is, the force is attractive. Successively, the Casimir force on the piston was studied for different dimensions, different fields and different boundary conditions[13]. The results indicate that the Casimir force on the piston can be attractive or repulsive for different cases. We discussed Casimir pistons for a massless scalar field with hybrid boundary conditions and obtained the repulsive Casimir force on the piston[14] and also gave the result for massive scalar fields considering the influence of the mass of the fields[15].

The three-dimensional piston is depicted in Fig. 1, where the boundary condition on the piston is Neumann and those on other surface are Dirichlet. For simplicity, we take the base as a square. The vacuum energy in cavity A is

$$E(a,b,b) = \frac{1}{2}\sum_{m=0}^{\infty}\sum_{n_1,n_2=1}^{\infty}\sqrt{\left(m+\frac{1}{2}\right)^2\left(\frac{\pi}{a}\right)^2+\left(\frac{n_1\pi}{b}\right)^2+\left(\frac{n_2\pi}{b}\right)^2} \tag{3}$$

When $a > b$, we get the regularized vacuum energy in cavity A as

$$\begin{aligned} E_R^A(a,b,b) = & -\frac{a\beta(2)}{48b^2}+\frac{\zeta(3)a}{16\pi b^2} \\ & +\frac{1}{2b}\sum_{m,n_1,n_2=1}^{\infty}\frac{\sqrt{n_1^2+n_2^2}}{m}\left[K_1\left(2\pi m\sqrt{n_1^2+n_2^2}\frac{a}{b}\right)\right. \\ & \left.-K_1\left(4\pi m\sqrt{n_1^2+n_2^2}\frac{a}{b}\right)\right] \end{aligned} \tag{4}$$

where $\zeta(s)$ is Riemann zeta function, $K_n(z)$ is the modified Bessel function and $\beta(2)$ is a Dirichlet series defined as $\beta(s) \equiv \sum_{n=0}^{\infty}(-1)^n(2n+1)^{-s}$ which comes from the relation $Z_2(1,1;s) = 4\zeta(s)\beta(s)$ [16] during the regularization. Substituting Eq. (4) and the corresponding expression for the regularized vacuum energy in cavity B into the following expression for Casimir force on the piston

$$F = -\frac{\partial}{\partial a}\left[E_R^A(a,b,b) + E_R^B(L-a,b,b)\right] \tag{5}$$

and taking $L \to \infty$, we obtain the force on the piston as

$$\lim_{L\to\infty} F = \frac{\pi}{b^2}\sum_{m,n_1,n_2=1}^{\infty}(n_1^2+n_2^2)\left[2K_1'\left(4\pi m\sqrt{n_1^2+n_2^2}\frac{a}{b}\right)\right.$$
$$\left.-K_1'\left(2\pi m\sqrt{n_1^2+n_2^2}\frac{a}{b}\right)\right] \tag{6}$$

The force is positive (i. e. repulsive) from the result of numerical calculation and it approaches zero with the ratio of a/b approaching infinity.

In the case that $a < b$, the regularized vacuum energy in cavity A can be reexpressed as

$$E_R^A(a,b,b) = \frac{7\pi^2 b^2}{11720a^3} - \frac{3\zeta(3)b}{64\pi a^2} + \frac{\pi}{192a}$$
$$+\frac{1}{4a}\sum_{m,n=1}^{\infty}\frac{m}{n}\left[K_1(\pi mn\frac{b}{a}) - 2K_1(2\pi mn\frac{b}{a})\right]$$
$$+\frac{b^{1/2}}{2a^{3/2}}\sum_{m,n=1}^{\infty}\left(\frac{m}{n}\right)^{3/2}\left[K_{3/2}(2\pi mn\frac{b}{a}) - \frac{\sqrt{2}}{4}K_{3/2}(2\pi mn\frac{b}{a})\right]$$
$$+\frac{b^{1/2}}{2a^{3/2}}\sum_{m,n_1,n_2=1}^{\infty}\left(\frac{m}{\sqrt{n_1^2+n_2^2}}\right)^{3/2}\left[K_{3/2}(2\pi m\sqrt{n_1^2+n_2^2}\frac{b}{a})\right.$$
$$\left.-\frac{\sqrt{2}}{4}K_{3/2}(2\pi m\sqrt{n_1^2+n_2^2}\frac{b}{a})\right] \tag{7}$$

Then the force on the piston is

$$\lim_{L\to\infty} F = \frac{7\pi^2 b^2}{3840a^4} - \frac{3\zeta(3)b}{32\pi a^3} + \frac{\pi}{192a^2} - \frac{\beta(2)}{48b^2} + \frac{\zeta(3)}{16\pi b^2}$$
$$-\frac{\pi b}{4a^3}\sum_{m,n=1}^{\infty} m^2\left[K_0(\pi mn\frac{b}{a}) - 4K_0(2\pi mn\frac{b}{a})\right]$$
$$-\frac{\pi b^{3/2}}{a^{7/2}}\sum_{m,n=1}^{\infty}\frac{m^{5/2}}{n^{1/2}}\left[K_{1/2}(2\pi mn\frac{b}{a}) - \frac{\sqrt{2}}{8}K_{1/2}(\pi mn\frac{b}{a})\right]$$
$$-\frac{\pi b^{3/2}}{a^{7/2}}\sum_{m,n_1,n_2=1}^{\infty}\frac{m^{5/2}}{(n_1^2+n_2^2)^{1/4}}\left[K_{1/2}(2\pi m\sqrt{n_1^2+n_2^2}\frac{b}{a})\right.$$
$$\left.-\frac{\sqrt{2}}{8}K_{1/2}(\pi m\sqrt{n_1^2+n_2^2}\frac{b}{a})\right] \tag{8}$$

The force is again repulsive and decreases with the ratio a/b increasing (see Fig. 2).

For the special case that $a = b$, which means cavity A is a cube, we find from both Eq. (6) and Eq. (8) that the force on the piston is (in unit $\hbar c$) $F = \frac{0.00041244}{b^2}$.

The piston model is the simple generalization of the single-cavity problem. The calculation is rigorous and exact, but it is not the purely internal vacuum pressure on the side of rectangular cavity. The recent study [17] pointed out that the attraction (or repulsion for a piston with Neumann boundary conditions) of a piston to the nearest face of the box does not negate the Casimir repulsion of boxes without a piston that have some appropriate ratio of the sides because the cases with an empty space outside the box and that with another section of the larger box outside the piston are physically quite different. The result indicates that the electromagnetic Casimir force between the opposite faces of a cube is repulsive for cubes of any size and it is increases with increasing temperature.

Although there have been numerous papers on theoretical study of Casimir energy and force in rectangular boxes or extended models, no experiment can be performed for these configurations. Actually, experimentally verification can only be made at present for simple configurations of two rigid plates or between one plate and one spherical face. The measurement for Casimir force in any single body such as a ball or rectangular box is unrealizable so no one can tell whether the Casimir force in these configurations is repulsive or attractive.

3. Experiment for repulsive Casimir force

Evidence for repulsive van der Waals interactions between solids separated by a liquid has been presented[18]. So it is natural to search for the repulsive Casimir force. The research group of Harvard [10] did the first measurement of long-range repulsive forces between solids separated by a fluid and they showed that the results are consistent with Lifshitz's theory within the uncertainties of the optical properties of the materials.

Repulsive forces between macroscopic bodies can be qualitatively understood by considering their material polarizabilities. The interaction of material 1 with material 2 across medium 3 goes as a summation of terms with differences in material permittivities $-(\epsilon_1-\epsilon_3)(\epsilon_2-\epsilon_3)$ over frequencies that span the entire spectrum[19]. When $\epsilon_1 = \epsilon_2$, $-(\epsilon_1 - \epsilon_3)(\epsilon_2 - \epsilon_3) < 0$, the force is attractive. However, when $\epsilon_1 > \epsilon_3 > \epsilon_2$, $-(\epsilon_1 - \epsilon_3)(\epsilon_2 - \epsilon_3) > 0$, the force is repulsive. Examples of material systems that satisfy the requirement $\epsilon_1 > \epsilon_3 > \epsilon_2$ are rare but do exist. The set of materials (solid-liquid-solid) they chose in their experiment that obeys the above inequality over a large frequency range is gold, bromobenzene and silica.

The detailed measurements show that the long-range quantum electrodynamics forces between solid bodies can become repulsive when the optical properties of the materials are properly chosen. For example, it might be possible to "tune" the liquid (possibly by mixing two or more liquids) so that the force becomes attractive at large separation, but remains repulsive at short range. this would provide the means for quantum levitation of an object and could lead to the suppression of stiction and to ultra-low friction devices and sensors.

4. Summary

Theoretically, a lot of research has been done on repulsive Casmir force, including closed geometry and boundary materials. Repulsive Casimir force has been found experimentally for two solid material boundaries immersed in a liquid whose permittivities satisfy certain conditions. The applications of the repulsive Casimir force to nanodevices remain to be investigated, but the prospects look exciting.

As a famous and important discovery, after more than 60 years development, Casimir effect still left a lot of mysteries for us to investigate. So we expect there will be new breakthroughs both in theory and practical application in the future, especially for repulsive Casimir force.

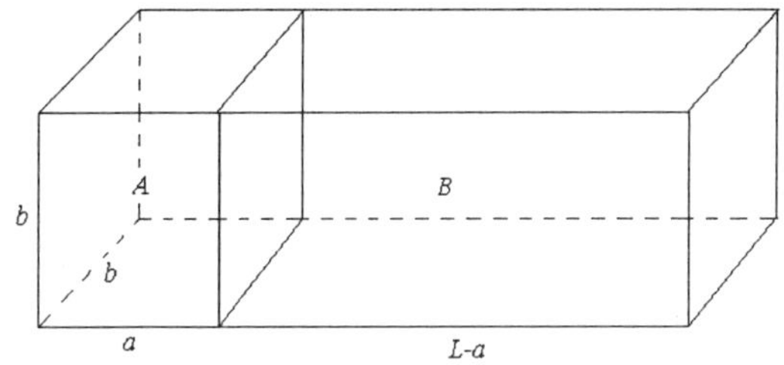

Fig. 1. Casimir piston in three dimensions.

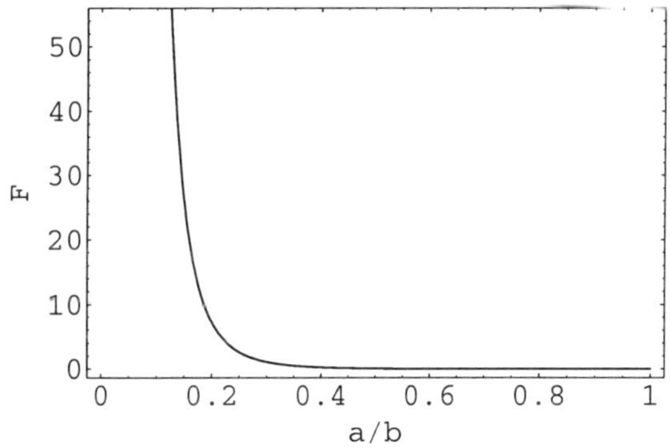

Fig. 2. Casimir force F (in units hc/b^2) on a three-dimensional piston versus a/b where a is the plate separation and b is the length of the sides of the square base.

Acknowledgments

This work is supported by National Nature Science Foundation of China under Grant No. 10671128 .

References

1. H. B. G. Casimir, *Proc. K. Ned. Akad. Wet.* **51**, 793 (1948).
2. S. K. Lamoreaux, *Phys. Rev. Lett.* **78**, 5 (1997); U. Mohideen and A. Roy, *Phys. Rev. Lett.* **81**, 4549 (1998); A. Roy, C. Y. Lin and U. Mohideen, *Phys. Rev. D* **60**, 111101 (1999); G. Bressi, G. Garugno, R. Onofrio and G. Ruoso, *Phys. Rev. Lett.* **88**, 041804 (2002); R. S. Decca, D. Lopez, E. Fischbach, G. L. Klimchitskaya, D. E. Krause, V. M. Mostepanenko, *Ann. Phys. (N. Y.)* **318**, 37 (2005); R. S. Decca, D. López, E. Fischbach, G. L. Klimchitskaya, D. E. Krause, V. M. Mostepanenko, *Phys. Rev. D* **75**, 077101 (2007).

3. M. Bordag, U. Mohideen and V. M. Mostepanenko, *Phys. Rep.* **353**, 1 (2001); H. Gies and K. Klingmuller, *Phys. Rev. Lett.* **96**, 220401 (2006).
4. F. M. Serry, D. Walliser and G. J. Maclay, *J. Microelectromech. Syst.* **4**, 193 (1995); H. B. Chan, V. A. Aksyuk, R. N. Kleiman, D. J. Bishop and F. Capasso, *Science* **291**, 1941 (2001).
5. T. H. Boyer, *Phys. Rev.***174**, 1764(1968).
6. W. Lukosz, *Physica* **56**, 109 (1971); F. Caruso, N. P. Neto, B. F. Svaiter and N. F. Svaiter, *Int. J. Mod. Phys. A* **14**, 2077 (1999); S. Hacyan, R. Jauregui and C. Villarreal, *Phys. Rev. A* **47**, 4204 (1993); X. Z. Li, H. B. Cheng, J. M. Li and X. H. Zhai, *Phys. Rev. D* **56**, 2155 (1997) ; X. Z. Li and X. H. Zhai, *J. Phys. A* **34**, 11053 (2001);A. Edery, *J. Math. Phys.* **44**, 599 (2003).
7. T. H. Boyer, *Phys. Rev. A* **9**, 2078 (1974); V. Hashwater, *Am. J. Phys.* **65**, 381 (1997).
8. C. G. Shao, D. L. Zhang and J. Luo, *Phys. Rev. A* **74**, 012103(2006); O. Kenneth, I. Klich, A. Mann and M. Revzen, *Phys. Rev. Lett.* **89**, 033001 (2002); I. G. Pirozhenko, A. Lambrecht, *J. Phys. A* **41**, 164015 (2008).
9. E. M. Lifshitz, *Sov. Phys. JETP* **2**, 73 (1956).
10. J. N. Munday, *Nature* **457**, 170 (2009).
11. G. Barton, *J. Phys. A* **34**, 4083(2001); N. Graham, R. L. Jaffe, V. Khemani, M. Quandt, M. Scandurra, and H. Weigel, *Phys. Lett. B* **572**, 196(2003); N. Graham, R. L. Jaffe, V. Khemani, M. Quandt, M. Scandurra, O. Schroeder, and H. Weigel, *Nucl. Phys. B* **677**, 379(2004).
12. R. M. Cavalcanti, *Phys. Rev. D* **69**, 065015(2004).
13. M. P. Hertzberg, R. L. Jaffe, M. Kardar, and A. Scardicchio, *Phys. Rev. Lett.* **95**, 250402(2005); G. Barton, *Phys. Rev. D* **73**, 065018(2006); V. N. Marachevsky, *Phys. Rev. D* **75**, 085019(2007); A. Edery, *Phys. Rev. D* **75**, 105012(2007); S. A. Fulling, L. Kaplan and J. H. Wilson, *Phys. Rev. A* **76**, 012118(2007); A. Edery and I. MacDonald, *JHEP* **09**, 005(2007); A. Rodriguez, M. Ibanescu, D. Iannuzzi, J. D. Joannopoulos, S. G. Johnson, *Phys. Rev. A* **76**, 032106 (2007).
14. X. H. Zhai and X. Z. Li, *Phys. Rev. D* **76**, 047704(2007).
15. X. H. Zhai, Y. Y. Zhang and X. Z. Li, *Mod. Phys. Lett.A* **24**, 393 (2009).
16. I. J. Zucker, *J. Phys. A* **9**, 499 (1976).
17. B. Geyer, G. L. Klimchitskaya, V. M. Mostepanenko, *Eur. Phys. J. C* **57**, 823 (2008).
18. A. A. Feiler, L. Bergstrom and M. W. Rutland, *Langmuir* **24**, 2274 (2008); S. Lee and W. M. Sigmund, *J. Colloid Inteface Sci.* **243**, 365 (2001).
19. V. A. Pasegian, *van der Waals Forces: A Handbook for Biologists, Chemists, Engineers, and Physicists* (Cambridge Univ. Press, 2006).

THE ROLE OF VARIATIONS OF CENTRAL DENSITY OF WHITE DWARF PROGENITORS UPON TYPE IA SUPERNOVAE

R. FISHER*, D. FALTA*, G. JORDAN**, and D. LAMB**

* *Department of Physics, University of Massachusetts Dartmouth, North Dartmouth, MA 02747-2300, United States*

** *Department of Astronomy & Astrophysics, University of Chicago, Chicago, IL. 60637, United States*

E-mail: robert.fisher@umassd.edu

Web : http://www.novastella.org

The discovery of the accelerated expansion of the universe using Type Ia supernovae (SNe Ia) has stimulated a tremendous amount of interest in the use of SNe Type Ia events as standard cosmological candles, and as a probe of the fundamental physics of dark energy. Recent observations of SNe Ia have indicated a significant population difference depending on the host galaxy. These observational findings are consistent with SNe Ia Ni-56 production in star-forming spiral galaxies some 0.1 solar masses higher – and therefore more luminous – than in elliptical galaxies. We present recent full-star, 3D simulations of Type Ia supernovae which may help explain the nature of this systematic variation in SNe Ia luminosities, as well as the nature of the Ia explosion mechanism. These insights may in turn eventually shed light on the mystery of dark energy itself.

Keywords: Type Ia supernovae; computational astrophysics.

1. Introduction : Observational Properties of Type Ia Supernovae

The discovery of the accelerated expansion of the universe using Type Ia supernovae[1,2] has stimulated a tremendous amount of interest in the use of SNe Type Ia events as standard cosmological candles, and as a probe of the fundamental physics of dark energy.

Supernovae come in different types, classified by their spectra. They are grouped into two main categories based on the presence or absence of Balmer absorption lines, indicating the presence or absence of atomic hydrogen. Type II contain hydrogen, whereas Type I lack hydrogen. These types are further subdivided by absorption line features and light curve shapes, see Table 1. In particular, Type Ia exhibit an absorption line of singly ionized silicon (Si^+) at 615 nm at peak light. Type Ib/c are further classified according to the absence of Si II lines, and the presence or absence of He lines, respectively.

Accordingly, Type Ia properties are consistent with white dwarf progenitors. The two leading models include white dwarfs in binary systems – either with a non-degenerate companion, the so-called "single-degenerate" scenario, or with another white dwarf – the so-called "double-degenerate" scenario. While both types of events are likely to occur in nature, and contribute to the diversity of explosion energies, there is still considerable debate over their relative frequencies, and even what the explosion mechanism is in each case. We outline some key theoretical and mounting observational evidence below based on nearby historic supernovae, which supports the identification of typical brightness "Branch normal" supernovae with single-degenerate events.

Table 1. Classification of Supernovae

Mass ⟶		⟵ Age
SN Ia	SN II (IIn, IIL, IIP, IIb)	SN Ib/c
No Hydrogen (Si^+ absorption) White dwarf Progenitor	Has Hydrogen Core collapse of a massive star	No Hydrogen (no Si^+) Core collapse (outer layers stripped by winds)

Tying the variable luminosities of Type Ia explosions to additional parameters is of key significance in any attempt to calibrate the Phillips relation. Analysis of light spectra demonstrates that Type Ia supernovae produce a relatively constant combined yield of stable nuclear statistical equilibrium nuclei (NSE) of ^{58}Ni and ^{54}Fe and intermediate mass elements (IME) Si-Ca of $1.05 \pm .09\ M_\odot$.[3] The approximately fixed burned mass over a wide range of SNe Ia luminosities suggests that SNe Ia share a single common explosion mechanism. However, recent observations of SNe Type Ia have indicated a significant population difference depending on the host galaxy. In particular, observations have found that SNe Ia in star-forming galaxies decline more slowly than those in ellipticals.[4] These findings are consistent with Ia SNe ^{56}Ni production in star-forming galaxies some $\sim 0.1\ M_\odot$ higher than that in ellipticals.[5]

Previous work has focused on the possible influence of the metallicity as a second parameter in explaining the luminosity variance of Type Ia events.[5–8] These results suggest that while the metallicity effect may indeed contribute to the observed variance in Type Ia luminosities, alone it accounts for only a portion of the total variance.[5,6] Consequently, we must look elsewhere to other physical effects – including both the central density and angular momentum profile of the white dwarf progenitors – which may explain the majority of the observed Ia luminosity variance. These progenitor properties may in turn be directly influenced by the surrounding environment of Ia event, in particular the accretion rate from the companion star.

To put the single- and double-degenerate white dwarf progenitor models into their proper astrophysical context, we briefly consider the observational evidence. A number of recent papers have lended strong support for the general viability of the single-degenerate channel for the origin of Branch normal SNe Type Ia, in which a progenitor white dwarf accretes material from a non-degenerate companion star.[9,10] Analysis of the spectrum from the Type Ia SN 2006X in Virgo suggests that the blast wave moved through the circumstellar medium and collided with the ejecta from an otherwise undetected red giant companion star.[11] Furthermore, a candidate G-type companion to SN1572 (Tycho) has been identified.[12] SN1572 has independently been identified as a Branch-normal Ia event based on analysis of historical light curves and light echo spectra.[13,14]

However, despite this string of recent successes, the single degenerate model faces significant theoretical challenges. In particular, models suggest that a non-rotating white dwarf can burn stably only over a relatively narrow mass accretion rate range.[10] The addition of differential rotation broadens this range.[15] However, when one introduces rotation to the white dwarf structure, one must make a number of assumptions regarding accretion and internal shear in order to understand the evolution of the angular momentum distribution of the white dwarf progenitor. The accretion rate in turn sets the central density of the white dwarf at ignition.

As a consequence of these challenges faced by theoretical descriptions of the single degenerate model, and the limited direct observational evidence constraining progenitors, there is a significant degree of uncertainty in the characterization of the degenerate progenitor. Here we focus on the role which the central density (or equivalently, the total mass) of a non-rotating white dwarf progenitor plays in determining the luminosity of Type Ia SNe.

2. Physics of Type Ia Supernovae

The energetics of the single-degenerate model of Type Ia supernovae can be simply estimated using nothing more than elementary physics. The internal energy of N electrons in a fully-degenerate white dwarf is N times the characteristic Fermi energy E_F. The Fermi energy can itself be estimated for a relativistic electron with momentum p as

$$E_F = pc \tag{1}$$

Applying the Heisenberg Uncertainty Principle, $p \sim \hbar n^{\frac{1}{3}}$, where n is the number density of electrons. Therefore,

$$E_F \sim \hbar n^{\frac{1}{3}} c \sim \frac{\hbar N^{\frac{1}{3}} c}{R} \tag{2}$$

Where in the second step we have estimated the mean number density within a white dwarf of radius R. Including the gravitational binding energy, the white dwarf has a total energy of

$$E = NE_F + E_G \sim \frac{\hbar N^{\frac{4}{3}} c}{R} - \frac{GN^2 m_p^2}{R} \tag{3}$$

here E_G is the binding energy, and m_p is the proton mass. Note that we focus here on the essential physics, and have therefore suppressed all factors of order unity, and neglected composition effects.

This elementary analysis reveals a remarkable feature of a fully degenerate star. In a fully degenerate gas, the degeneracy pressure is fundamentally independent of temperature. This immediately leads to the result that both the internal energy term and the gravitational binding energy term scale inversely with radius. Consequently, a fully-degenerate white dwarf cannot seek a lower total energy state by an adiabatic spherical compression or expansion. This stands in sharp contrast to stars supported by ordinary gas pressure, which can indeed minimize their total energy by spherical adiabatic compression or expansion. We are led to conclude that the stability of the white dwarf is therefore set solely by the sign of the total energy E. The critical case is where the total energy E equals zero; solving for the maximum total mass M_{Chandra} of the star then yields

$$M_{\text{Chandra}} \sim \left(\frac{\hbar c}{G}\right)^{\frac{3}{2}} \frac{1}{m_p^2} \sim m_{\text{Planck}} \left(\frac{m_{\text{Planck}}}{m_p}\right)^2 \sim 1.5\ M_\odot \tag{4}$$

This analysis shows the critical mass, which is known as the Chandrasekhar limit, to be a combination of fundamental physical constants. Indeed, it is fundamentally set by the Planck mass times a large dimensionless number, which is the ratio of the Planck mass to the proton mass, squared. This simple analysis, which neglects composition effects, demonstrates the Chandrasekhar mass is of order a solar mass; a more precise calculation demonstrates it to be 1.4 $M_\odot$ for a predominantly C/O white dwarf.

We take the progenitor to be a carbon-oxygen white dwarf. The mass, which is close to the Chandrasekhar limit, undergoes carbon burning, releasing roughly 10^{18} ergs/g. The characteristic nuclear energy available for a Type Ia is about $3 \cdot 10^{51}$ ergs $\sim$ 3 foe, where 1 foe is a convenient unit representing 10^{51} ergs, a typical Type Ia luminosity. Therefore, there is more than sufficient nuclear energy within the white dwarf progenitor to represent not only typical Type Ia events, but also the more luminous events observed.

In the single-degenerate model of a Type Ia supernova event, the system begins as a binary pair of main sequence stars. As time progresses, the more massive star evolves into a giant, accreting gas onto its companion, which expands and is eventually engulfed. The core of the giant and its companion begin to spiral inward inside a common envelope. Tidal torques cause the envelope to be ejected, and the binary separation to decrease. The core of the giant then collapses into a white dwarf that begins to accrete gas from its aging companion. The white dwarf continues to accrete until reaching a critical mass close to but not equal to the Chandrasekhar limit. At a critical mass set by the accretion rate from the companion, the central core of the white dwarf ignites a nuclear flame that causes the supernova explosion, and ejects the companion from the system.

The precise initial conditions leading to ignition are still poorly understood. The problem arises because of a large dynamic range between the long convection phase (on the order of a hundred years), and the brief deflagration/detonation phase following flame ignition, which is on the order of one to several seconds.[16] At some point during convection, a runaway nuclear burning occurs during which the burning exceeds the neutrino cooling. The runaway reaction ignites a deflagration flame slightly off-center from the white dwarf, giving rise to a buoyancy force. This buoyant flame bubble undergoes subsonic burning while rising toward the surface. The possibility of the bubble burning though the entire star in a pure deflagration[17] has generally lost support because it tends to burn inefficiently, leaving behind a significant amount of nuclear fuel, and producing events which are generally underluminous with respect to typical Branch normal Ia events. It is most likely that the bubble either undergoes a deflagration to detonation transition (DDT)[18] or leads to a gravitationally-confined detonation (GCD).[19] In the case of a GCD, the deflagration bubble breaks through surface of the star. Ash is launched into the atmosphere and wraps around the surface of the star under the force of gravity. The ash quickly reaches the opposite side of the star and collides with itself. This collision creates a supersonic detonation front that unbinds the star.

At present, both the DDT and GCD mechanisms have their advantages and disadvantages. The DDT model can be tuned to be consistent with observations. However, in the absence of a full first-principles understanding of the detonation transition, DDT simulations set the detonation transition as a parameter. The current belief is that detonation can occur when the flame is ripped apart by turbulence, and transitions into the distributed burning regime. However, the precise conditions under which this transition occurs remain a matter of intense research, so that the transition density is largely a free parameter in the simulations. On the other hand, the GCD mechanism can successfully produce detonations without fine-tuning, though initial models typically overproduced ^{56}Ni, underproduced intermediate mass elements (IME), and generated overluminuous events. We address one aspect of the luminosity problem in the following section, by varying the central density of the white dwarf progenitor models in the simulations.

3. Simulations of Type Ia Supernovae

We begin with a carbon-oxygen white dwarf with a pre-ignited flame bubble of initialized size and location. The model is run through detonation. The evolution of the white dwarf and bubble has a broad range of length scales. On the largest scales, one must be able to follow the expansion of the supernovae out to several tens of thousands of kilometers to capture the homologous expansion phase, and on the smallest scales, one must begin to capture the flame physics, which extends down to the laminar flame thickness on centimeter scales. We employ the Paramesh library within the FLASH code to implement an adaptive mesh refinement (AMR) mesh. Even the power of AMR still does not allow one to avoid the enormous dynamic range between the large-scale physics of the explosion and the flame scale – over 10^9 in linear dimension – so we incorporate a thickened model of the flame surface, which artificially thickens the flame over ~ 4 grid cells.

We evolve the simulation using a comprehensive multiphysics model, including the coupled equations of hydrodynamics, self-gravity, and nuclear combustion. Specifically, we incorporate the Euler equations of hydrodynamics :

$$\frac{\partial \rho}{\partial t} + \vec{\nabla} \cdot (\vec{v}\rho) = 0 \tag{5}$$

$$\frac{\partial \rho \vec{v}}{\partial t} + \vec{\nabla} \cdot (\vec{v} \otimes \rho \vec{v}) = -\vec{\nabla} P - \rho \vec{\nabla} \Phi \tag{6}$$

$$\frac{\partial \rho E}{\partial t} + \vec{\nabla} \cdot [\vec{v}(\rho E + P)] = \rho \vec{v} \cdot \vec{\nabla} \Phi + \rho \epsilon_{nuc} \tag{7}$$

with source terms for self-gravity and nuclear energy release. Here ρ is mass density, $\vec{v}$ is velocity, P is pressure, Φ is gravitational potential, ϵ_{nuc} is the specific nuclear energy release. E is the total energy density, given by

$$E = \rho \left(U + \frac{1}{2} v^2 \right) \tag{8}$$

Where U is the specific internal energy.

The Euler equations are coupled to Poisson's equation for self-gravity

$$\nabla^2 \Phi = 4\,\pi G \rho \tag{9}$$

and an advection-diffusion reaction model of the thickened combustion front

$$\frac{\partial \phi}{\partial t} + \vec{v} \cdot \vec{\nabla} \phi = \kappa \nabla^2 \phi + \frac{1}{\tau} R(\phi) \tag{10}$$

$$R(\phi) = \frac{f}{4} (\phi - \epsilon_0)(1 - \phi + \epsilon_1) \tag{11}$$

Here ϕ is a scalar progress variable which monitors the advancement of the flame surface, and sets the nuclear energy release function ϵ_{nuc} above. κ is a parameter representing the diffusivity of the thickened flame, τ is the reaction timescale, R is the reaction term, which is set by a lowered Kolmogorov-Petrovski-Piscounov (KPP) binomial, specified by the three parameters, f, ϵ_0, and ϵ_1.

Our first sucessful 3-D simulation of a Type Ia detonation was a cold white dwarf model in initial equilibrium with initial mass 1.36 $M_\odot$. The nuclear bubble was ignited within a spherical region slightly offset from the center of the white dwarf. The supercomputer simulation "marches" this condition forward in time in 3-D, using full equations describing the flame, hydrodynamics, and self-gravity. An numerical analysis yields a critical conditions for initiation.[20] We found that the critical conditions for degenerate white dwarf matter are robustly met in our 3-D simulations. We have confirmed that detonation arises independent of the resolution in the detonation region, and also for a wide variety of initial bubble sizes and offsets. Simulations of the GCD model produce intermediate mass elements

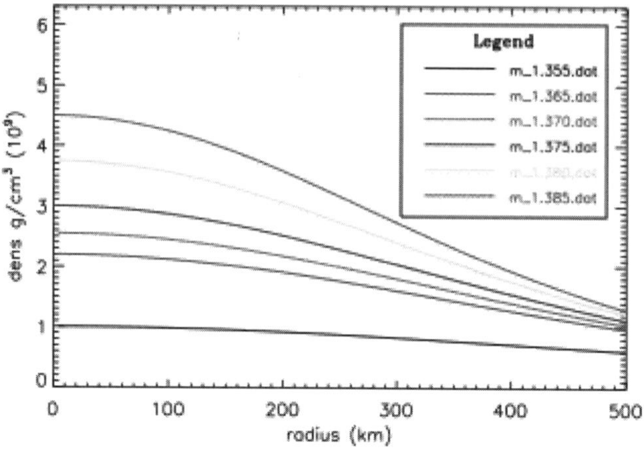

Fig. 1. Density versus radius for a series of cold white dwarf models, ranging from 1.355 $M_\odot$ to 1.385 $M_\odot$. The plot focuses on the innermost region of the white dwarf to emphasize the sensitive dependence of central density on total mass.

at a velocity coordinate $\sim 11,000$ km/s, creating a layered structure of IME and Fe peak (NSE) products similar to observation.[3]

However, despite these successes, initial simulated 3D GCD models generally underproduced intermediate mass elements and overproduced Ni. Consequently, the models are generally too luminous in comparison to Branch normal Ia events. The underabundance of IME suggest that additional pre-expansion is required to produce Ni and IME abundances consistent with observation. Moreover, other 3D simulations were found to pre-expand significantly and lead to failed explosions.[21]

One factor influencing this outcome is the uncertainty in the central density of the progenitor model, which is fundamentally set by the accretion rate in the single degenerate scenario. As shown in figure 1, a small change in the total mass of a cold 50/50 C/O white dwarf progenitor model from 1.355 $M_\odot$ to 1.385 $M_\odot$ leads to more than a factor of four change in central density : from 10^9 gm/cm^3 to $4.4 \cdot 10^9$ gm/cm^3. This increase in central density in turn directly impacts the nuclear energy release during the deflagration phase. Fundamentally, this is due to two reasons. First, the critical wavelength for Rayleigh-Taylor instability $\lambda_c \propto S_l^2/g$ is sensitively dependent on the density through the dependence on the laminar speed S_l. Here g is the gravitational acceleration. Higher central densities lead to a larger critical wavelength for the development of the Rayleigh-Taylor instability, which in turn suppresses the growth of turbulence during the early stages of the burning, and leads to an enhancement in the burnt mass. Second, the increased total mass of the progenitor with increased central density causes it to be closer to the Chandrasekhar mass, and therefore more easily pre-expanded than models with lower central density.

The net effect on the energy release during the deflagration phase in the GCD model is dramatic, as shown in figure 2. An increase in total mass by just one percent, and an increase in central density by a factor of two, leads to a four-fold increase in the total fractional energy release during the deflagration phase. This leads to a significantly enhanced pre-expansion rate in the case of the higher central density model.

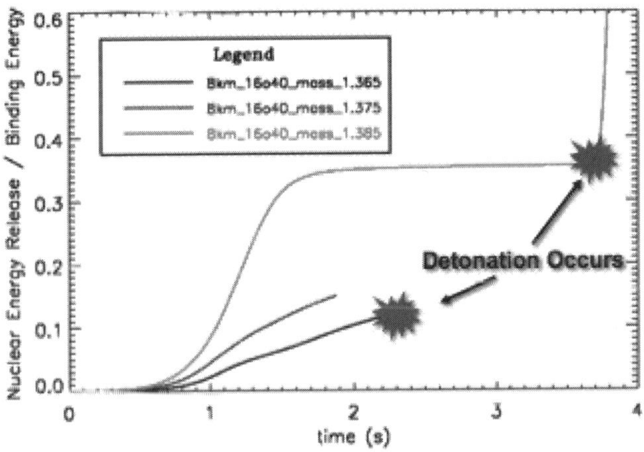

Fig. 2. A plot of the nuclear energy released as a function of time, shown for a range of stellar masses, from 1.365 $M_{\odot}$ to 1.385 $M_{\odot}$.

4. Conclusion

The GCD model simulations demonstrate that successful self-consistent detonations of Type Ia supernovae in 3-D are possible without having to be initiated by hand. The nuclear energy release during the deflagration phase is found to be sensitively dependent upon the central density of the white dwarf progenitor model. With a greater pre-expansion afforded by the greater nuclear energy release in higher central density models, the GCD mechanism can lead to observed Branch normal luminosities, and typical levels of intermediate mass elements (Si -Ca).

Acknowledgments

The software used in this work was in part developed by the DOE-supported ASC / Alliance Center for Astrophysical Thermonuclear Flashes at the University of Chicago. The PARAMESH software used in this work was developed at the NASA Goddard Space Flight Center and Drexel University under NASA's HPCC and ESTO/CT projects and under grant NNG04GP79G from the NASA/AISR project.

References

1. A. G. Riess *et al*, *Astronomical Journal* **116**, 1009 (1998).
2. S. Perlmutter *et al*, *Astrophysical Journal* **517**, 565 (1999).
3. P. A. Mazzali, F. K. Röpke, S. Benetti and W. Hillebrandt, *Science* **315**, 825 (2007).
4. D. A. Howell, M. Sullivan, A. Conley and R. Carlberg, *Astrophysical Journal Letters* **667**, L37 (2007).
5. A. L. Piro and L. Bildsten, *Astrophysical Journal* **673**, 1009 (2008).
6. F. X. Timmes, E. F. Brown and J. W. Truran, *Astrophysical Journal Letters* **590**, L83 (2003).
7. D. A. Chamulak, E. F. Brown, F. X. Timmes and K. Dupczak, *Astrophysical Journal* **677**, 160 (2008).
8. X. Meng, X. Chen and Z. Han, *Astronomy and Astrophysics* **487**, 625 (2008).
9. J. Whelan and I. J. Iben, *Astrophysical Journal* **186**, 1007 (1973).
10. K. Nomoto, *Astrophysical Journal* **253**, 798 (1982).
11. F. Patat *et al*, *Science* **317**, 924 (2007).
12. P. Ruiz-Lapuente *et al*, *Nature* **431**, 1069 (2004).
13. P. Ruiz-Lapuente, *Astrophysical Journal* **612**, 357 (2004).
14. O. Krause, M. Tanaka, T. Usuda, T. Hattori, M. Goto, S. Birkmann and K. Nomoto, *Nature* **456**, 617 (2008).
15. S.-C. Yoon and N. Langer, *Astronomy and Astrophysics* **419**, 623 (2004).
16. K. Nomoto, F.-K. Thielemann and K. Yokoi, *Astrophysical Journal* **286**, 644 (1984).
17. J. C. Niemeyer, W. Hillebrandt and S. E. Woosley *Astrophysical Journal* **471**, 903 (1996).
18. A. M. Khokhlov, *Astronomy and Astrophysics* **245**, 114 (1991).
19. T. Plewa, A. C. Calder and D. Q. Lamb, *Astrophysical Journal Letters* **612**, L37 (2004).
20. J. C. Niemeyer and S. E. Woosley, *Astrophysical Journal* **475**, 740 (1997).
21. F. K. Röpke and S. E. Woosley, *Journal of Physics Conference Series* **46**, 413 (2006).

International Organizing Committee

David Blair (Australia)
Mario Novello (Brazil)
Valery P. Frolov, Don N. Page (Canada)
Yong-Jiu Wang, Jun Luo, K. S. Cheng, Wei-Tou Ni, Liao Liu, Rong-Gen Cai (China)
Jean-Michel Alimi, Roland Triay (France)
Spiros Cotsakis (Greece)
Varun Sahni (India)
V. de Sabbata (Italy)
Masakatsu Kenmoku, Hideo Kodama, Kazuaki Kuroda, Misao Sasaki, Katsuhiko Sato (Japan)
Yong-Min Cho, Sung-Won Kim (Chair of IOC) (Korea)
A. Garcia (Mexico)
Dmitri V. Gal'tsov, Vladimir N. Lukash, G. S. Bisnovaty-Kogan, Vitaly N. Melnikov, V. K. Milyukov Valentin N. Rudenko, Alexei A. Starobinsky (Russia)
Igor D. Novikov (Denmark)
C.-M. Chen, James M. Nester (Taiwan)
Jong-Ping Hsu, F. Everitt, H. J. Paik (USA)

Local Organizing Committee

Zhong-Kun Hu (Huazhong University of Science and Technology)
Run-Qiu Lau (Academy of Mathematics and Systems Science, Chinese Academy of Science)
Fang-Yu Li (Chong Qing University)
Tan Lu (Purple Mountain Observatory,Chinese Academy of Science)
Jun Luo (Huazhong University of Science and Technology) (Chair of LOC)
Qiu-He Peng (Nanjing University)
Bin Wang (Fudan University)
Yue-Liang Wu (Institute of Theoretical Physics, Chinese Academy of Science)
Ren-Xin Xu (Peking University)
Hsien-Chi Yeh (Huazhong University of Science and Technology)
Ye-Fei Yuan (University of Science and Technology of China)
Shuang-Nan Zhang (Tsinghua University)
Yuan-Zhong Zhang (Institute of Theoretical Physics, Chinese Academy of Science)
Ze-Bing Zhou (Huazhong University of Science and Technology)

Supported by
Asia Pacific Center for Theoretical Physics (APCTP)
National Natural Science Foundation of China (NSFC)
The Ministry of Education of P. R. China
Huazhong University of Science and Technology (HUST)

List of Participants

1.	P.P. Antonio	Purple Mt. Observatory, CAS	antonio@pmo.ac.cn
2.	P. Bender	U. of Colorado, USA	PBENDER@jila.colorado.edu
3.	G.S.Bisnova-tyi-Kogan	Space Research Inst. RUS	gkogan@iki.rssi.ru
4.	R.-G Cai	Inst. of Theore. Phys., CAS	cairg@itp.ac.cn
5.	C.-M. Chen	Nat. Central U, Taiwan	chiangmeichen@gmail.com
6.	J.-W Cao	Tsinghua U.	jcao@mail.tsinghua.edu.cn
7.	Z.-J. Cao	Inst. of Math., CAS	zjcao@amt.ac.cn
8.	X.-S. Chen	Huazhong U of Sci. and Tech.	cxs@scu.edu.cn
9.	R. Fisher	U. of Mass. Dartmouth, USA	rfisher1@umassd.edu
10.	D. Galtsov	U. of Moscow, RUS	galtsov@phys.msu.ru
11.	Z.-M. Gan	Huazhong U of Sci. and Tech.	ganzhaoming@smail.hust.edu.cn
12.	C.-Q. Geng	Nat. Tsinghua U, Taiwan	geng@phys.nthu.edu.tw
13.	B.-P. Gong	Huazhong U of Sci. and Tech.	bpgong@mail.hust.edu.cn
14.	S. Hayward	Shanghai Normal U	sean_a_hayward@yahoo.co.uk
15.	Feng He	Hunan U of Sci. and Tech.	fhedoc@163.com
16.	J. P. Hsu	U of Mass. Dartmouth, USA	jhsu@umassd.edu
17.	D.-Z. Hu	Huabei U of Sci. and Tech.	Hudezhi_502@163.com
18.	W.-R. Hu	Inst. of Mechanics, CAS	wrhu@imech.ac.cn
19.	Z.-K. Hu	Huazhong U of Sci. and Tech.	ggg@mail.hust.edu.cn
20.	C.-G. Huang	Inst. of High Energy Phys, CAS	huangcg@ihep.ac.cn
21.	M. Kenmoku	Nara Women's U, Japan	kenmoku@asuka.phys.nara-wu.ac.jp
22.	S. P. Kim	Kunsan Nat. U, Korea	sangkim@kunsan.ac.kr
23.	S. W. Kim	Ewha Womans U, Korea	sungwon@ewha.ac.kr
24.	K. Kuroda	U of Tokyo, Japan	kuroda@icrr.u-tokyo.ac.jp
25.	Y.-K. Lau	Inst. of Mathematics, CAS	lau@amss.ac.cn
26.	F.-Y. Li	Chongqing U	fangyuli@cqu.edu.cn
27.	G.-Y. Li	Purple Mt. Observatory, CAS	gyl@pmo.ac.cn

28.	Yi Ling	Nanchang U	yling@ncu.edu.cn
29.	D.-J. Liu	Shanghai Normal U	djliu@shnu.edu.cn
30.	Qian Liu	Beijing Normal U	liuqian0814@mail.bnu.edu.cn
31.	Yu-Xiao Liu	Lanzhou University	liuyx@lzu.edu.cn
32.	Z.-Z. Liu	Huazhong U of Sci. and Tech.	zzl129@mail.hust.edu.cn
33.	Tan Lu	Purple Mt. Observatory, CAS	t.lu@pmo.ac.cn
34.	Jun Luo	Huazhong U of Sci. and Tech.	junluo@mail.hust.edu.cn
35.	Yi-Qiu Ma	U of Sci. and Tech. of China	myqphy@mail.ustc.edu.
36.	V. Melnikov	Peoples' Friendship U of Russia	melnikov@phys.msu.ru
37.	V. Milyukov	U of Moscow, RUS	milyukov@sai.msu.ru
38.	H Motohashi	U of Tokyo, Japan	motohashi@resceu.s.u-tokyo.ac.jp
39.	J. Nester	Nati. Central U, Taiwan	nester@phy.ncu.edu.tw
40.	W.-T. Ni	Purple Mt. Observatory, CAS	wtni@pmo.ac.cn
41.	J. Ovalle	U of Simon Bolivar, Venezuela	jovalle@usb.ve
42.	H. J. Paik	U of Maryland, USA	hpaik@umd.edu
43.	J.-J. Peng	Central China Normal U	pengjjph@163.com
44.	J.-S. Ping	Shanghai Ast. Observatory, CAS	pjs@shao.ac.cn
45.	A. Ruediger	A. Einstein Inst, Hannover, Gm.	albrecht.ruediger@aei.mpg.de
46.	C.-G. Shao	Huazhong U of Sci. and Tech.	cgshao@mail.hust.edu.cn
47.	J.-X. Tian	Dalian U of Technology	lanmanhuayu@yahoo.com.cn
48.	Hao Tong	Nanjing U	haotong@nju.edu.cn
49.	R. Triay	Centre Phys. Theorique - CNRS	triay@cpt.univ-mrs.fr
50.	L.-C. Tu	Huazhong U of Sci. and Tech.	tlc@mail.hust.edu.cn
51.	C. Wang	Zhejiang Forestry U	wangchong@zjfc.edu.cn
52.	D.-X. Wang	Huazhong U of Sci. and Tech.	dxwang@mail.hust.edu.cn
53.	Y. Wang	Nanjing U	wangyan7677000@gmail.com
54.	P.-X.Wu	Hunan Normal U	wpx0227@gmail.com
55.	S.-Q. Wu	Central China Normal U	sqwu@phy.ccnu.edu.cn
56.	S.-C. Wu	Huazhong U of Sci. and Tech.	scwu@mail.hust.edu.cn
57.	X-N. Wu	Inst. of Mathematics, CAS	wuxn@amss.ac.cn
58.	Y. Xia	Purple Mt. Observatory, CAS	yxia@pmo.ac.cn

59.	R.-X. Xu	Peking U	r.x.xu@pku.edu.cn
60.	J.-H. Yang	Hunan U Arts Sci.	yjianghe@163.com
61.	J. Yang	Lanzhou U	yangjiev@lzu.edu.cn
62.	H.-C. Yeh	Huazhong U of Sci. and Tech.	yexianji@mail.hust.edu.cn
63.	H-W. Yu	Hunan Normal U	hwyu@hunnu.edu.cn
64.	Y.-Fe.Yuan	U Sci. and Tech. of China	yfyuan@ustc.edu.cn
65.	X.-H. Zhai	Shanghai Normal U	zhaixh@shnu.edu.cn
66.	C.-M. Zhang	Nat. Ast. Observatories, CAS	zhangcm@bao.ac.cn
67.	T.-J. Zhang	Beijing Normal U	tjzhang@bnu.edu.cn
68.	Y. Zhang	U Sci. Tech. of China	yzh@ustc.deu.cn
69.	Y-Z. Zhang	Inst. of Theoretical Phys., CAS	zyz@itp.ac.cn
70.	Z-B. Zhou	Huazhong U of Sci. and Tech.	zhouzb@mail.hust.edu.cn
71.	Z.-H. Zhu	Beijing Normal U	zhuzh@bnu.edu.cn

HUST, Wuhan, P. R. China 28 June - 2 July, 2009
ICGA-9

2009年全国研究生暑期学校——广义相对论理论和实验
湖北·武汉 华中科技大学物理学院 2009.6.20—2009.7.9

ALBERTA

Wei Tou Ni
Chairman

朗学校的老师和学员！